Managing the Smart Revolution in Tourism Firms

Innovation and Value Creation in the Era of Data

This book is dedicated to my wife María José,
and my daughters Sofía and Alicia.

Managing the Smart Revolution in Tourism Firms

Innovation and Value Creation in the Era of Data

Francisco J. Navarro Meneses

Universidad Antonio de Nebrija, Madrid (Spain)

CABI

CABI is a trading name of CAB International

CABI
Nosworthy Way
Wallingford
Oxfordshire OX10 8DE
UK

Tel: +44 (0)1491 832111
E-mail: info@cabi.org
Website: www.cabi.org

CABI
200 Portland Street
Boston
MA 02114
USA

Tel: +1 (617)682-9015
E-mail: cabi-nao@cabi.org

A catalogue record for this book is available from the British Library, London, UK.

ISBN-13: 9781789249309 (hardback)
9781789249316 (paperback)
9781800620254 (ePDF)
9781800620261 (ePub)

DOI: 10.1079/9781800620261.0000

Commissioning Editor: Claire Parfitt
Editorial Assistant: Emma McCann
Production Editor: Marta Patiño

Typeset by: SPi, Pondicherry, India
Printed and bound in the USA by Integrated Books International, Dulles, Virginia

Books From California
51 W. Easy St
Simi Valley, CA 93065
UNITED STATES
customerservice@booksfromca.com

Books From California
51 W. Easy St
Simi Valley, CA 93065
UNITED STATES

To: Alibris APEX DC 76906266-30 - APEX
800 Avondale Ave.

Grandview Heights, OH 43212-3473
UNITED STATES

If this order is shipping from the US to an international buyer, remember to attach proper documentation. Complete and attach USPS Form CN-22 for all Priority Mail International Envelopes or First-Class Mail International shipments (available at your post office).

Order Number:	33969804
Ship Method:	Standard
Customer Name:	Alibris APEX DC 76906266-30 - APEX
Order Date:	9/11/2026
Alibris Order #:	76906266-30
Email:	

Items:

Qty	Item	Locator	Item ID	Condition
1	Managing the Smart Revolution in Tourism Firms: Innovation a...Creation in the Era of Data Navarro-Meneses, Francisco Javier SKU: mon0003375436 ISBN: 1789249309 - Books	C -2-01-734-001-1053	76906266-30	Like New

Notes:

Fine As New Condition-Over 500, 000 Internet Orders Shipped. hardcover

alibris

Alibris guarantees the condition of every item as it is described on our Web site. If you are dissatisfied for any reason, return your purchase within 60 days of receipt for a refund of the item price. We'll also refund shipping costs if the return was a result of our error. All return requests must be submitted via the Alibris Web site. Any item returned without accompanying paperwork and/or more than 60 days after its shipment date will be discarded and you will not be eligible to receive a refund.

If you have any questions or concerns regarding this order, please contact us at customerservice@booksfromca.com
Thanks for your order!

Contents

Foreword

Tourism firms are at a crossroads when it comes to creating value. The dramatic effects of the COVID-19 pandemic and destabilizing geopolitical events in Eastern Europe have not only significantly damaged the ability of tourism firms to sustain value creation, but their potential to emerge stronger from crises raises serious questions, especially for small and medium-sized businesses. The world is not going to be the same again and the advent of smart technologies also has a lot to do with it. Everything around us seems to have become smarter: our phone is getting smarter, our TV, our running shoes, and even our home is now smarter. Consumers are accustomed to things, products, services and even experiences becoming "smart", which means that humans are starting to be replaced when we buy something or order a service in favor of machines and "intelligence" embedded in things. Smart technologies have become the perfect allies for consumers when they travel to a destination for the first time, from which they obtain tailored information, access personalized services, and are in contact with other users and friends. Up to this point there is nothing wrong. After all, we like to be treated in a personalized way and to make our lives easier.

However, solving the equation of value creation in an environment as complex and competitive as that of today's tourism is a great challenge for all organizations. From the perspective of the owner and manager of the tourism firm, this requires, first of all, putting together the pieces of the puzzle to begin to understand the true dimension of the moment in which we live. Of course, this is not the first time that consumers and firms have experienced disruption, where ways of thinking and acting stop making sense. Let us simply remember what tourism experienced with the internet revolution and when all started to go digital. What is different this time is that the speed, depth, and scope of the opportunities offered by the Smart Revolution are of a magnitude never seen before. And this makes a huge difference!

For the most daring firms, the Smart Revolution offers enormous opportunities to create long-standing sustainable competitive advantages. For all the others, the probability of falling behind and dying is equally enormous. Addressing the Smart Revolution requires starting to think and behave differently at all levels of the organization and embracing new ways of managing and leading organizations that traditional firms may not be used to. Above all, it means raising data to the category of critical asset of the organization on which to build a new agile and flexible work culture capable of responding to the challenges of the environment with speed and precision. And here is where this book comes in.

The book is divided into four main parts and contains 23 chapters that shed light on the challenges that the Smart Revolution brings to tourism firms in two ways. On the one hand, it provides updated knowledge obtained from the literature and state-of-the-art research about the critical variables that frame the phenomenon of the Smart Revolution and the impact on the tourism firm. This knowledge is essential to be able to operate technically and functionally in the world of smartization. On the other hand, the book aims to guide business leaders, and those who want to become one soon, on how to get it right when making their decisions about smart transformation.

Ultimately, the owners and managers of tourism firms have at their fingertips with this book an orderly, systematic, but never complete or exhaustive framework of some of the latest and most relevant advances in the field of smart transformation that can be put into practice to continue creating value in the years to come.

Francisco J. Navarro Meneses
Torrejón de Ardoz, Spain
June 2022

Acknowledgements

This book is a tribute to all owners and managers of tourism firms who struggle daily to win customers, keep them loyal, and stay afloat. Over the years, and there are already many, I have had the privilege of working with many business leaders, both in tourism and other service industries, from large international corporations to small and medium-sized firms. I have consulted them on organizational transformation issues and trained them and their teams. Even I myself have spent years being an entrepreneur in tourism. The talent and perseverance with which these leaders work day and night to make tourism that great "experience industry" has always inspired me. I have also been amazed many times by how much they achieve despite not always having the best tools and the most up-to-date knowledge to tackle the enormous challenges that lie ahead. For that I feel indebted to them, so this book is my humble contribution to their efforts at such a decisive time for their business. They are many and it would not be possible to name them all, but I want to express grateful recognition to all of them. Last but not least, I would like to acknowledge the dedication of the CABI editorial team, especially in the names of Claire Parfitt, Emma McCann, and Marta Patiño for their constant assistance and support in the hard work of getting this book off the ground, and which took this author longer than initially planned. It goes without saying that the book would not have been possible without the love and support of my dearest family.

Introduction

Tourism business owners and managers are at a critical tipping point. The business environment in which they operate has not only become intrinsically more competitive and complex but keeping up with innovation and disruption, while still managing day-to-day operations effectively and efficiently, has become a real challenge. With the advent of the 21st century, people and organizations have realized that we all live in an increasingly VUCA world, an acronym that all tourism professionals should be familiar with and that describes the increasingly volatile, uncertain, complex, and ambiguous world in which we live. Scholars and professionals now accept that firms are dynamic in nature, that change can happen at any time and come from anywhere, and that we have lost the ability to predict what is going to happen next in business and in our lives. Furthermore, the speed at which change occurs is so high that it is extremely difficult to make the right decision about the direction that our organizations should take. In this context, the unexpected is king and the rules that govern how firms operate are unequivocally different.

There are many reasons that can help us understand how we have arrived at this stage, including globalization, digitalization, geopolitical balances, as well as countless social and economic interconnections that operate at the macro and micro levels. Notwithstanding the foregoing, there is still a key factor that is driving tourism firms towards this new era of rapid and accelerating change: technology. Advances in information and communication technologies (ICTs) are raising consumer expectations and disrupting tourism ecosystems at a breakneck pace, to the point where a new generation of non-traditional data- and technology-focused competitors are driving a radical transformation of tourism. Technology is now embedded in the things we normally see and touch, but also in the things we don't perceive (e.g., factory production lines, supply and delivery chains, etc.) having become ubiquitous. In fact, many analysts believe

DOI: 10.1079/9781800620261.intr

that we are already immersed in the Fourth Industrial Revolution, a new era characterized by cyber-physical systems that are creating radical breakthroughs in various fields of society at an extraordinary fast speed (Schwab, 2017).

Today, everyone carries a smartphone in their pocket, is in constant communication with their friends, family, and co-workers in real time, and may even live in a house or drive a car that is "intelligent" (Kaivo-Oja et al., 2015). Consumer use of social media has reached unimaginable levels with the support of personal mobile and connected technologies, shaping the way people communicate with each other and with firms. The power of networks and crowds has increased substantially relative to major market leaders, causing changes in people's social behavior, which, in turn, contain the seeds of a new system of values and beliefs. All these changes are even more noticeable in generations of people who were born between the 1980s and mid-1990s, the so-called Generation Y, whose communication and work habits have forever changed the way they interact with firms and how they make decisions (Burchardt & Maisch, 2019).

The impact of all these developments for tourism small and medium-sized enterprises (SMEs) is materializing in a hyper-competitive business environment in which competitive advantages are only temporary (D'Aveni et al., 2010). Organizations need to constantly innovate to stay ahead of the competition and make a profit that vanishes all too quickly (Mariani et al., 2016). Even more worrying is that owners and managers can only expect these changes to continue to intensify further in the future as internet penetration consolidates and the entire population ends up connected via smart devices and high-speed broadband. Driven by these transformative conditions, digital and smart technologies are pushing tourism SMEs in new and somewhat unpredictable directions, providing unprecedented opportunities to develop new markets, develop new tourism products and services, and improve their competitive edge. No doubt these powerful trends towards accelerating change, technology adoption, and growing interconnectedness are fundamentally redefining the preferences and needs of the so-called "new tourist" (Pasca et al., 2021). The "new tourist" favors the use of mobile technologies and digital platforms, is much more informed, knowledgeable, and connected than ever, extraordinarily demanding, and expects services to be provided when and how she wants them, not how tourism firms want them to be (Zine et al., 2014).

However, it is not all good news. Along with the new opportunities and promised big benefits, there are new risks and threats on the horizon for tourism firms. Smart technologies can create an uneven playing field in the business ecosystem, exacerbated by a widening gap between globally connected, data- and technology-driven tourism firms, and traditional micro and small firms whose business models are grounded in the low use and low investment in technology. This disparity is further accentuated as the world's most innovative and highest-value digital companies (e.g., Airbnb, Booking, HomeAway, Uber, etc.) coexist with the larger number of traditional tourism SMEs still struggling to understand the opportunities and reap the benefits of digitalization (OECD, 2019).

Furthermore, reality is much more complex than anybody imagines, and business owners and managers should not consider technology as the only transformative factor that is pushing change hard. Although much of the attention of tourism leaders to date has focused on figuring out how to implement digital technologies, deciding which technology is best, or developing the skills needed to excel in the digital world, time has shown that technology is less effective in improving productivity and driving innovation in an increasingly global and competitive market. Owners and managers should be aware that people are just as important as technology. The leaders who mobilize their organizations to embrace change and provide innovative and effective responses to challenges, and the frontline employees who do their best to satisfy their demanding customers, are essential parts of the smart transformation puzzle.

This is the new normal. A time of profound and transformative change that no tourism firm can escape, no matter its size, location, or sector. All tourism firms are affected equally and strongly. So, you may be wondering right now, how can my business fully grasp the true dimension of the Smart Revolution and what it needs to do to address it effectively, given the myriad challenges it faces? This is exactly what this book is about. In it, owners and managers will find a support guide to understand the phenomenon and the true dimension of the Smart Revolution that tourism firms must tackle and will learn how to navigate these turbulent times by transforming the organization in the most effective and efficient way possible.

Let me welcome you to this wonderful journey and make yourself comfortable. I hope that by reading this book you will discover a better future for yourself and for everyone in your organization.

1 A Call to Action

Tourism firms, like many in other industries, are not immune to digital disruptions and the deep structural and behavioral changes that digitalization is bringing about in the business ecosystem and in consumers. These disruptions manifest themselves in many different ways and to varying degrees, including the way tourism firms create value, the competitive landscape of tourism activities, or how firms communicate and engage with partners, suppliers, and customers. The variety, extent, and speed at which these developments are taking place are leaving tourism business owners and managers out of options, as the decision to tackle transformation is no longer a matter of choice, but a must that requires vision, determination, resources, and leadership among key members of the organization.

Firms that do not invest enough in transformation will find it difficult to survive in the future (OECD, 2020), and those who do not achieve a constant pace of technology adoption and renewal will inevitably lose their competitive edge and the big opportunities ahead (i.e., for innovation and value creation). When digital technologies are used wisely, the potential for growth and improvement for tourism firms is phenomenal (Reddy & Reinartz, 2017). On the contrary, when digital technologies are used without intelligence, everything can end disastrously (Kaivo-Oja et al., 2015).

In this cumbersome context, SMEs are at high risk of being left behind, so it is essential that they start by re-examining their business end-to-end and redesigning it if they are to leverage on the promises and gain the benefits of transformation in terms of efficiency, agility, and competitiveness (Busulwa et al., 2020). Doing so requires not only that business owners and managers understand digital technologies and their implications (e.g., the impact of privacy, cybersecurity, business models, interoperability, scalability, ethics, ecosystems, etc.). It fundamentally requires firms to effectively lead transformation

DOI: 10.1079/9781800620261.0001

internally as a process that spans the entire organization, from senior management, through middle management, to front-line employees. This is no doubt a big ask for tourism SMEs, which surely explains why so many firms fall short of expectations and lag behind in this endurance race of transformation (Lv et al., 2021).

It is obvious that a large number of SMEs do not have anywhere near the resources, talented people, skills, and knowledge necessary to successfully tackle transformation. Although the latest innovations in the field of connectivity, networks, artificial intelligence, Big Data, analytics, and automation, to name just a few examples, are opening new windows of opportunity for SMEs, they are also much more sophisticated and far from the capacity of most SMEs to activate change (Earley, 2014). Furthermore, there are still deeply rooted beliefs in many SMEs about whether it is worth investing the time and money in transformation and turn the organization upside down, since it is hard to predict and quantify the true benefits that transformation can bring and when and where these benefits will emerge. Quite often bad past experiences in the execution of technological projects and mismatches in the realization of expectations have historically contributed to this disbelief.

It cannot be denied that there may be a good justification behind these attitudes. No one can attest today, be they consultants, scholars, or technologists, how much or at what precise moment in time transformation will improve customer satisfaction, or the service experience. Notwithstanding the above, it is equally true that no tourism firm can afford today to lose customers, manage its resources inefficiently, or assume failure simply because it is unable to decide in which direction its organization should point in today's hyper-competitive and technology-centric environment (Nagle et al., 2013). The lack of a clear understanding of the strategic direction that the firm must follow and not having a tourism-specific roadmap focused on creating value (for the organization and the customer) are critical factors that very often prevent firms from advancing in the process of transformation that is needed (Udovita, 2020). Those running a tourism business must understand the fundamental nature of transformation and the implications that technological disruptions are having for organizations. Only after owners and managers are clear on what needs to change to close the gap can it become clear which strategy needs to be executed and which parts of the organization need to take the lead in the new strategic direction. With so much at stake, business leaders need to be aware that failure to act on the opportunities offered by transformation and digital technologies will have lethal consequences for the organization's ability to survive in the future.

For all of the above reasons, this book is fundamentally a call to action for business leaders and organizations alike to get down to work as soon as possible and seize the vast pool of opportunities that exist and grow tourism firms in times of the Smart Revolution. To achieve this there are no shortcuts; only the knowledge, attitude, and resources to implement an action plan can help. That is why you have this book in your hands, so that those who run a tourism SME, or intend to do so, equip themselves with the most relevant knowledge and management practices to devise and implement a smart tourism transformation plan. Firms around the world are looking for leaders who are able to understand this complex world and who know how to address the opportunities of smartization. Now here is your chance to stand out from the crowd.

2 The Need for This Book

The increasingly influential and sophisticated field of transformation and technology uses a specialized terminology that is often difficult for non-specialists to understand. Stop for a moment and think about all those acronyms, abbreviations, and jargon that are constantly heard and read, such as IoT, AI, Analytics, Big Data, or 5G. It is often difficult to know what they really mean. For many of the owners and managers who are approaching this growing field of knowledge and practice for the first time, or who are still at a beginning level, the language itself becomes a barrier to understanding the key issues. This not only happens with those who do not have prior technical training, but also with the most experienced practitioners due to constant innovation and the new practices and technologies that emerge every day.

Despite these obvious difficulties, a large part of the reference books and training available on the subject are highly technical or academic and focus primarily on information technology or systems management. By doing so, most authors and trainers miss out on helping business owners and managers understand the terminology and fail to guide them through the process of business transformation and organizational improvement. In addition, owners and managers have grown accustomed to hearing about the phenomenal benefits that some "magic" technologies are bringing to companies, consumers, and society in general, but few people tell them about the failures and bad experiences of those who have tried it before and have not experienced the expected returns despite having invested large amounts of time and money in the process.

To fill this gap the book provides readers with an up-to-date framework to understand the key concepts, technologies, and management practices that are crucial to tackle the challenges of smartization and the underlying organizational transformation. Always using an eminently practical approach attached to reality and an perspective independent of the technology industry and consultants,

DOI: 10.1079/9781800620261.0002

the book highlights both the good things that smart technologies can bring to the tourism firm and the not-so-good things. Special attention is also devoted to the barriers and limitations that owners and managers will encounter in the transformation process and the mistakes that should be avoided.

The book delves into the world of digital and smart technologies from a managerial perspective, keeping technical details to the minimum necessary to facilitate understanding of concepts and ideas. It also highlights the opportunities that may arise from the implementation of more than ten key smart technologies, including Big Data, analytics, artificial intelligence, Internet of Things, augmented and virtual reality, etc. and how they can shape the tourism firm's strategy and operations. Furthermore, the book emphasizes the crucial role that people and organizational leaders play in harnessing smart technologies and enabling the organization to pursue efficiency, improve the tourism experience, and gain agility for sustainable competitive advantage. To conclude, an implementation framework that draws on best practices is provided that owners and managers can use as a roadmap to take the first steps in smart transformation and start reaping its benefits. Finally, the author offers his personal vision of what could be the next wave of innovation and disruption that is yet to come in the field of smartization and transformation and that could mark the future of tourism firms.

A variety of discussion questions are provided throughout the chapters of the book to allow current and future business leaders to engage with the concepts and internalize the importance of the ideas presented. In cases where the book is used by groups of undergraduate or graduate students, this material can help stimulate class discussion about the relevance of the ideas presented by the instructor and thus achieve a more collaborative and experiential learning.

After reading the book, anyone running a tourism SME should have a broader understanding of the relationships between organizational transformation, smart technologies, and the management practices that can unlock the wealth of opportunities that they offer. Whoever reads the book carefully should also be prepared to set a plan that optimizes the firm's strategy and its operations in the context of the Smart Revolution and lead the organization towards the technological developments that are to come. Those readers will also understand that a successful smart transformation process does not depend solely on technology, but that the role that leadership and people play in mobilizing the organization and aligning it with common goals is essential. Ultimately, owners and managers who read the book will be more confident the next time they find themselves in a position to lead a transformation project within their organization or are asked by senior management to lend their support on transformation initiatives. They will surely be better equipped than others to serve as agents of change and contribute to increasing the firm's performance by keeping up with the accelerated pace of change and innovation that characterizes the era of smart tourism.

3 Who Should Read This Book

This book aims to serve as a reference and support book for owners and managers of tourism SMEs in the development and implementation of their strategies for smart transformation and organizational change, regardless of the size, geographic location, and sector in which they operate. Likewise, the book is aimed at graduate students in any of the branches of business management, change management, and digital transformation with a specialty in tourism, as well as graduate studies in tourism, hospitality, travel, and leisure management. The book may be useful to advanced undergraduate students who are approaching digital and smart technologies for the first time and who are willing to specialize in the latest trends of tourism and learn how to manage them. The materials in the book may also be suitable as part of the curriculum for courses in strategic business management, Big Data and analytics, digital transformation, and ICT management in the degrees of tourism, hospitality, travel, and leisure management.

Although this is a book in which smart technologies are prominent protagonists of the story, it has been written trying to make the language of technology and business management accessible to those who are not specialists in any of these fields. It must be said that this is not a book written for specialists in ICT management or systems engineers, nor is it meant for those who want to delve into the operation of specific technologies or become experts in their functionalities. There are very well-written works available out there that cover these subjects and to which the interested reader should turn. However, technology plays an instrumental role in the book insofar as it has an impact on organizational transformation and value creation in tourism firms. Therefore, a holistic and interdisciplinary perspective focused on management prevails in the book, rather than one specialized in technology.

DOI: 10.1079/9781800620261.0003

For Owners and Managers

The book is perfect for those owners and managers who are approaching the fundamentals of the Smart Revolution for the first time and want to acquire a comprehensive and holistic view of the consequences on the tourism firm from the perspective of organizational transformation and strategic management. In this case the book provides the reader an understanding of the key concepts that are shaping the Smart Revolution in tourism firms, as well as the factors that affect the creation and destruction of value in organizations and how to manage them.

For those owners and managers with a track record of managing technology disruption, change initiatives, or who have already embarked on digital/smart transformation processes, the book provides valuable guidance to further explore the management framework that is applicable to the tourism firm to succeed in implementing the smart transformation and unlock its benefits.

Although each chapter has been written to be read and understood on its own, it is highly recommended that beginning and experienced owners and managers alike first read the book from cover to cover before deciding to skip to specific chapters. This will allow them to get a more detailed insight into the key concepts, the relationships between them, and the consequences that arise for the management of the smart transformation in the tourism firm. Experienced owners and managers will thus be able to update their understanding of the main building blocks that shape modern smart transformation and the new management practices that have emerged to harness them. However, if owners and managers feel that they have enough knowledge and experience or are already familiar with the key ideas in this book, they are free to skip directly to the parts or chapters of the book that are of most help or interest to them.

For Instructors

Instructors can use this book as a reference book in an introductory graduate, undergraduate, or professional course on digital/smart transformation, Big Data and analytics management, organizational change management, or IT management in tourism studies, hospitality, travel, and leisure. The book contains a sequence of parts and chapters with increasing complexity and depth that can be used as stand-alone or connected topics for specialized course sessions, whether they last a few weeks or a full semester. In some cases, instructors may prefer to use the book as supplemental material to a course for prospective tourism business owners, entrepreneurs, or managers, to encourage engaging class discussions, broaden the knowledge of more advanced students, and help students focus on their final coursework.

For Students

Graduate and undergraduate students who are approaching smart technologies, smart transformation, organizational change, tourism trends, or any of the branches related to ICT management for the first time, will find in this book a valuable reference work that will accompany them both in their last years of studies and during the first steps in their professional career. The book offers them a broad and complex overview of the fundamental concepts that govern the Smart Revolution in tourism firms, as well as the key issues and applicable management practices that tourism firms can implement from a manager's perspective. Those students who are more proactive or interested in these subjects will find numerous references, case studies, and discussion questions in the book that will guide them to continue deepening their knowledge and broaden their vision of the sector in which they wish to work or, eventually, specialize.

4 How to Use This Book

The book is structured in four major parts:

1. Introduction to Smart firms;
2. Smart Technologies;
3. Smart Management Practices;
4. What's Next.

Part 1 includes four chapters that introduce the key ideas and concepts that are shaping business transformation and smartization in organizations around the world, and especially in tourism firms. Part 2 focuses on understanding the technologies that are the basis of the Smart Revolution, not from a deep technical or functional perspective, but from the positive (and negative) impacts that their implementation is starting to have, and the transformation needs for the tourism firm. Part 3 contains six chapters that examine some of the main management practices that tourism organizations should consider implementing in order to make a successful transition from conventional to smart firm. Finally, Part 4 aims to guide business leaders in the first steps of the smart business transformation process, as well as to suggest some of the trends that will continue to disrupt tourism firms in the future.

Each of the 23 chapters of the book is self-contained and can be studied on its own. However, the author recommends that the reader make a first complete reading of the book from beginning to end and follow the established order of the parts and chapters. In this way, the reader will better comprehend the key concepts covered in the book, the relationships between them, and the consequences for the smart management of the tourism firm. After a first reading of the book, readers will realize the true dimension of the Smart Revolution and will feel more comfortable to later delve into specific chapters to review or search for ideas, learn more through the references provided or use the discussion questions at the end of the chapters to stimulate debate and reflection.

DOI: 10.1079/9781800620261.0004

Part 1 Introduction to Smart Firms

Part 1 of the book introduces the key concepts and ideas that are shaping the Smart Revolution in tourism, as the by-product of a longstanding process of deep change that has been impacting the business ecosystem, consumers, and tourism firms for years. Over the next four chapters, business owners and managers are going to learn about the main components that qualify the Smart Revolution as a substantial step forward in creating innovative opportunities and addressing new challenges for tourism management, making clear the main drivers that will guide the transformation of their organizations in the coming years. Emphasis has been placed on using clear and concise language that allows business leaders to get past the confusion surrounding the smartization process once and for all. The aim is to provide the reader with an up-to-date stock of knowledge about the far-reaching consequences that the Smart Revolution is having on tourism firms, stressing the need to address this process through intense data-centric organizational transformation. Given the complexity and length of the topics covered, an approach focused more on breadth than on depth has been chosen in which those factors that are essential to consolidate sustainable competitive advantages and reap the benefits of the Smart Revolution are highlighted.

5 Fundamentals of the Smart Firm

Tourism is transforming at an accelerated pace that has never been seen before, impacting consumers, businesses, and destinations alike. However, the reasons that are driving this moment of profound change are neither new nor exclusive to tourism. Throughout history, both society and businesses have witnessed other moments of dramatic change that paved the way for new structures, cultures, and behaviors to reach what we are today.

The First Industrial Revolution brought advances in the production of goods. The introduction of a new source of energy such as steam meant that many of the tasks that were previously carried out manually could be mechanized, leading to the emergence of production lines for the first time. At the end of the 19th century, the changes initiated in the First Revolution accelerated, giving rise to the Second Industrial Revolution, which led to the introduction of mass production. Since then, millions of consumers began to have access for the first time to thousands of products that were of good quality and affordable. Not many years passed until the Third Industrial Revolution, also called the Scientific-Technological Revolution, introduced the massive adoption of information and communication technologies (ICT) in the means of production, thus facilitating the emergence of new distribution channels that broke the limits of trade at a geographical level and expanded it to all corners of the planet through e-commerce and advances in transport networks (Gaurav & Kongar, 2021). Our time in history is sustained by the unstoppable digital advance that has caused a drastic reduction in employment in the agricultural and manufacturing sectors, while employment in the service sector has skyrocketed.

Has the history of human progress ceased since then? Of course not. History has not stopped being written for a single moment and has run its course inexorably, to the point that today we take for granted that the Third

DOI: 10.1079/9781800620261.0005

Industrial Revolution has been far surpassed and that we are living the dawn of a new revolution: the Fourth Revolution.

What makes our time different from other past times? Our time is distinguished by the massive incorporation of information and knowledge in everything we do, buy, or produce. We individuals have become accustomed to others doing the work for us, or at least making it easier for us – with someone recommending us those products and services that best fit our personality, state of mind, or moment in life. We make decisions based on the experience that others before us have had when they bought a product or service, and we expect firms to do everything possible to adapt their services to our preferences and demands.

Have we all become spoiled children? Surely not. Still, we'd better examine how we got to this stage, which certainly hasn't been a bed of roses.

5.1 The Origins of Smart Tourism

The road to where we are today has been littered with disruptive changes that previous generations had to respond to and adapt to, no doubt. Those people had to struggle to find the right mix of entrepreneurship, innovation, and technology adoption that would allow them to leap forward. Is it something different now? Not at all. Now it is our turn to do the same because we have entered a comparable historical moment.

In tourism, the rapid development of ICT has led to similar phases of progress that nonetheless have their own particularities. Until now, traditional tourism activities were based on close personal relationships, mutual trust, and personal attention. At a business level, the tourism ecosystem was dominated by a few large companies that influenced tourists' decisions almost at will, monopolizing value offers in the world's main tourism markets. On the other hand, consumers did not usually complain and accepted the conditions imposed by the big operators, grateful to be able to enjoy a vacation without unpleasant surprises and at prices adjusted to their purchasing power.

The advent of Google changed the rules of the game. The swift popularization of search engines allowed prospective tourists to access information from hundreds of tourism service providers and compare them in a matter of seconds. That was only the beginning of a story of sweeping changes to come in the way tourism consumers and businesses interacted, which continues to this day. Unimaginable new business models began to emerge that replaced many of the traditional models and sought to satisfy entirely new market niches and facilitate consumer decision-making. It was in these times that travel blogs, recommendation sites, and price comparison systems started to appear (Zsarnoczky, 2018).

The accelerated adoption of digital and online media by consumers meant that tourism firms had to start thinking very seriously about the need to change and adapt their "analog" products and services, if not to create them from scratch. Designing products from the perspective of the "new tourist" began to become a necessity. Since then, getting to know the "new tourist" in depth has become an integral part of the design of business models, and of products and services. Nowadays it would be simply unthinkable for a tourism firm to survive in the market without having this knowledge.

Challenged by this avalanche of sweeping changes in the way tourism firms are required to operate, many firms have simply been unable to keep up with the pace of change and have been left out of the market. This would explain the renewal that the tourism ecosystem has experienced in recent years at the business level. Figure 5.1 illustrates the evolution described, from traditional tour operators with a low level of digitalization to the present day, characterized by the digital transformation of operators and the growing but unstoppable integration of smart technologies in tourism firms.

Phase 1: Traditional tourism

This phase is characterized by the pervasive influence of tourism service providers on consumer decision-making, as they dominate the tourism ecosystem at a global level. The adoption of ICT by tourism firms is still very low and most are largely unaware of the individual tastes and preferences of their customers. Instead, management's focus is more on continuous quality improvement, the efficient operation of processes, and the generation of economies of scale to improve commercial margins.

Phase 2: e-Tourism

This second phase (1990–2000) is marked by the integration of the internet into the daily lives of consumers and tourism firms. Firms realize the need to be on the internet to serve a growing number of potential customers and

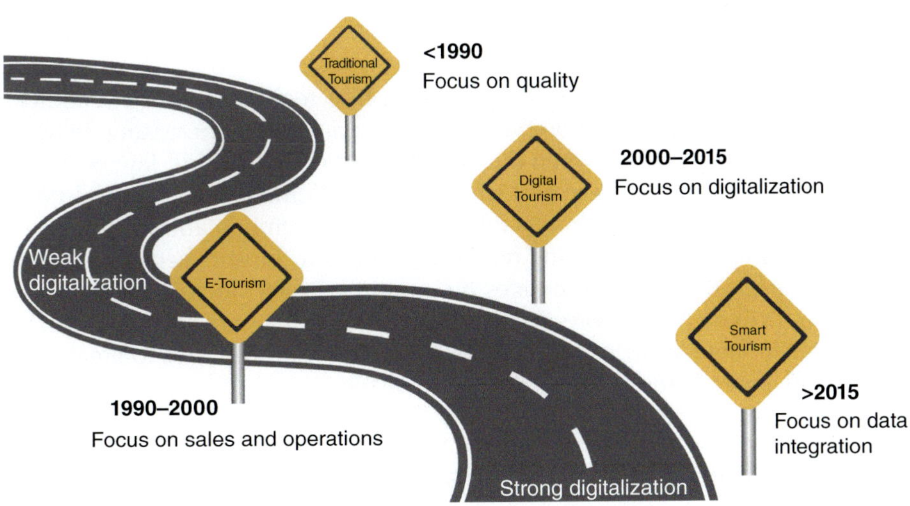

Fig. 5.1. The roadmap towards smart tourism. Source: own elaboration based on Dredge et al. (2019).

the great potential of the World Wide Web as a sales and marketing tool. Nonetheless, communication with customers remains rudimentary and mostly one-way (e.g., via phone, email, and electronic forms). The websites of tourism firms become shop windows open 24 hours a day to any customer in the world regardless of their physical location, and owners and managers begin to see the electronic channel as a promising source of income generation. Paper is replaced as the main support in communications with customers and tourism firms starting to mass-produce digital content, aware that it can be delivered instantly to the people. New electronic and online marketing tools emerge that contribute to improving business performance.

Phase 3: Digital tourism

In the next decade (years 2000–2010), the internet consolidates as the main source of information for tourists, and the main channel through which information is shared between consumers and firms. New internet technologies openup the possibility for consumers to search and compare information for their trips in a matter of seconds and for firms to create transactional gateways for their products and services. New intermediaries appear that develop business models unknown until this moment. The consumer becomes a co-creator of content and has a much more influential role in decision-making. Traditional operators are forced to transform their business models if they want to survive, meaning that they need to establish much closer ties with their customers and be willing to personalize their product and service offerings. On the other hand, customers begin to think of tourism as an experience that they themselves are willing to create, and the traditional view of the product or service that cannot be controlled is surpassed.

Phase 4: Smart tourism

Starting from 2010, advances in mobile technologies, communication networks, and cloud computing, accompanied by increasing integration and interoperability between the physical and digital worlds, lay the foundation for the transition to the era of smart tourism. Combined with social networks and Web 2.0 platforms that facilitate two-way communication with consumers, the possibility of co-creating new experiences with users becomes a reality in this phase. All the above creates the opportunity for firms to obtain an amount of data like never before. From now on, the ability to exploit data and turn it into actionable knowledge becomes a crucial factor for firms to create competitive advantages and thrive in the marketplace.

5.2 Definition of Smart Tourism

No one doubts that the term smart has become a buzzword nowadays. It is not only used as an adjective to describe people or animals, but we can practically find it in anything that incorporates some technology with which we can interact in a more "intelligent" way, that is, faster, automatic, intuitive, or more efficient (e.g., a television, a sports watch, a motor vehicle, etc.) (Poslad, 2011). The *Oxford English Dictionary* defines the term smart as something "smart" "controlled by a computer, so that it seems to act intelligently". So we can ask ourselves, how is the term intelligent defined? In the case of a person, it is the ability to understand and learn things. In the case of a computer or program, it is the ability to store information and use it in new situations.

Although these definitions give us a rough idea of what the term smart means, the truth is that intelligent and smart have some differences that should be considered, especially when we are in a business context. "Smart" usually refers to the ability to adapt to changing situations caused by different customer needs, past experiences, or particular environmental conditions. In this context, technology is one of the key drivers that would enable organizations to speed up response to changing conditions. In "intelligent tourism", tourists become passive consumers who are offered various options among which they must "intelligently" choose the one that best suits their needs. Different levels of intelligence would thus lead to different types of services and experiences.

In this book, the term smart is used to refer to state-of-the-art information systems and processes that provide tourists and firms with relevant information, facilitate better decision-making, greater mobility, and a more pleasurable experience in accordance with the preferences and needs of the tourist at all times (Pinheiro et al., 2021). Smart information systems encompass a wide range of technologies, which can be very useful for the tourism firm depending on the environment and the stage of the experience cycle in which they are used. Some examples include recommendation systems, environmental sensing systems, and ambient intelligence systems. Notwithstanding the foregoing, owners and managers must be aware that the term smart not only covers information systems and digital technologies, but also denotes a process that we will call "smartization", which transcends the mere technological dimension and implicitly affects other core organizational components such as innovation, leadership, and human capital (Boes et al., 2015). In a similar fashion, when we talk about the adoption of a higher degree of "smartness" through the transformation of the organization, we will be referring to a process that is intrinsically different from the implementation of conventional technologies, and that has a strong relationship with the management practices that must be used to accommodate change within the organization.

So, now that we have an idea of what smart means in this book, what do we mean by smart tourism? The notion of smart tourism is a reflection of the profound impact that smart technologies are having on the tourism consumer and on the products and services (Mehraliyev et al., 2020). The importance of smart technologies is increasing as the role of tourism consumers and firms is changing, and with it the boundaries of the tourism ecosystem. Smart tourism is a much broader notion than intelligent tourism, as it emphasizes the results that technology can offer tourists and that require a large amount of information and data to materialize. In smart tourism firms need tourists to be active all the time so that they can provide a huge amount of information that can be collected, processed, analyzed, and transformed into individualized knowledge by firms to deliver accurate products and services to tourists. Whereas smart tourism relies heavily on the availability of data and data processing capabilities to anticipate the needs of tourists, intelligent tourism provides a much more basic utility from the available data and information. As an example, the process of generating a travel route automatically from a small number of inputs provided by the user through a mobile application to which an optimization algorithm is applied, is intelligent. If in the same process information is added about the user's experience, information about the location where the user is or wants to go, or other information about the user's buyer profile, personality, etc., this is smart (Y. Li et al., 2017).

In summary, smart tourism involves the use of technologies to collect and exploit large amounts of data and provide real-time service delivery to tourists. These so-called smart technologies are not only the internet and mobile devices used to communicate, but also include the infrastructure that integrates hardware, software, and communications, as well as networks that provide real-time data to facilitate faster and more efficient decision-making for stakeholders (Dorcic et al., 2019; Gretzel et al., 2015a). Ultimately, the use of smart technologies by tourism firms largely depends on their ability to collect data in large quantities that can then be transformed into useful knowledge.

5.3 What Is Smart Tourism For?

The debate in academic and professional forums is still ongoing in trying to reach a definitive and widely accepted conclusion of what smart tourism is and what its main components are. Unfortunately, on many occasions these discussions lead to ideas of little use to owners and managers, who are more concerned with understanding the phenomenon of smartization and how to benefit from it than with theoretical discussions that contribute little or nothing to improving the performance of their business.

The fundamentals of smart tourism lie in the tourist. The new applications of smart tourism are activated by the behavior of tourists, so changes in their needs and demands are the real driving force of smart tourism. Any benefits derived from smart tourism will arise from the needs of tourists and changes in their behavior (Y. Li et al., 2017). Consequently, the role of the tourist as an informational asset is essential, as evidenced by the fact that the value of information for tourists, tourism firms, and destinations has not stopped growing in recent years. Without tourist information there simply cannot be smart tourism.

For tourists

Smart tourism provides tourists with access to more comprehensive, portable, and personalized information based on their needs, allowing them to organize and fine-tune the tourism experience more quickly and accurately. For more and better information to be accessible, tourists must provide relevant connected data about their behavior, demands, or personality, so that tourism firms can deliver accurate digital informational resources at their fingertips (e.g., text, pictures, video) through a variety of devices and technologies. The type of value that tourists obtain in exchange for the data they provide is independent of the way in which tourism firms collect and process that data about their behavior and needs.

For owners and managers

Smart tourism is about building an integrated framework of services and technologies that is comprehensive and provides the tourist with a range of services that are accurate, convenient, and ubiquitous. From a technical perspective, such a framework allows for continuous interaction between the assets of the firm (more specifically between the digital representation of those assets) and the tourist needs to create more meaningful relationships with tourists. In this way, smart tourism provides tourists with entirely new ways of consuming tourism services, and tourism firms with new possibilities to create value through innovative combinations of tourism information, services, and technologies.

Readers should note that not all tourist information falls within the scope of smart tourism. Indeed, tourists can obtain valuable information from very different sources. Sometimes the information will come from websites; other times from social networks, or even from the tourist information offices of a destination. However, only the information that is ubiquitous and that is delivered to the tourist according to their individual requirements is what we will call smart tourism.

5.4 Components of Smart Tourism

It is not an easy job trying to summarize in a few pages the most important components that characterize smart tourism and that distinguish it from conventional tourism. Nor could it be exhaustive because any attempt to do so would require much more extensive work. Most likely, it would not be of much value to the reader either since the very idea of smart tourism is not yet fully developed. Even so, the components with which every owner and manager should be familiar and that most directly affect the tourism firm are described below.

One of the main components of smart tourism is tourist information. In smart tourism, information is no longer that old static information aimed at poorly differentiated groups of consumers that was delivered through a few channels enabled by firms in a rather unstructured way. In contrast, the information in smart tourism is ubiquitous, and is created to meet the individual needs of the tourist. The ubiquity of information means that tourists can easily access information from anywhere, at any time, and through any type of media on the internet. Access to tourist information is provided through devices that are part of everyday life for most people and that provide tourists with a large portable computing capacity, such as mobile devices, portable devices, and other non-conventional means. The very idea of ubiquity implies that, through smart tourism, people have overcome the limitation of having to sit in front of a desktop computer to access tourist information and services. Instead, it involves a radically new way of shaping space and time to the individual needs of the tourist.

Ubiquity, however, not only encompasses information, but also the tourism firm itself, to the point that we could speak of the "ubiquitous firm". These are firms that have the ability to provide, optimize, and improve services as they are delivered, using ubiquitous marketing, ubiquitous supply chain, and ubiquitous management. The target of the ubiquitous firm is no longer the group of tourists (understood as a set of consumers who share certain attributes that make them similar in the eyes of the firm and for whom generic products and services are created), but the tourist as an individual. This means that smart tourism firms remain focused on individuals with special needs, who have an autonomous behavior that makes them different from other tourists and with the ability to establish their own preferences through the (ubiquitous and bidirectional) communication channels made available by the firm.

Smart tourism firms devote their resources (e.g., people, money, infrastructure, etc.) to create and capture value from the data generated by the behavior of tourists at any given time. This is what we call business development the smart way. In other words, smart tourism firms' development depends to a great extent on Big Data and their ability to extract value from it through more relevant relationships with tourists and improved products/services. To create more relevant relationships between tourists and firms another key component must be considered: the smart tourism ecosystem.

Smart tourism firms operate within an ecosystem in which tourism products/services (and experiences) are created, managed, and delivered through smart technologies. These technologies are very special because they require the intensive use and constant exchange of information resources to, from, and between tourists, tourism firms, and destinations, to create shared value from the Big Data (Gretzel et al., 2015b). Smart technologies are basically those ICTs that allow constant communication and interaction with tourists, from whose behavior and demands data is extracted that can later be exploited by the tourism firm to generate useful knowledge for the business. Note that the smart tourism ecosystem would simply not work if the actors involved were isolated or separated from each other. They must be connected through (smart) technologies, with the tourist located in the center so that the smart ecosystem can work properly.

This tourist-centric approach that characterizes smart tourism is part of its idiosyncrasy and is another of the key components that differentiate it from conventional tourism. Business owners and managers should not forget that smart tourism is created with the aim of satisfying the individual travel, mobility, accommodation, catering, and leisure needs of tourists based on their personal preferences. The tourist-centric approach of tourism firms is made possible thanks to the rapid advances and intensive use of smart technologies (e.g., Big Data, analytics, artificial intelligence, cloud computing, mobile and network technologies with high computing capacity and connectivity). The adoption of smart technologies into the consumption habits of tourists has forever transformed the way they consume tourism products/services and enjoy tourism experiences.

Tourism firms' use of smart (customer-centric) technologies puts them in a privileged position to meet the individual demands of tourists and improve satisfaction levels. It also provides tourism firms with the opportunity to radically improve their products/services and revamp tourism experiences through the implementation of new and innovative management practices that optimize the use of internal and external resources and are aimed at co-creating value with tourists, destinations, and the ICT industry. In short, smart tourism with all its components constitutes a framework for action to improve the management of tourism firms that can contribute very significantly to improving their competitiveness and performance.

5.5 Impact of Smart Tourism

The process of change towards smartization is a breakthrough that confirms the transition of the tourism sector towards a modern service industry. This process brings highly relevant impacts that will undoubtedly be the catalysts for the future developments of tourism in the coming years. As we have seen, smart tourism builds on top of earlier phases of social and business development, when everything was more focused on the informatization and digitization of the tourism business. That is why it is worth examining the most notorious impacts that smart tourism is having on consumers and firms.

Impacts on tourism firms

One of the key benefits of smart tourism for firms is that it provides a two-way communication channel between tourism firms, tourists and the rest of the actors involved in the value chain. "Smart channels" are created that operate as high-speed highways through which a constant flow of data and information circulates and is disseminated in a rather personalized and efficient way among all tourism stakeholders. In this way, smart tourism strongly impacts the commercialization and marketing of tourism products/services, given the variety of media that can be used (e.g., text, images, videos, and hybrid combinations), the ability to monitor the demand and quality of the services provided, and the possibility of quantitatively measuring the performance of campaigns.

From a management perspective, smart tourism offers tourism firms the opportunity to configure their own value offer to the customer in a quite flexible way, activating a much deeper knowledge about the customer that translates into delivering information, products, and services in accordance with the location and the requirements of the customer. However, achieving "smartness" is not straightforward. It requires the firm to redesign its internal processes, so that integration with smart technologies is seamless and contributes to improving the operational efficiency of the firm's business. For example, a hotel can have room check-in and check-out done via radio-frequency identification (RFID) systems, or the tourist can read a menu, book a table, and have access to entertainment services in the room all through the same mobile device.

One consequence of the above is that the tourism firm can end up optimizing itself thanks to smart tourism while changing the way it relates to the customer. The customer, in turn, may change behavior when searching for information and consuming tourism products/services to enjoy all the advantages that smart tourism promises.

Impacts on the tourist

Smart tourism and its components impact the tourist in several ways, including the experience, satisfaction, behavior, and decision-making (Mehraliyev et al., 2020). Smart tourism can enhance the tourist experience by fostering greater engagement and interaction with and between tourists, and promoting the exchange of experiences (Buonincontri & Micera, 2016). For example, today it is easy for tourists to share their travel experience in many different ways: they can post photos and videos of their trip, or write a review to share information with their family and friends on social media. Geocaching is another example that shows how the experiences of younger tourists can be enhanced by diversifying and gamifying the activities they enjoy in a destination (Skinner et al., 2018).

Smart tourism facilitates tourist planning and decision-making before and during travel, e.g., "smart travel guides" that allow interaction between content and the tourist in a highly dynamic way (Chuang, 2020; Ruíz et al., 2017). By using smart technologies, tourists are less constrained by the arrangements they make before they leave on a trip, as destination information is readily available wherever they are during their trip. Moreover, tourists can obtain relevant information for their trip through websites or mobile applications based on their previous behavior, consumption patterns, and browsing records on the networks and the internet. This allows tourists to adapt their consumption throughout the trip, as their needs change or they discover new information on an on-going basis (Y. Li et al., 2017). The result is that tourists can end up becoming much more unpredictable consumers than before. One consequence of the above is that the efforts made by tourism firms to improve tourist satisfaction (through smart technologies and process reengineering) are increasing day by day, for example, by implementing automatic billing processes in self-service restaurants through the use of neural networks (Aguilar et al., 2018), by installing lighting control systems in hotel rooms to improve the quality of service (Feng et al., 2017), by developing dynamic software tools that allow the characteristics of hotel rooms to be adjusted according to the activities carried out by the customer (Cetina et al., 2013), or by promoting the use of virtual and augmented reality systems to experience tourist attractions in a more enjoyable way.

Last but not least, another important impact is related to consumer preferences when it comes to adopting smart technologies and smart information. Although this is an area that needs more research, there are some studies that highlight that ease of use and perceived usefulness are key factors influencing consumer attitudes towards smart technologies (Bae et al., 2017; Chung et al., 2015). Other factors, such as informativeness, interactivity, accessibility, and personalization are also highly relevant (Chung & Koo, 2015). In conclusion, it is important that tourism firms are aware of the factors that affect their customers and users when adopting smart tourism, and that they understand how tourists prefer to enjoy tourism experiences (e.g., "more smart", "less smart"). This is a field of knowledge that will gradually gain in importance and that is very likely to change the way in which tourism firms segment and interact with customers.

5.6 Criticism of Smart Tourism

The tacit consensus reached among most academics, consultants, and leaders from the technology industry reveals an overly optimistic view of the effects of smart tourism, both for tourism firms and tourists. It often happens that many of the benefits that are continually pointed out by industry and academia are poorly supported by empirical data, which advises owners and managers to be cautious in trusting the promised benefits and the ease of obtaining them. This does not mean that owners and managers should distrust smart tourism recommendations by default, but rather that many slogans have been widely criticized due to the lack of studies to support them.

The first criticism has to do with the limited empirical evidence that academic and industry research has provided in recent years on the benefits of smart tourism, beyond highly specific case studies. In addition, the little practical research that has been carried out in this respect has provided very limited theoretical advances, which prevents establishing broad generalizations that cover all tourism activities and firms, since they have no validity beyond the isolated case studies investigated.

A second criticism is that the study on smart tourism has been carried out on very few fronts (e.g., destinations, specific smart technologies, the tourist experience, combined effects with other technologies). This means that there is neither a solid body of empirical evidence nor an elaborate theoretical framework that is exclusive to tourism and on which we can base conclusions about the real impacts that smart tourism has on tourism firms. This lack of understanding about the business perspective of smart tourism means that the answer to the question of how smart tourism actually affects tourism firms, and what factors limit its adoption, remains open.

The third criticism relates to the lack of understanding about tourist preferences, as it is still hard to know for sure which aspects of smart tourism are most (or least) preferred by tourists, or which segments of tourists are those most/least likely to choose smart tourism. Ultimately, what owners and managers should note is that there is a general absence of solidly grounded discourse on the impacts and consequences of smart tourism. This leads us to point out the significant limitations that exist in determining the drivers, drawbacks, and costs associated with implementing a smart tourism strategy in the tourism firm. Therefore, business owners and managers should take a critical standpoint, aware that not all the effects of the Smart Revolution are positive for the tourism firm. There is still a long way to go in identifying, measuring, and mitigating the negative effects and drawbacks of smart tourism in real life, and only more and better empirical work will help clarify the pros and cons.

e-Lienation

One of the most studied negative aspects of smart tourism is that of e-lienation, which highlights the negative impact that ICTs have on the consumer's own experience (Tribe & Mkono, 2017). According to these authors, there is sufficient evidence to affirm that the use of smart technologies can promote isolation and addiction on the part of the consumer, as well as self-marginalization, loss of routines, and even the breaking of social ties and misconduct when complying with rules, uses, and customs. These behavioral alterations that can affect the consumer, both in an individual and social context, can even pose a threat to the restorative purpose sought by the tourist with a trip or vacation. Circumstances like this are more frequent than most think and they happen when tourists have a hard time disconnecting from their work obligations and take their computer and mobile devices on a getaway or on vacation, or when they feel obsessed with chatting and sending emails to friends and family at home, or keep tweeting or posting on Facebook and Instagram to show the world their "perfect self in the perfect place".

These behaviors are a reflection of e-lienation and can negatively influence the quest for authenticity of tourist experiences to the point of reinforcing anxiety, narcissism, and addiction. Of course, not all the uses that tourists make of ICT have to end in e-lienation, but it seems evident that the more we use ICT in our leisure and tourism time, the more opportunities arise to suffer from e-lienation (Tribe & Mkono, 2017). Business owners and managers need to be aware of these circumstances and the negativities they can bring to the tourism experience. This means being able to identify the critical factors that can lead to e-lineation before deciding to undertake a transformation strategy that involves the implementation of smart technologies and organizational change.

Discussion Questions

- What kind of problems is smart tourism helping to solve related to your tasks and responsibilities? What types of real problems of tourism firms are the most suitable to be addressed by smart tourism?
- What is the balance between the costs and benefits of implementing a smart transformation strategy in the tourism firm?

- What factors specific to tourism (and non-tourism) are contributing most to making the term smart a buzzword?
- How do business owners and managers in your local area conceptualize smart tourism? To what extent does it coincide with or differ from the definition of smart tourism provided in this chapter?
- What benefits apart from those specifically examined in this chapter does smart tourism have?
- Is smart tourism for all tourism firms? Are there certain types of tourism firms (depending on size, location, business model, etc.) that are more likely to succeed with a smart transformation strategy?
- What negative effects other than those described in this chapter can smart tourism have for the tourism firm?

6 Smartization and Digital Transformation

Undertaking the process of smartization in a tourism firm necessarily implies that digital transformation has taken root in the firm or is being carried out in parallel. Effective smartization involves a large-scale migration of the firm's operational and organizational model to another model based on data and knowledge. This requires that firms enforce management practices that involve data-driven processes and technologies and a more agile and efficient customer-centric organizational design that sustains activities over time (Dahlström *et al.*, 2017).

So, you may be asking yourself, is smartization the same or is it different from digital transformation? Is there any true difference between the two? Smartization represents a big step forward beyond digital transformation. Nonetheless, smartization is not possible without digital transformation, and if you are not first familiar with the essentials of digital transformation, it will be more challenging for you to continue to dive into the basics of smartization. For this reason, this chapter examines the basic elements of digital transformation that every business owner and manager should know to successfully tackle the process of smartization in the tourism firm.

6.1 Defining Digital Transformation

Just as happened when we defined the terms smart and smart tourism, defining digital transformation is not straightforward. The literature on digital transformation is vast and many authors and organizations have contributed their particular vision of the phenomenon in recent years, generally emphasizing those dimensions that most affected or concerned them. Although it is not within the scope of this chapter to theorize or try to establish a universal definition of what digital transformation is, it is worth stopping for a moment a make clear how

DOI: 10.1079/9781800620261.0006

the notion of digital transformation has been used in this book, so that we can share a common vocabulary.

The term "digital transformation" refers in this book to a process of profound and evolutionary change that affects all aspects of the organization, led by digital capabilities and technologies aimed at creating value. The fact that digital transformation is an evolutionary process indicates that the phenomenon of transformation is continuous; that is, it is maintained over time (Morakanyane et al., 2017). There are some authors, however, who argue that digital transformation involves more radical or disruptive change than simply an evolutionary process. This is a challenging academic discussion that has been going on for a long time in which the author cannot take sides because, in the end, both streams of thought are adequate lenses to tackle digital transformation (Riasanow et al., 2019). The idea of an evolutionary process accurately describes what happens most of the time in the digital transformation of businesses in real life, although it is also true that tourism firms must be prepared for disruptions in their routine activities and business processes when faced with digital transformation (Balaji & Lavanya, 2018).

Digital technologies play a vital role and are the foundation of digital transformation initiatives. However, digital transformation is not connected to any specific technology (e.g., cloud computing, mobile phones, social media, etc.); nor is it driven by any particular digital innovation. Even those who think that using digital technologies is enough to drive change and achieve transformation are wrong. Organizations fundamentally require a set of new skills, as well as a distinctive mindset and culture, to successfully tackle digital transformation. It is precisely into this set of new capabilities that digital technologies must be incorporated if the organization intends to reap the benefits of digital transformation.

Digital technologies, coupled with organizational capabilities, can create endless opportunities with the potential to transform tourism firms, especially when they are related to new business models, operating processes, and products and services. In the end, what owners and managers should seek is to create value both for the organization and for customers by generating operational efficiencies, improving the customer experience, reducing costs, improving the relationship with stakeholders, etc. In view of the above, there are three different but closely related dimensions that shape digital transformation (Reis et al., 2018):

- **The technological dimension,** since digital transformation is based around the use of digital technologies that include infrastructures, networks, devices, as well as the software that makes digital applications possible.
- **The organizational dimension,** since digital transformation itself entails a significant change in the organizational processes, and even in the culture and business model of the tourism firm.
- **The social dimension,** since digital transformation is a process that affects consumer behavior and beliefs, both individually and in society (i.e., the way tourists consume tourism products and services, the reasons why tourists prefer some channels to others, the way tourists interact with friends and family, or how tourists enjoy a travel experience).

When it comes to digital transformation, the mere implementation of digital technologies in the organization is not enough. Rather, digital transformation encompasses far-reaching changes that extend to organizational management, the firm's operational processes, and its own organizational culture. This large-scale change process affects all members of the organization and their way of relating to the customer, leading to improvements in the firm's performance and the creation of competitive advantages.

6.2 Essentials of Digital Transformation

Digital transformation is a challenging undertaking that takes time to complete and is often more complex than expected. It makes products and services obsolete faster, while encouraging the emergence of new ones that are accessible through a wide variety of channels, such as websites and social applications (Brown & Brown, 2019). Digital transformation redesigns the role of people in business organizations (Bala, 2018), and prompts business leaders to rethink strategy, leadership, and organizational culture (Mahraz et al., 2019).

These substantive transformational changes differ from the typical organizational changes of earlier stages in several respects. For example, the technologies involved today, such as Big Data, analytics, artificial intelligence, etc., differ substantially from previous technologies. In addition, the complexity of the environment in which tourism firms operate is much greater than it was before. Digital technologies provide more information, computing, and connectivity capacity than ever, thus enabling new patterns of collaboration between distributed networks of actors that are more diverse. New dependencies also emerge between actors whose interests may not be fully aligned (Vial, 2019). This means that digital transformation is not limited to a specific type of business or industry, or to the implementation of particular ICTs, but also encompasses much larger and more complex ecosystems and has major organizational implications (Högberg, 2021).

Managing digital transformation requires orchestrating work within the firm to achieve better products/services and build business models that harness the benefits offered by digital technologies. It is thus important that tourism firms consider their own digital transformation very seriously, since those firms that fail to adapt to the digital environment will fall victim to "digital darwinism". In other words, firms that resist digital transformation will surely end up disappearing and only those that adapt and respond to technological trends will survive and remain in the competitive arena (Schwartz, 2002).

Tourism firms face increasing pressures from customers, competitors, and employees to initiate or accelerate their pace of digital transformation, and constantly redefine their value proposition. Connectivity, automation, new data, and new digital customer interfaces are challenging traditional value chains as we knew them. Therefore, tourism firms need to carefully assess their product and service portfolios, and the skill set they have. This means starting to make progress on their digital maturity (if they haven't already started) to seize opportunities, develop better offerings, and bring them to market quickly. How and in what direction the firm can take effective steps remains unclear. In the end, everybody seems to agree that digital is better, deeper, and faster (Bala, 2018), but to benefit from digital transformation tourism firms must be able to put together a plan that is coherent and consistent.

The accelerated pace in the development of digital technologies and their wide-ranging effects on tourism firms make digital transformation processes so uncertain that it is virtually impossible to predict which businesses will succeed in their transformation. Even so, it is becoming increasingly clear that competitiveness has long since ceased to depend solely on physical products, and now depends more on the integration of organizations in ecosystems of smart technologies and services. The new possibilities offered by digital technologies are ultimately pushing tourism firms to evaluate the potential to innovate their business models and constantly reinvent themselves. Digital transformation is now an inevitable and irreversible process, at full throttle in the economy and in society as you read these lines (Mahraz et al., 2019). But for digital transformation to gain momentum in tourism ecosystems, the heroic efforts of a few firms, however important they may be, are not enough. It is thus essential to organize concerted actions between all actors in the tourism ecosystem and policy makers to achieve a harmonized regulatory framework, powerful communications, fluid information infrastructures, and an environment that encourages investment to give a new impetus to the competitiveness of tourism firms through digital transformation and smartization.

6.3 Digitization, Digitalization, and Digital Transformation

Although academics and practitioners generally take it for granted that digital transformation is affecting and challenging business owners and managers in all industries and contexts, the truth is that there is an overwhelming lack of clarity when it comes to understanding it. One of the main challenges is that there is no reconciled definition of the core elements that make up digital transformation. In fact, many studies see digital transformation as entirely different things: for some authors it means only a slight change enabled by technology (e.g., when a new enterprise resource planning (ERP) system is implemented in a department of a firm); for others it means the impact that technology has in the strategy or business model of an organization; while for others it implies a more profound and evolutionary approach that develops over time (Morakanyane et al., 2017). It is also quite worrying that there are several key terms, such as digitization and digitalization, whose differences are often unclear, and which add even more confusion and ambiguity when owners and managers use them (Bala, 2018; Henriette et al., 2015; Mahraz et al., 2019). Therefore, we are going to pause for a moment along the way to clarify some terminological differences and learn how to distinguish between them.

Digitization

The term digitization means the conversion of information from an analog format to a digital one. It is a pure technical process that seeks to convert something that is analog into a binary language (composed of "0s" and "1s"), so that it can be processed by digital information systems (e.g., a computer, a smart phone, etc.) (Castaldi et al., 2018; Mahraz et al., 2019; Pesonen, 2020; Trenkle, 2019; Udovita, 2020). An example of digitization is converting a paper photo to a digital photo to store it on a flash drive, using a scanning process.

Digitization involves significant improvements in the use that organizations make of ICT and in guiding technology implementation strategies. This explains why technology adoption models and the criteria used to select ICT tools in a business context, as well as how data can be integrated into information systems, are common themes pertaining to digitization. In the end, digitization is the first and foremost step in advancing digitalization processes and, subsequently, in digital and smart transformation.

Digitalization

Digitalization involves the use of data and digital technologies to optimize or automate management processes in the organization, including those related to the collection and analysis of data itself. Digitalization encompasses change processes that are much deeper than those of digitization, since it involves the use of technologies in a much broader organizational context (Castaldi et al., 2018; OECD, 2020; Trenkle, 2019). The terms computerization and informatization are concepts closely related to the notion of digitalization and are often used as synonyms.

Digital transformation

The term digital transformation is used by academics and professionals to refer to changes of all kinds that occur in human activity as a result of digitalization. For business firms, it embodies a socio-technological process whereby the application of digitalization is raised to a systemic level within the organization.

Digital transformation emphasizes aspects related to cultural change, the way in which the organization is managed, the roles of people, as well as the changes that occur in the operating environment as a result of the adoption of digital technologies. In other words, digital transformation implies a strategic change process focused on advancing the systematic digitalization of the firm while keeping the different business strategies in sync with the IT strategy. Therefore, the concept of digital transformation is different from that of digitization and their meanings are not interchangeable (Riasanow et al., 2019; Udovita, 2020).

Meanwhile, both digitalization and digital transformation make use of digitization and digital innovation with the intention of improving the firm's products and services (Yoo et al., 2012). This means that, for many authors (including consultants and ICT industry leaders), these terms mean the same thing and are used as synonyms. However, both differ in the different nature of the objectives pursued by the organization: systemic objectives in the case of digital transformation and objectives limited to specific processes in digitalization, and in the depth of the changes that both entail (much deeper in the case of digital transformation). Figure 6.1 describes this conceptual journey from digitization to smart transformation.

6.4 Components of Digital Transformation

Digital transformation is essentially made effective through the following four drivers of change supported by enablers and value propositions, as shown in Figure 6.2 (Bala, 2018; Berger, 2015; Castaldi et al., 2018):

- **Digital data:** The capture, processing, and analysis of digital data to make decisions and improve the forecasting capabilities of the firm.
- **Automation:** The combination of conventional technologies with artificial intelligence capabilities to enable systems to work autonomously and to be self-organized. Through automation, firms increase the speed of processes, reduce error rates, and reduce operating costs.
- **Connectivity:** The integration of the firm value network through mobile or fixed broadband telecommunications networks to synchronize the components of the supply chain and shorten both production times and innovation cycles.
- **Access to the digital consumer:** The use of the internet to provide direct access to digital consumers, substantially increasing the transparency of markets and choice options.

Within the framework of digital transformation, digital technologies are key drivers that provide the firm with greater interactivity and the possibility for stakeholders to participate in the creation of new value for the business, in the design of products/services, in the generation and exchange of knowledge, and even in the contribution of new experiences. This requires firms to thoroughly review their business management practices, increasingly determined by the ability to act in the new value networks that emerge from their integration into ecosystems capable of self-learning and self-organization.

6.5 Benefits of Digital Transformation

Digital transformation is happening all around us all the time. In fact, there is almost certainly not a single aspect of our lives that has not been affected

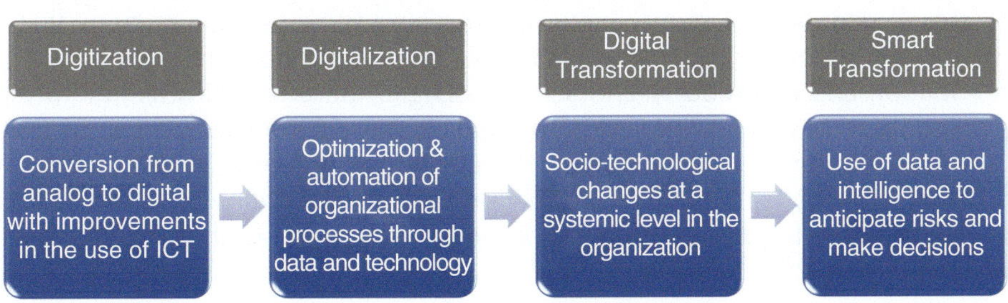

Fig. 6.1. The journey from digitization to smart transformation. Source: own elaboration.

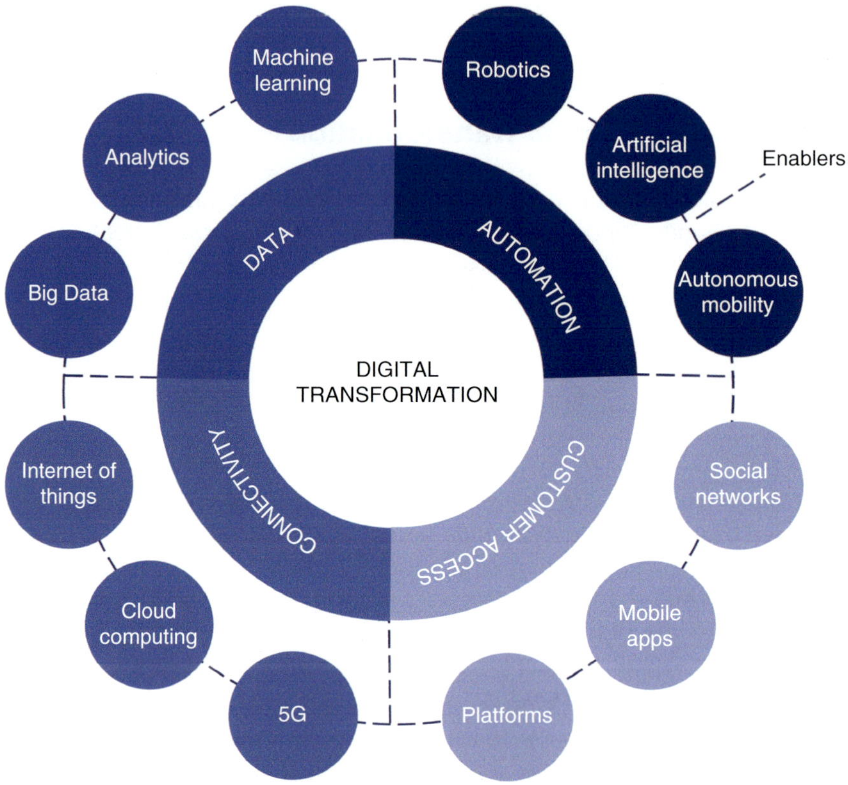

Fig. 6.2. Framework of digital transformation. Source: Own elaboration based on Roland Berger (2015).

by it. The effects of digital transformation are considerable and have a lasting impact on firms and economic systems, but also on the lives of individuals and society in general. If there is still someone who needs to be convinced of the importance of digital transformation for their organization, here are some of the most widely accepted benefits (Reddy & Reinartz, 2017).

Interaction costs

In all economic systems, including business firms, the interactions that take place in the markets generate costs that are the result of the exchange of information and of products and services. Digital transformation is key to reducing these costs and, therefore, to creating value for organizations, so that the more exchanges that take place, the greater the potential benefits for the firm. Furthermore, as digital transformation makes information more accessible to all, asymmetry between market participants is reduced, thus markets become more fluid and competitive.

Information and value exchange

Digital transformation pushes firms to move from unilateral and bilateral exchange models to one based on networks. In other words, consumers and businesses are evolving from linear chains of exchange (and value creation) to becoming part of networks in which multiple exchanges take place at the same time. In addition, these networks emerge, grow, and vanish constantly. As a result, the number of connections that arise between the actors of the tourism ecosystem is growing exponentially, as well as the speed at which value is created and exchanged between them.

Massive amounts of data

The ever-increasing computing power and its highly distributed nature around the new value networks lead to the so-called Big Data. Cisco Systems' predictions suggest that global IP traffic will reach 396 exabytes per month by 2022, up from 122 exabytes per month in 2017. That is 4.8 zettabytes of traffic per year by 2022. In this data-flooded landscape, data can be accessed from everywhere and in real time, leading to a huge number of data handling operations. Therefore, a key challenge for stakeholders in the tourism ecosystem is how to analyze and interpret the patterns that are hidden in huge volumes of data and how to obtain valuable insights to make better informed decisions with them. It should not be a surprise that, in the era of smart transformation, the competitiveness of tourism firms no longer depends on whether they are digital or not, but on whether they are able to learn from their own data (Castaldi et al., 2018).

Internationalization

With the advent of digital technologies, the role of distance is undermined, as knowledge and intelligence can be generated and distributed on a global scale. This leads to new ways of producing and using knowledge assets, which in turn transforms the role and importance of geographical proximity (Castellani et al., 2017). Local assets, however, do not lose importance immediately, but their role changes since they now have the possibility of integrating into global channels that connect geographically dispersed sources of knowledge. By virtue of digital transformation, an empowerment effect can also be triggered that spurs SMEs to access digital networks and interact with distant counterparties, thus facilitating internationalized business models that extend their presence globally without involving significant amounts of direct investment. The ability of modern firms to access international markets with a smaller asset footprint has led to micro-multinationals and born global firms, which are those that with minimal cross-border investment rapidly achieve global reach (Gestrin & Staudt, 2018).

Irreversibility

Digital transformation is permeating and fundamentally transforming our daily lives, and it is doing so in an irreversible way. The main driver of these changes is the accelerated reduction in the cost of computing. According to Yale Professor William Nordhaus, while the average annual decrease in costs between 1945 and 1980 was 37%, in the 1980s and 1990s these costs decreased at an average annual rate of 64%. In the coming decades, as computing power continues to grow exponentially, the forces driving digital transformation will continue to be highly relevant.

Digital transformation is expected to drive the creation of more tangible and intangible value for tourism firms in the future. However, it is worth noting that all these benefits also come with often unforeseen costs and risks. Therefore, it is important that the owners and managers of tourism firms have a good understanding of the opportunities that are presented to them, but also of the potential challenges that surround the creation of value in digital environments. We are all going to be affected to one degree or another and no stone will be left unturned.

Accor and Digital Transformation

Accor looks to the future of digitalization by focusing on the customer experience and providing value propositions for all stakeholders in the hospitality value network. Increasingly, the hospitality industry is transforming into a digitally enabled end-to-end provider that works with different tools and technological applications that are increasingly powerful and integrated (e.g., mobile, cloud, artificial intelligence, blockchain, Internet of Things, etc). Hotels have the potential to deeply engage with guests at every possible touch point in the customer journey and provide them with new opportunities to create personalized experiences. However, just focusing on the customer is not enough. It is also crucial to reach agreements with other players in the ecosystem that allow hospitality firms to remain competitive and create new revenue streams (e.g., airlines, cruise lines, travel agents, etc.). Another important challenge is ensuring that the organization's workforce has the right digital skills to manage new customer-centric strategies. In the end, hospitality firms must understand how to apply digital technologies to their daily operations to improve guest satisfaction, employee awareness, and higher collaboration with partners. That is why the three main pillars of Accor's digital experience are: customers, employees, and partners.

Source: Own elaboration based on Damnjanović et al. (2020).

6.6 Challenges of Digital Transformation

Too often, when someone from the ICT industry or academia delivers a speech or masterclass on digital transformation, they do so by showing the most optimistic side of the story, that is, ignoring the challenges and pitfalls that are common. However, figures speak for themselves. In 2019 alone, US$1.3 trillion was spent on digital transformation efforts with unsatisfactory rates of return (Tabrizi et al., 2019). What's more, according to a recent Bain & Company report, only 5% of digital transformation initiatives met or exceeded expectations, 20% produced less than half of expected benefits, and 75% underperformed (Baculard et al., 2017). Despite these disappointing success rates, most firms still believe that the potential payoff of digital transformation is significant and that it can help stay competitive in the marketplace (Gaurav & Kongar, 2021).

Some of the main challenges involved in the execution of digital transformation are examined below. Although this is not an exhaustive account of all the challenges that may arise, it does reflect how the effects of digital transformation encompass both the organization itself, its resources, and the relationship with the customer (Bala, 2018).

Customer experience

The transformation of the customer experience is evidenced through the different ways in which consumers use the products and services of tourism firms, as well as by changes in consumer behavior and loyalty. It is also reflected in the interactive communication that occurs with customers during the sales process and the different digital touchpoints made available between the firm and customers.

Customers in the digital environment are extremely demanding and expect firms to listen, understand, and be more flexible regarding changes in their needs. They are already used to using digital technologies in their daily lives and expect to continue to use them in all other facets of life such as when shopping, working, or traveling. This is even more apparent with the new generations known as digital natives, who were born surrounded by technologies and who now demand that their service experiences have a digital dimension. It should therefore come as no surprise that firms that want to build satisfying relationships with customers are forced to rethink the consistency of their digital communication channels and integrate new customer tracking systems, which ultimately requires investing in new infrastructure, processes and skills (Mahraz et al., 2019).

Processes and business models

The advance of digital transformation challenges firms to transform their main operational and business processes, encouraging them to be more agile and efficient through digital technologies. This includes, for example, the automation of product/service delivery, distribution, and marketing processes (Vial, 2019), the creation of intelligent products and services that incorporate artificial intelligence and allow algorithm-based decision-making, or the ability of workers to perform their functions in different functional areas of the same organization, all of which help to generate cost savings and increase the performance of firms (Mahraz et al., 2019). Furthermore, new business models may emerge by adding digital content to existing products and services, or by creating entirely new digital solutions (Schwertner, 2017).

Legacy systems

One of the main challenges posed by digital transformation in business firms is to overcome the limitations of their old legacy systems, since they lack the capabilities and functionalities necessary to respond to the demands of the new digital environments. Overcoming this situation is no simple task. Quite often, a portion of the workforce will be very comfortable working with the old legacy systems despite their limitations. In these cases, it is common to find resistance from workers who fear failure, or the negative consequences of the changes derived from the implementation of new digital systems.

Change management

The lack of an effective change management method is a considerable obstacle to successfully implementing digital transformation initiatives in the firm, which is greater the more complex the organizational environment. Moreover, getting all areas of the organization to agree on a common course of action can certainly be a challenge. This challenge is frequently aggravated not only by the technical difficulty involved in migrating to a new digital environment, but also by the incompatibility that usually exists between the technologies available to each department of the firm. To deal with this situation, it is necessary for firms to develop visionary and committed leadership, capable of mobilizing the critical resources of the organization around plausible and measurable objectives, leadership that is able to partner with IT teams to drive innovation. Unfortunately, it is still very common today to see how leadership models are linked to organizational models that are too hierarchical, which hinders innovation and prevents digital transformation from developing at the speed it should.

Commitment to transform

There are many occasions when digital transformation initiatives do not have the appropriate level of commitment and leadership from senior management to be carried out successfully. This is one of the key reasons why a large number of digital transformation initiatives do not achieve the expected results. This lack of commitment from senior management can become even more apparent when firms try to move forward from developing digital initiatives to implementing them. Business owners and managers often spend their time talking about the need to innovate in increasingly competitive markets. However, this discourse often ends up turning into a deeply ingrained attitude of resistance to organizational learning and innovation. In other words, very often there is a disconnect between what business leaders say they value and the values they actually display through their actions and behaviors. It is still all too easy to find leaders for whom learning is valuable, but their organizations punish failure for daring to learn and innovate, which means they really don't want to risk change.

Skilled labor

For digital transformation to take root in the organization, it is essential to have people who are well trained and prepared to work in a digital environment. This need involves both training in digital business and digital technologies and should be aimed at both junior profiles and senior profiles with greater responsibility in the management of the organization. Furthermore, training programs must be well supported by appropriate investment and a business culture consistent with the challenge (Sestino et al., 2020). At the same time, leaders must find a way to trust those employees who have specific digital transformation skills, so that creativity and innovation can be fueled within the organization. At the end of the day, smart organizations must have a workforce that knows how to adapt to change through an organizational culture that is itself change-oriented and capable of responding to challenges. Organizational culture is very often a determining factor in how quickly firms embrace digital transformation.

Infrastructure

Modern infrastructures of hardware, software, networks, connectivity, facilities, etc. hold significant value potential to successfully implement a digital transformation strategy. Given the wide variety of devices and technologies that customers use (e.g., mobile phones, PCs, websites, social networks, wearables, etc.) and the intelligent systems that tourist destinations and competitors are already implementing, new challenges arise for the management of infrastructures in firms. These challenges affect both the availability of digital services and the need for the infrastructure to be flexible enough to facilitate the continual addition of new requirements. To tackle this challenge, tourism firms should react in an agile way to the changing requirements that are presented to them.

Financing

Digital transformation requires significant funding to meet the investment needs in infrastructure, digital tools, innovation, training, and new systems development. Therefore, it is highly recommended that tourism firms have a budget at their disposal with sufficient resources dedicated to the digital transformation program and that they avoid diverting to other types of operational policies within the organization.

Discussion Questions

- Is digital transformation the solution to the emerging challenges of tourism firms in the digital age, or is it just another marketing buzzword (Morakanyane et al., 2017)?

- What positive and negative effects characteristic of digital transformation should draw more attention from the owners and managers of tourism firms?
- What do you think is the optimal degree of digitalization that a tourism firm should achieve? What factors does this decision depend on?
- When implementing a digital transformation strategy, what effect can a B2B or B2C business model have on the results?
- What is the recommended duration of a digital transformation process in the tourism firm?
- When and where should a tourism firm start executing a digital transformation strategy? Which areas are the first in which changes should be made and to what extent?
- What aspects differentiate digital transformation from smart transformation in a tourism firm?

7 Tourism and Digitalization

Tourism was one of the most important economic sectors in the world in 2019, contributing 10.14% to global GDP and 10.6% of workers worldwide (Juc & Misrahi, 2021). With the advent of the COVID-19 pandemic, the contribution of tourism to GDP has practically been reduced by half in 2020. However, the estimates of the main international economic organizations (UNWTO, 2021) point to a strong recovery for 2022 as the epidemiological situation improves. Be that as it may, tourism has proved to be a major growth engine for the economies of developed and developing countries around the world. Furthermore, as more and more parts of the world's population travel, whether for pleasure or work, tourism has become a great source of opportunities that contribute to reducing poverty and narrowing the income gap between citizens.

Tourism is undergoing a radical transformation as a result of advances in information technology and digitalization. The internet has been a before and after in this process, allowing customers to search for and book travel tickets, book their accommodation online, or share their experience with friends and family through social media. It is not surprising that tourism is one of the sectors that is leading the development of digital technologies and platforms in the world, driven by strong global demand, the growing middle class in emerging markets, and the increasing relevance of experiences in modern societies.

Today no one doubts that technological advances are having a profound impact on tourist services. These developments span a wide variety of areas, from business management technologies (e.g., mobile technologies, robotics, blockchain, cloud computing), technologies for product and service delivery (e.g., virtual and augmented reality, Internet of Things), to technologies that are capable of generating deeper knowledge about customers and markets (e.g., Big Data, analytics, artificial intelligence) (OECD, 2017b). This suggests that digitalization has become an unstoppable phenomenon that is vital for the firm to meet consumer

 Managing the Smart Revolution in Tourism Firms: Innovation and Value Creation in the Era of Data (F.J. Navarro Meneses)
DOI: 10.1079/9781800620261.0007

expectations in the future (Tajeddini et al., 2019; Weinelt & Moavenzadeh, 2017). Ordinary tourism firms have had no choice but to start incorporating more digital elements into their business models, at a time when native digital giants (e.g., Airbnb, Booking.com, Tripadvisor) are consolidating and shaking up the traditional tourism value chain (Ozdemir et al., 2020; Weinelt & Moavenzadeh, 2017)

This convergence of industry, customer, and technology trends is redefining business and operating models in the whole tourism ecosystem, with tourism SMEs being particularly affected as digitalization moves tourism activities from linear value chains to highly complex value networks (Kelly, 2015). Understanding how value is created and captured among the multiple stakeholders operating in emerging tourism ecosystems is now critical to unlocking the full potential of digitalization. For example, the benefits obtained from the expansion of markets, the productivity gains in firms, or the substantial changes in the way in which customers and firms interact and communicate (OECD, 2020; Tajeddini et al., 2019) are greatly increasing the complexity of the tourism ecosystem, posing significant challenges for tourism SMEs.

Whilst much emphasis has been placed on the positive effects of digital transformation, it is also important to be aware of the negative effects that need to be addressed with speed and vigour. Tourism firms are not usually at the forefront of digitalization efforts and, when they are, they often do not have the will or the necessary capacities to embark on the deep transformation process that the tourism organization needs. That is why it is very important that those organizations that are falling behind change their mentality and the way they work to start reaping the benefits promised by digital and smart transformation.

7.1 What Makes Tourism Unique?

Although many of the challenges and opportunities that digital transformation represents for tourism firms are similar to those of other industries, tourism has some particularities that make it unique. The tourism ecosystem is highly fragmented and involves a wide variety of actors that are heterogeneous and cover a broad range of services. It is a sector characterized by the overwhelming presence of SMEs (around 85% of the firms with a relevant role in the provision of tourist services according to the OECD) that coexist with a small group of large companies and especially with micro-enterprises.

Tourism is, by definition, information-intensive, which means that many tourism services are eligible for digitalization and smartization. However, since the average size of tourism firms is small, there is a significant challenge when it comes to integrating technologies and investing in customer relations through digital means. Another peculiarity is that tourism firms operate in a global marketplace, although they deliver locally. Furthermore, they are part of unique tourist destinations, with which they are intimately connected in terms of service delivery and consumer decision-making. The attractiveness and accessibility of destinations is thus a factor that greatly influences the ability of tourism firms to generate business and stay competitive. There are tourism firms that can overcome the role reserved for destinations and attract visitors due to their special service attributes, but the general rule is that the choice of destination is an inherent part of tourist decision-making and influences business. On the other hand, digital tourism firms, i.e., those firms whose business model is totally digital, such as online travel agencies (OTAs) and accommodation platforms, are playing a key role and helping to transform tourism by connecting tourism products and services with customers from anywhere in the world in real time. The good news for tourism SMEs is that they can significantly increase their visibility in the market at relatively low costs. The downside is that many tourism SMEs now depend, at least in part, on large intermediaries who dominate online consumer traffic and impose their business terms often unilaterally.

These are not the only challenges facing tourism firms. In addition to the strict conditions imposed by the intermediaries on the platforms, the customers also demand that firms improve the quality of their services (i.e., the best possible service at the lowest possible price) and deliver experiences that always meet their needs. Traditional tourism firms are risk averse and far more concerned with their business survival than with the appetite for innovation and technology. This will have to change drastically if they want to be alive in a few years.

7.2 The State of Digitalization in Tourism

Digitalization represents a major opportunity for tourism firms with the potential to unlock approximately US$1 trillion of value for the industry and society over the next decade (Weinelt & Moavenzadeh, 2017). The World Economic Forum (2017) has estimated that by 2025, digitalization will create up to US$305 billion of additional value only for the tourism sector thanks to higher profitability. It is worth noting that a significant part of this value will be transferred from traditional tourism firms to digital businesses. Digital transformation is also expected to generate US$700 billion in benefits by reducing the environmental footprint, improving safety, and saving time and costs for consumers. Notwithstanding these somewhat optimistic estimates, the reality shows some imbalances between aspirations and achievements. Recent research by the European Commission found significant imbalances in the adoption of digital technologies in tourism across Europe. The geographical location of the firms and their size are factors that could explain these differences; for example, firms located in the Nordic countries show a higher level of acceptance of digital technologies than those located in Eastern and Southern European countries.

SMEs are also lagging behind large companies (e.g., hotel chains, airlines, big tour operators), especially in those technologies that are more sophisticated or that require greater integration at an organizational or technological level, such as Big Data, analytics, artificial intelligence, or cloud computing (Dredge et al., 2019). According to the European Commission, tourism SMEs are characterized by a low level of digitalization, with most of the technologies they use focused on enabling e-commerce and social media. Furthermore, SMEs fall short in mid- and high-level digitalization technologies, and there is little evidence that major steps are being taken towards the implementation of smart technologies and their interoperability. The European Commission also highlights some political-institutional and policy-making factors that have an impact on the digitalization of tourism firms. For example, the higher the level of education, the greater the access to digital technologies, and the higher the public sector spending, the higher the level of digitalization is reached, as is the case of the Nordic countries, compared to the countries of Eastern Europe (Dredge et al., 2019). Despite this unequal assimilation of digital technologies by tourism SMEs, digital transformation is having and will continue to have a crucial role in transforming tourism, driven by the cumulative nature of technological change, the convergence of technologies, and the progressive price reduction (UNCTAD, 2018).

The shift to a digital economy offers great opportunities for tourism firms of all sizes wherever they are. These opportunities translate into access to new markets and knowledge networks, the provision of new tourism services to consumers on a global scale at a relatively low cost, the improvement of competitiveness, and higher performance and improvement of business productivity (OECD, 2017a). However, capturing these benefits will depend on achieving a balanced combination of investment in digital infrastructure, development of human capital skills, and innovation in business models (OECD, 2019). But this is not a rapid or easy task. This balance is key to starting a transition towards the digitalization of tourism SMEs and then continuing to smartization. Unfortunately, in some countries there are still significant gaps between the investment of tourism firms in digital technologies and those made by firms in other sectors.

7.3 Drivers of Digitalization in Tourism

The use of ICTs in the tourism industry has a long tradition that becomes more evident from the year 2000 when many of the repetitive tasks of the tourism business begin to be automated (e.g., buying tickets, booking accommodation, etc.), and electronic commerce emerges strongly introducing new technological capabilities that allow completely new ways of distributing and marketing tourism products and services. Similarly, as consumers become more accustomed to searching, planning, and booking travel online, digital technologies emerge powerfully, driving new advanced capabilities for tourism firms.

But what really are the key drivers that motivate tourism firms to make the transition to digitalization? This section summarizes some of the main drivers of digitalization that business owners and managers should be aware of (Fig. 7.1).

Consumer demand

The appetite for travel is growing strongly due to the demographic evolution of the population on a global scale. The increasing incorporation of millennials

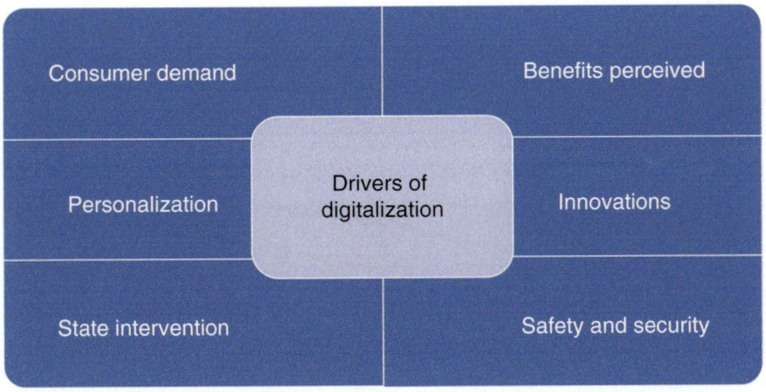

Fig. 7.1. Drivers of digitalization in tourism firms. Source: own elaboration.

into travel, the expansion of the size of the middle class in the regions with the highest demographic potential, and the general increase in disposable income are making this growth in tourism demand quite robust everywhere. The International Air Transport Association (IATA) forecast sees the total number of travelers reaching 4 billion in 2024, surpassing pre-COVID-19 levels (103% of the 2019 total) (IATA, 2022). A little more time will be needed for international passenger traffic to reach pre-COVID-19 levels, as it will not do so until a year later, in 2025, according to the IATA forecast (Fig. 7.2).

Growing demand certainly brings great opportunities, but it also challenges stakeholders in the tourism ecosystem to quickly adapt their strategies to capture growth. For example, tourism firms are increasingly seeking stronger and more stable interactions with their customers, while trying to adapt operations according to the better knowledge they have about customer demands and preferences. Connected wearable devices and artificial intelligence are already providing opportunities to make operations efficient and enable collaboration by sharing knowledge assets between firms. Ultimately, digital technologies are also having an impact on the workforce, allowing employees to have real-time information to speed up decision-making and focus on their strengths (Weinelt & Moavenzadeh, 2017).

Benefits perceived by firms

Dredge et al. (2019) provide the following reasons why tourism SMEs perceive significant benefits from digitalization based on their survey of thousands of European SMEs.

1. Growth: Two of the top benefits perceived by SMEs include "improving online presence" and "improving growth". Both are indicators of how tourism firms are confident that digitalization can be instrumental in achieving their growth ambitions.

2. Future opportunities: Tourism SMEs rely on the opportunities that digitalization offers for their businesses. Interestingly, this situation contrasts with the observations made by other studies according to which SMEs lack awareness and knowledge about the opportunities of digitalization.

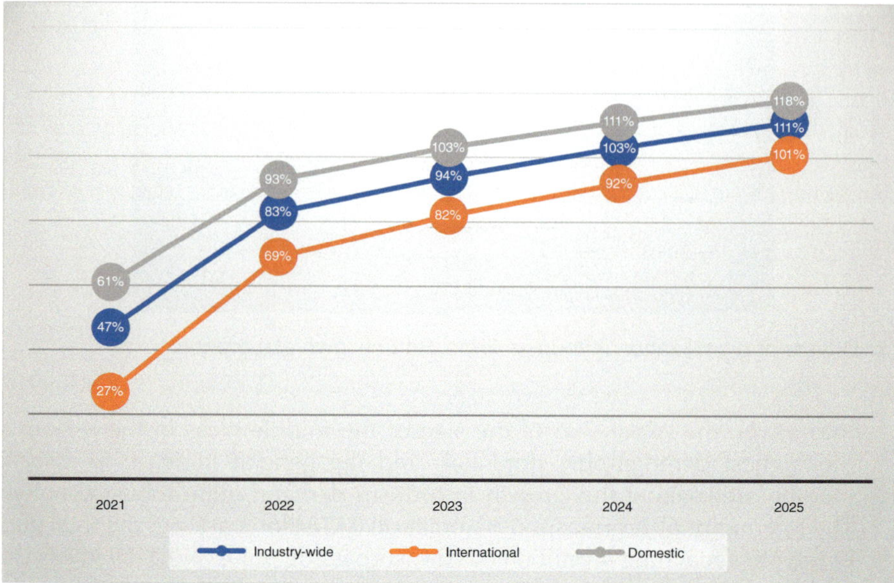

Fig. 7.2. Passenger numbers, share of 2019. Source: own elaboration with data from IATA/Tourism Economics Air Passenger Forecast, March 2022.

3. **Cost savings:** For many SMEs, digitalization is nothing more than a way to reduce costs and improve access to markets when dealing with fluctuations in demand, such as those caused by seasonality, economic crisis or pandemic.
4. **Interactivity with the ecosystem:** Tourism SMEs also perceive digitalization as a source of opportunities to strengthen connections between the different actors that make up the tourism ecosystem. Through digitalization, tourism SMEs can improve their links with other SMEs, with destinations, with consumers, and with other subsectors (e.g., transport, accommodation, attractions, etc.). This logic also suggests the perception on the part of SMEs that digitalization can help make supply chains more flexible and create new value through participation in expanded digital business ecosystems.

Personalization of experiences

With the rise of the digital consumer and the emerging new trends in tourism, there is a growing demand for tailor-made products and services beyond mass tourism, as consumers expect increasingly personalized solutions that meet their individual habits and preferences (Zsarnoczky, 2018). This phenomenon is more noticeable among younger tourists, who are individualists and seek experiences that respond to their own ideas and preferences.

The emergence of individualism creates numerous opportunities, which may explain why a good number of new start-ups in the tourism sector are led by young people. Individualism also leads to customer fragmentation, which in turn leads to market fragmentation. In such a disintegrated market, consumers tend to behave selfishly, paving the way for personalized services and tailored solutions. As a result, tourism firms are driven to optimize the use of the resources at their disposal to provide flexible experiences to customers, which means they must continually collect and exchange data and extract useful insights from it.

Today, most tourism firms have access to extensive information about their customers and are able to monitor and track consumer behavior and changes in preferences in a cost-effective manner. Moreover, basic capabilities for customizable experiences are available through various tools, such as customer relationship management (CRM) customer databases. For example, cloud-based CRM systems can configure effective value propositions by analyzing past sales records and demographics (Zsarnoczky, 2018). Leading firms in the tourism sector are able to analyze huge datasets and use scalable Big Data analytics methods to search for patterns of behavior. This process of extracting value from data is one of the critical foundations of smartization.

Disruptive innovations

Innovations in tourism manifest in many different ways. Some of the most powerful are the result of disruptive changes in customer behavior and expectations, shifts in the competitive landscape, and the availability of data (Högberg, 2021). Today's tourists are highly knowledgeable thanks to digital technologies. They can easily move from one digital channel to another (e.g., from hotel websites to OTAs) and between different devices (e.g., from mobile phones to tablets, and to PCs). With the use of digital technologies and mobile devices, hotel guests can manage their reservation and search for information on their smart phones before their arrival at a destination or at their convenience during their stay. Social media technologies, such as mobile phone applications and real-time instant communication, are also part of this new context that enables customers to become co-creators of tourist experiences (Tajeddini et al., 2019). Consequently, the expectations of the guests are greatly affected and become much broader and more demanding.

This new panorama of ever-greater and always unsatisfied customer expectations is proving to be an ideal breeding ground for the emergence of new non-traditional players (e.g., Airbnb, Tripadvisor, OTAs such as Expedia, Booking.com, etc.) that compete to establish new alliances with customers through highly differentiated digital ecosystems. Tourism firms become dependent on the one hand, on these platforms that provide them with the visibility to attract customers and the trust to generate new leads (i.e., through their pricing policy, reviews made by other guests); on the other hand, firms compete with the platforms themselves in attracting customers and avoiding the intermediation commissions that they charge for each reservation made through the platform. The engagement of tourism firms in digital platforms has drastically altered the power relationship between firms and their customers, as firms now have much more information available about their customers than ever before, allowing them to act in a more informed and accurate way.

Furthermore, some of the predictions made about the future of product development strategies in various industries indicate that within a short period of time almost all of our everyday objects and equipment will be accessible via the internet. The devices of the future, unlike the devices of today, will communicate bidirectionally, providing the user experience with new capabilities linked to personalized differentiation and autonomy of decision. New emerging demands for continuous real-time connectivity will mean continuous data collection while using smart devices. Eventually, all relevant information will combine together into a centralized real-time data and service management system known as the Internet of Things (IoT), which will provide robust and secure data handling for all connected devices (Zsarnoczky, 2018). However, a major problem is that firms need Big Data and artificial intelligence-based computing solutions to store, manage, and organize all this enormous amount of information, and these new innovations affect the speed, efficiency, and current computing capabilities of most firms to deliver consistent tourism experiences.

State intervention

States intervene in the tourism ecosystem through policies of different scope at the national, regional, and local level with the aim of accelerating the adoption of digital technologies. These policies are conditioned by a wide variety of factors, including the business structure of the ecosystem, the political and institutional context, the legal framework, the geospatial conditions, the social and demographic characteristics, the climate of confidence for investment, the competitive environment, etc. (Dredge et al., 2019). These influences manifest themselves with different degrees of intensity in the different states and in the components of the tourism ecosystem, leading to digitalization processes at different speeds. The way states design and implement their digitalization policies affects the likelihood that tourism firms will embrace digitalization, so it is important that states consider the aforementioned factors to improve their success rate in digital initiatives.

Safety and security

As digital acquires a more relevant role in the management of tourism firms, a greater collaborative effort to promote cybersecurity and protect the privacy of tourist data becomes imperative (Weinelt & Moavenzadeh, 2017). Security concerns are relevant both for physical environments (e.g., airport controls, city mobility systems, hotel facilities)) and for digital environments (e.g., data privacy, veracity of electronic transactions). Security breaches and accidents, whether in the physical or digital realm, can lead to severe financial and reputational damage to tourism firms. Similar risks exist when geopolitical tensions or terrorist activity occur, all of which fuels an unprecedented interest in safety and security. The real challenge for tourism stakeholders today is how to support the growing demand for seamless tourism experiences, while improving safety standards. As tourism firms continue to move down the digital path, security and cybersecurity-focused digital technologies (e.g., biometrics, crowd analytics, artificial intelligence applied to video monitoring, etc.) will continue to evolve to create a safe and secure ubiquitous environment. After all, higher levels of cybersecurity are critical to maintaining customer trust and public safety.

7.4 Opportunities of Digitalization

Digital technologies are bringing substantial transformations to the tourism business, revolutionizing tourism firms, business ecosystems, consumers and destinations. Digitalization is changing the traditional roles of tourism firms and consumers, giving rise to new relationship models between them and changing the way institutions manage tourism public policies. The spread of smart devices has transformed the way tourism firms manage their operations and has opened up new horizons for innovation in sales and marketing throughout the tourism industry. As a result, new business models are emerging, many of them driven by new digital platforms that build alliances and link the objectives of stakeholders in the tourism ecosystem, increasing the quantity and variety of products, services, and experiences.

In the digital era, digital platforms become new sources of value creation, since they allow expanding connections with many external actors and facilitate innovation from abroad. This becomes key for tourism firms to remain relevant in the digital economy. Nonetheless, there are tradeoffs. The emergence of the digital platforms is also aggravating the fragmentation of the tourism market, and the competition between traditional and non-traditional competitors is intensifying, not to mention that tourism SMEs are very often subject to harsh commercial conditions imposed by the platforms given the control of the market they have. All these changes are creating many opportunities, as well as challenges, for tourism firms, the customer and society in general. Figure 7.3 shows a comparison between the market capitalization of OTAs and that of traditional players in the tourism industry, which justifies the true dimension of this phenomenon.

Tourism firms can fundamentally benefit at the operational level through lower unit costs thanks to improvements in the efficiency of customer relationship processes and a higher volume of demand generated by personalized products and services and better experiences provided. Intelligent automation (e.g., artificial intelligence, Internet of Things, robotics) is making it possible for tourism firms to produce products and services at reduced cost and with a smaller environmental footprint. Powered by smart technologies, the next wave of solutions will collect massive amounts of data from all kinds of systems across multiple touchpoints between customers and tourism service providers. By combining data, people, and things, the customer experience will be transformed, leading to an increased likelihood of new bookings.

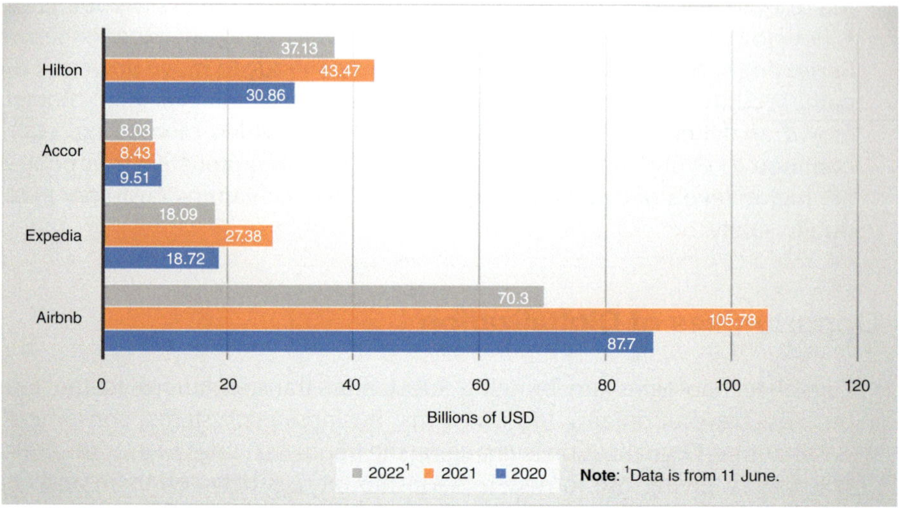

Fig. 7.3. Market capitalization of OTAs and traditional players. Source: own elaboration with data from https://companiesmarketcap.com.

The benefits for customers are also apparent, such as the opportunity to enjoy seamless experiences tailored to individual preferences and demands across the entire tourism ecosystem. However, the extent to which customers realize these benefits will depend on their willingness to provide data to tourism service providers.

On a societal level, digitalization can help generate higher income for those participating in the sharing economy using their own assets (e.g., houses, vehicles, bicycles, etc.), as well as contribute to a lower environmental impact through improvements in the efficiency of production processes and the optimized use of assets involved in tourism experiences (Weinelt & Moavenzadeh, 2017). Now that digital technologies have become more affordable and portable, there are no longer barriers preventing people and things from connecting. Digital trust has become a cornerstone of the tourism industry, since without digital trust firms cannot use and share the data that enhance their operations. Fortunately, tourism consumers are likely to share more information when they feel it is necessary for safety reasons or when they are looking for more personalized experiences.

Very often tourism SMEs perceive the opportunities of digitalization as short-term market-focused advantage. This is far from recognizing the potential that digitalization has for innovation and the development of new products and services, or for building a more efficient ecosystem. Less digitally mature SMEs are often more interested in digital technologies as tools that can help them improve service quality, increase customer retention and streamline processes, while SMEs with a medium level of digitalization are looking more for ways to expand their commercial reach internationally (Dredge et al., 2019).

Moreover, the particularities of digitalization in tourism SMEs still need to be nuanced when we look at the different subsectors of tourism. For example, in the accommodation subsector, SMEs with a lower level of digitalization are more concerned about the international outreach of their operations, perhaps due to the important role that online booking and accommodation sharing platforms play in it. On the other hand, more mature SMEs prefer that digitalization facilitates the entry of new intermediaries into the market, thus increasing the opportunities to create new value through the expansion of the tourism ecosystem. In contrast, in the travel subsector, less mature SMEs are more interested in using digitalization to generate profitability and acquire customer insights. All in all, digitalization provides a vast range of opportunities that are multidirectional and that can help tourism firms create new sources of value, both inside and outside of tourism.

7.5 Challenges to the Digitalization of Tourism Firms

Due to the traditional nature of most tourism firms, digitalization brings with it several challenges that have a significant impact on SMEs. We have just seen in the previous section how the tourism ecosystem is fragmented into several subsectors (e.g., transport, accommodation, restaurants, personal services, etc.), which in turn are subject to different challenges (and opportunities) in terms of digitalization. Each subsector has its own particularities in human resources, digital skills and abilities, different levels of access to resources (financial and non-financial), and different levels of awareness. These differences in turn are magnified or attenuated according to the particularities of the ecosystem and the destination in which firms operate. Weinelt and Moavenzadeh (2017) have identified the following five challenges that digital transformation is expected to have for tourism.

1. **Regulation:** Regulation has a significant impact on the speed of digital transformation and can either be a great enabler of transformation or pose serious obstacles. In some cases, regulatory regimes may deter tourism firms from implementing new technologies because they are unable to capitalize on them. It is also quite common for innovation to advance at a faster rate than regulation and supporting policies, urging institutions to formulate new regulations for new technological developments.
2. **Infrastructure:** As digital business models gain more and more prominence, physical assets are beginning to lose the central role they used to have. This means that tourism firms must pay more attention to the quality and level of service provided by their digital assets and the investments they make in digital infrastructures (without neglecting their physical assets, which will continue to be important). The costs of implementing new digital technologies are a major concern, especially if we compare SMEs with more limited financial resources and large enterprises. Furthermore, SMEs located in urban settings benefit from modern fiber broadband and wireless connections, while in more remote or rural areas the infrastructure is usually underprovided.
3. **Legacy systems:** Non-digital native firms pursuing digital transformation need their legacy systems to continue operating for some time while

new systems are developed and implemented. Transitioning from old to new systems reduces the organization's agility to transform quickly, so finding a way to manage the process of transformation with different speeds in mind is one way to curb the negative impact of this challenge.

4. Jobs and skills: Digital transformation requires a skill set that differs from what workers have today. This represents a great challenge for tourism firms and society in general, as it suggests a hard transition for one in every 11 jobs worldwide by 2025. Digitalization is also transforming the tourism ecosystem by creating new types of jobs. Addressing this challenge requires a concerted effort among tourism leaders, government, educational institutions, and civil society to reskill and upskill human capital, thereby mitigating digitalization potential negative impacts.

5. Demographics: Demographic trends have an important influence on the speed of adoption of digital technologies and on the world tourism market. Although the growing middle classes of Asia and Latin America will lead the growth of travel worldwide in the coming decades, technology adoption will continue to be led by markets in developed countries rather than those in emerging countries.

It is important that owners and managers consider digitalization as a means to improve the customer experience, but never to complicate it. In other words, tourism firms should make digitalization work seamlessly and transparently for the customer, without bothering them or requiring them to go the extra mile (Seigneur, 2018). Otherwise, digitalization can easily become a barrier that the customer rejects.

Digitalization should also seek to increase the income level of local communities in destinations, thus capturing a part of the value that is appropriated by GAFA companies (Google, Apple, Facebook, Amazon) and the big digital platforms (e.g., Booking.com, Tripadvisor, etc.) through their service fees. In this vein, the implementation of smart technologies in destinations and local firms that are based on the Internet of Things could create new data lakes beyond the scope of GAFA companies and lead to new ways of capturing value that can be transferred from the giants to the locals.

Other key challenges facing tourism SMEs include difficulties in accessing financial resources, the high costs of training and high-speed broadband connectivity, the availability of technology and skilled human resources, and the rapid pace of technological change, thus making technology investments obsolete very quickly. In addition to these "tangible" challenges, there are other "intangible" challenges, such as those related to corporate culture that lead to a lack of understanding of digital opportunities and to considering the current levels of digitalization sufficient and questioning the strategies that are most suitable for transformation in the short and long term (Dredge et al., 2019).

Some of the main challenges outlined above reflect a conspicuous lack of understanding of digitalization, its opportunities and challenges by tourism SMEs. This is a major constraint that prevents firms from making informed decisions and acting decisively in the face of changing customer needs and threats from new market entrants, limiting themselves to being reactive in their responses and missing opportunities that arise.

Discussion Questions

- How are tourism firms affected by digital innovations and what patterns of response do they typically use for their adoption?
- What innovative formulas could be implemented in the tourism ecosystem so that tourism firms have access to human and financial resources?
- How can tourism firms encourage customers to share personal data in exchange for tangible benefits (i.e., highly personalized experiences) while ensuring the customer's right to privacy?
- What collaboration mechanisms with other actors in the tourism ecosystem could be established so that tourism firms can increase their chances of success in digital transformation?
- How should the operating models of tourism firms change in an increasingly connected business ecosystem where the boundaries between online and offline are blurring? How will these changes affect the behavior of the tourism consumer?
- To what extent is the regulatory framework in your country or region contributing to accelerating the digital transformation process of tourism firms?
- Through what mechanisms can local tourism firms capture more value that is now appropriated by the digital giants?

8 Towards the Data-Driven Organization

Tourism is an extraordinarily information-rich industry. The shift in focus to the tourism "experience", coupled with the highly competitive environment, pushes tourism firms to focus more on information and data. The collection, storage, organization, processing, and dissemination of information thus become crucial parts of the performance of tourism firms. In this context, the success of tourism firms depends on the creation of new knowledge from the data and information that is available and, more specifically, on the rapid and timely integration of relevant knowledge for decision-making (Kaivo-Oja et al., 2015; Nikolskaya et al., 2021).

The use of data by firms is nothing new. It has always been at the root of successful business management. What has changed today is that data, which used to be expensive and scarce, is now everywhere (users and consumers generate data all the time), has new properties (since it is meta-tagged), it is produced in real time, and it comes from many different sources and in many varied formats (e.g., texts, images, videos, audios). Its size and relevance for business and society has reached such relevance that many call this moment the "data revolution" or the "Big Data Era", and include the set of innovative technologies and techniques focused on collecting, storing, managing, and analyzing large data sets that could not otherwise be dealt with by conventional data management methods (Bashkirova, 2016).

Industries around the world collect all sorts of data and information from customers, suppliers, partners, and even from competitors. There is also a growing need at the organizational level for new capacities to tackle the growing complexity of the tourism ecosystem and the needs arising from digitalization. Today, firms' digital touch points with customers are many and varied, allowing the entire customer life cycle (before, during, and after the purchase) to be recorded for later analysis. Distributed sensors can also record what is happening in a physical environment, transmitting information wirelessly or over

DOI: 10.1079/9781800620261.0008

a fixed broadband network wherever there is the ability to use it. Additionally, renewed interest in artificial intelligence and machine learning tools is accelerating business transformation due to their promise of stronger analytics capabilities and systems that are more powerful and can operate in a more automated way (Berntsson Svensson & Taghavianfar, 2020).

The result is a tremendously complex landscape with an exponential amount of data flowing through hundreds or even thousands of applications, which in turn is leading to an accelerated growth in the capabilities of the firms to do something with it, such as storing, processing, and presenting it (Andersen et al., 2018). For organizations this massive flow of data is proving very difficult to manage and extracting real value from it has become a daunting challenge, even more so for organizations with decades of history in IT management that have never managed data as an asset.

In this regard, the successful transformation to a data-driven organization involves the widespread application of the latest technologies in data storage, processing, and knowledge creation/integration. Furthermore, any information-related process in a modern organization relies on dozens or more applications that typically have their own internal data model, their own interfaces, and even require their own expertise. This very often far exceeds the capabilities of what any individual in an organization can comprehend. For this reason, many organizations have appointed chief data officers (CDOs), in charge of managing the data in the organization, and data scientists, who receive, examine, and analyze data of all kinds in search of value for the business.

Far from diminishing, the sheer volume of data will continue to grow in the coming years, regardless of the size and type of organization. Data calls for more data, thus forming an upward spiral in which data is constantly increasing and less understood (Chessell et al., 2018). Converting all this data, the vast majority of which is unstructured, into structured knowledge will be one of the main challenges for the aspiring smart firm. In addition, firms that want to gain market share should become familiar with knowledge management methods and systems (Bashkirova, 2016) that can help improve firm performance. Indeed, some authors have reported that there is a correlation between business performance and the organization's use of data analytics and knowledge, whether for efficiency, growth, or competitive differentiation. High-performing organizations are using data to make decisions more than twice as often as low-performing organizations, and for a wide variety of decisions, big and small (LaValle et al., 2011).

However, tourism SMEs still do not use data much to create value and there is still a long way to go in this regard (Del Vecchio et al., 2018). Nowadays, tourism firms seem more concerned with implementing digital technologies and infrastructures focused on their daily operations (e.g., reservation systems, websites, information management systems), despite the relevance that data is acquiring to improve the competitiveness of any tourism business. The penetration of these technologies is also very different within and between the different tourism subsectors, as well as their impact on the growth of tourism firms. This makes data a central issue that tourism firms need to start acting upon to continue creating value for stakeholders. Business owners and managers have no more time to lose and should start planning for the transition to a data-driven organization.

8.1 The Value of Data

The transformation of traditional tourism firms into data-driven organizations mainly affects the organizational processes of the organization, the allocation of internal and external resources, and even the long-standing customs and culture of the people. Ultimately, these transformational changes aim to make the business more agile and responsive to customer demands, that is, more customer-oriented in every way.

Smart technologies can provide a major boost to the transformational effort of tourism firms if supported by proven standards embedded in technology platforms and everyday data-driven tools (e.g., tools and platforms that automate data storage, classification, and analysis). Furthermore, new technologies, such as artificial intelligence, real-time customer intelligence, analytics for prediction and personalization of services, to name a few, are already key to creating value and driving major changes throughout the tourism ecosystem.

But what exactly is "data" and why is it so valuable? Data is a collection of observations, measurements, facts, or even raw representations of phenomena that are not organized in any particular way and do not contain any apparent meaning. Data is the basic unit used to register events, activities, operations, or transactions. Data differs from typical organizational resources as it enables replicable services at almost zero marginal cost. It also provides great flexibility of use regardless of the device it runs on (Veit et al., 2014). However, data alone is useless unless it can be interpreted and made meaningful, and to make this possible, organizations must integrate data with analytics technologies. Data analysis thus activates a process of transforming the value of the data by which data pass through three different value states: "raw data", "information", and "knowledge" (Fig. 8.1). This is a sequential process where raw data is first collected, organized, and transformed into a stream of interpreted data called information, and subsequently transformed into usable knowledge through intellectual operations that provide more value to information (Monino, 2021). Through data analytics, organizations have the potential to harness data to uncover previously unseen market insights and business opportunities.

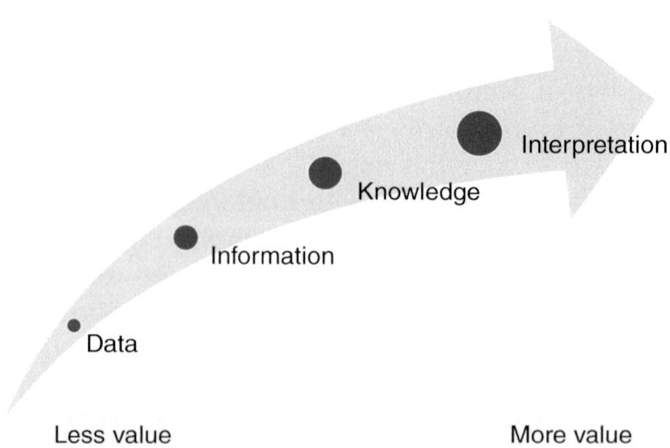

Fig. 8.1. Data value transformation process. Source: own elaboration.

Tourism firms have used data, even big data sets, for a long time, and they have a natural tendency to store all kinds of data. What distinguishes the present moment from any previous one is the ability that firms now have to combine these vast amounts of data with innovative manipulation and knowledge generation techniques (e.g., Big Data, analytics, artificial intelligence, etc.) and consequently boost innovation and business performance (Tsaih & Hsu, 2018). Forecasts predict that the amount of data stored worldwide will increase five-fold, from 33 zettabytes in 2018 to 175 zettabytes in 2025 (Reinsel et al., 2018). The aftermath is a market for Big Data technology that is projected to grow from US$41.3 billion to US$116.1 billion in 2027 (Fortune Business Insights, 2020). Both are clear indicators that show tourism firms how the economic value of data is a key factor that cannot be ignored and that firms are experiencing an unequivocal transition from an economy based on tangible assets to one based on intangibles (Mihet & Philippon, 2018). Tourism firms should therefore not delay in adopting this process of change and start integrating data into all new and existing processes.

8.2 Fundamentals of Data-Driven Organizations

A data-driven organization is one whose decision-making relies on data from a combination of sources in order to gain competitive advantage and create value (Berntsson Svensson & Taghavianfar, 2020). The three main components of a data-driven organization are as follows (Kiron, 2017).

1. A high-quality, well-governed database.
2. A culture where data is viewed (and treated) as a key organizational asset and continuously analyzed in search of insights that inform business strategy, new business value offerings, and increase customer and employee engagement.
3. Data-driven capabilities that enable the use of analytics to support decision-making and strengthen the business model.

One good reason to become a data-driven organization is the need to do something with the constant influx of data generated by mobile phones, information systems, machines, the Internet of Things, and all kinds of applications. But also because of the constant requests from many owners and managers to be able to make decisions based on evidence that, ultimately, improve strategies and the effectiveness of actions (Hume & West, 2020).

In theory, a data-driven organization can use data for all types of decisions (e.g., strategic, tactical, operational) and for all types of analysis (e.g., descriptive, predictive, prescriptive). In practice, organizations use a combination of types of analysis and decisions based on their goals and the sort of value they expect from each decision. A good handful of organizations have already begun to use data in a highly systematic and calculated way to address strategic challenges and react to, and even anticipate, changes in the marketplace. What these organizations have in common is that they view data as a critical asset.

To create value from data, it is not enough that the right person receives the right information at the right moment. The information must also be relevant to the person making the decision. The problem is that in an age of information overload like ours, the ability to discern the correct information is undermined. Most decisions in tourism firms are still made based on the experiences, opinions, and intuitions (or a combination of them) of different stakeholders. Moreover, many decisions about which products or service to innovate, with which partners to ally, or how much to invest in a new market are influenced by politics and individual agendas rather than by the true value expected for the organization or for the customer. Even when data is available, too much information or lack of relevance of the data can puzzle the decision maker and ruin decisions (Berntsson Svensson & Taghavianfar, 2020).

In organizations where managers are used to making decisions based on their own experience and intuition, it is normal to think that they do so because they are the most appropriate people given their qualifications and the relevance of the information they possess. However, in data-driven organizations, business leaders are convinced that data analytics can help their organization improve its competitive edge and are therefore willing to use data systematically to support decision-making. Data analytics is not only used to prepare reports or make operational or tactical decisions, but to decide on key issues such as budget allocation, the creation of new products and services, and even the business model (Kiron, 2017). Furthermore, leaders pay close attention to the quality of data collection, data processing, and analysis techniques, aware that the quality of their decisions depends heavily on the quality of data sources and visualizations, as well as their ability to interpret the processed data (Berntsson Svensson et al., 2019).

All the above could lead business owners and managers to think that in the data-driven organization there is no place for experience or intuition, but the truth is that data cannot replace experience, nor experience data. The way data is created, processed, and analyzed is the result of many individual decisions that introduce personal biases. Data-driven evidence will continue to be weighted with experience, and as analytics becomes a more widespread tool for creating value for organizations, experience will continue to play an important role in determining what value can be extracted from data.

8.3 Key Elements of the Data-Driven Organization

Although the potential benefits to be gained from data are huge, the number of tourism firms using data to successfully transform into data-driven organizations remains very low. Smart tourism firms are data-driven organization which rely on the efficient management of their data assets. For smart firms, data management and analytics become decisive success factors in developing the ability to create, acquire, classify, and generate knowledge, as well as apply it where it can create value and generate competitive advantage. However, there is a conspicuous lack of information on how an organization can manage to transform itself into a smart (data-driven) organization and what are the key elements to consider when transitioning towards a data-driven business model (Ghorbani et al., 2019). Some recent studies have tried to synthesize the high-level dimensions that describe an organization based on data, which are these five (Fig. 8.2) (Hupperz et al., 2021).

Digital transformation and data-driven culture

For new and existing processes in the organization to integrate data and take a step forward in smartization, it is necessary that the organization has adopted a digital transformation strategy. This requires that the organization has established a clear action plan that specifies the goals and stages to be followed for an effective transition from the current situation to one based on data. Formulating a digital transformation strategy is always challenging, but implementing the strategy is even more so, as time and resource constraints play a crucial role. The most effective way to succeed, though not always the fastest and easiest, is to build a data-driven culture that creates employee awareness of the vision and strategy to follow, and to integrate the organization into an ecosystem where the firm can engage with other stakeholders to address challenges in an open and innovative way.

Data science and analytics

Data science brings together advanced knowledge and techniques to analyze and extract value from data. In essence, data science can add value to the

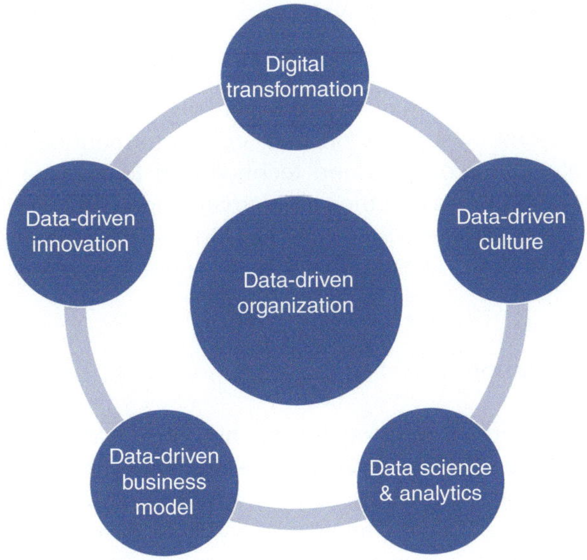

Fig. 8.2. Key elements of the data-driven organization. Source: own elaboration based on Hupperz et al. (2021).

organization by bringing greater transparency to firm processes and decisions, uncovering new needs, and identifying new opportunities to innovate. It can also enable more efficient management of the firm's resources and operations given its ability to make connections between various factors and predict what might happen. This requires the firm to have professionals with the knowledge and skills necessary to produce useful and enlightening insights for the business, which is a preliminary step for the organization to create competitive advantages.

Data analytics is the set of knowledge, techniques, and tools that drive the generation of business insights and the creation of value through data for the organization. It can be classified into the following three categories, which we will examine in more detail in Chapter 11 of this book (Berndtsson et al., 2018).

- **Descriptive analytics:** collects and generates data through data repositories to explain what happened in the past.
- **Predictive analytics:** uses data mining techniques (e.g., classification, clustering, association) to find patterns in data that were previously unknown to predict what is going to happen.
- **Prescriptive analytics:** analyzes both descriptive and predictive business information (e.g., data on customers, sales forecasts, etc.) to help decide what can be done.

Data-driven business model

To create true economic value for the firm and its stakeholders, the business insights produced by analytics must be transferred to the firm's business model. There are many possible ways to develop a data-driven business model, but perhaps one of the most compelling is to pay close attention to market demand and figure out how the organization is going to discover, create, and capture value, especially from its key resource: digital data. With that purpose in mind, organizations must focus on their core informational resources and capabilities, integrate processes and technologies to explore and exploit their data resources, and measure the value created and captured. In addition, organizations must continue to fine-tune the business model using internal and external operational data to ensure that it will continue to generate value in the future.

Data-driven innovation

Data-driven organizations generate and collect vast amounts of data that should not only be used to transform business activities and make them more efficient, but also to innovate. To use data for innovation purposes, organizations need to constantly discover data, analyze it, and try to predict and optimize future events using algorithms. Once new business insights have been generated through the exploration and exploitation of data, the organization must have mechanisms to distribute all this information and knowledge to the different departments of the firm so that they can start creating value from them. The new capabilities that data brings to organizational innovation make research and development activities increasingly relevant to discovering business opportunities that might not otherwise be realized.

8.4 Benefits

Some studies on non-data-driven organizations have come to identify as many as 23 potential benefits of migrating to a data-driven organizational model. The more data-driven an organization is, the more productive it becomes, specifically up to 5% more productive and 6% more profitable compared to the competition (McAfee et al., 2012). Berntsson Svensson & Taghavianfar (2020) have grouped these benefits into the following six categories.

- Decisions
- Understanding the customer and user
- Creativity and innovation
- Productivity
- Market position
- Growth opportunities.

Among the above benefits, the most relevant for the study participants were those related to decisions, customer/user understanding, and productivity. According to firms, organizational decisions improve significantly as a result of a more data-driven approach, and decisions become more accurate by combining internal data with external data (e.g., competitor data with market trends and sales). In addition, for some firms, decision-making also becomes faster and, therefore, they can react much sooner to changes in their environment.

Another key benefit noted by firms is improved customer satisfaction, as the data-driven organization develops insights that enable it to offer customers the products and services they need. In this way, firms can better understand customer/user behavior and where customer churn occurs. Improved productivity is another potential benefit identified by firms. Data-driven organizations can increase efficiency in product and service development and reduce time to market. Furthermore, because data processing enables automated operations, it's easier to identify problems, provide faster solutions, and get more done with fewer people.

8.5 Challenges and Enablers

Some field studies have attempted to explain in recent years the main challenges that organizations face in their journey to become data-driven. For example, in a survey of over 3000 business executives, LaValle and colleagues reported that organizations have more data today than they can use effectively, and that organizational leaders want more computational power and analytics to exploit their growing data and become smarter. In other words, business leaders are willing to run their organizations on data-driven decisions (LaValle et al., 2011). They want to be able to simulate scenarios that provide advance guidance on what are the best actions to take when unexpected or disruptive events occur (e.g., unexpected competitors, earthquakes, pandemics). In the end, what owners and managers demand are ways to understand the optimal course of action to take, based on complex business parameters and scant information. For these expectations to be met, a lesson must be learned that the information generated through data analytics must be aligned with the business strategy.

Organizations should not collect data and analyze something simply because it is easy to do. They need to determine what their goals are and what kinds of questions need to be answered beforehand. In addition, the answers must be easy for end users to understand and make them available when needed. Ultimately, the above requires that knowledge be instilled in everything the organization does (i.e., development of new products and services, hiring of new staff, financing decisions, etc.).

Challenges

Realizing that data-driven opportunities are critical to business growth, many business owners and managers are looking for the best place to start. However, often that entry point is elusive and the challenges too daunting. All in all, the firm must address three main challenges from the outset, regardless of whether everything else has been handled correctly (Berntsson Svensson & Taghavianfar, 2020):

- **Data vs. Intuition:** Many organizations on the path to becoming data-driven struggle to figure out how to move from subjective decisions based on intuition, experience, and opinion, to data-driven decisions. Owners and managers must be aware that organizations do not make this transition overnight but go through a maturation process in which intuition and feelings gradually lose weight in decisions. Contrary to popular belief, the ultimate goal is not to suppress opinion-based decisions, but rather to strike a balance between data and intuition that minimizes errors and inaccuracies in decision-making and focuses on value creation for the business.
- **Trust:** When the organization decides to focus on data, trust in the quality of the data and the insights gained from data processing is of paramount importance to decision makers. Without being sure that the data is reliable and relevant, and that the findings are interpretable and make sense, decision makers are likely to ignore the data or simply be unwilling to implement data-driven decision-making.
- **Culture:** It is critical to transforming into a data-driven organization. Indeed, all organizational changes to becoming data-driven should be accompanied by an organizational culture aligned with those higher goals. Creating a data-driven culture is a process that cuts across all functions within the organization. It is never an easy job, and much less in organizations that are not very agile and have deep-rooted hierarchical leadership styles,

as in data-driven organizations culture means that people are open to sharing data, are aware that data is a critical asset, and they understand why they do this.

Other authors have identified even more challenges to becoming data-driven, such as difficulty accessing reliable and relevant data, lack of a strategy, lack of adoption and understanding by middle management, insufficient organizational alignment, and employee resistance. Addressing the challenges posed by the data-driven model can also vary significantly by business type and size, as well by the type of activities (e.g., transport, accommodation, personal services). It also depends on the access to data sources, digital technologies available, financial resources, and qualified personnel. Other factors that may moderately affect progress towards a data-driven business include the location of the firm and the maturity of the tourism ecosystem and/or the destination in which the business operates, as well as leadership skills, which determine the way in which owners and managers perceive the opportunities in their environment and struggle to benefit from them.

Enablers

Among the key enablers driving the transition to a data-driven organization, technology features prominently. As an example, data-driven and analytics-driven organizational transformation simply would not be possible without powerful computing power that is easily accessible and affordable (e.g., the internet, communication networks, and resourceful software and algorithms) (Kiron, 2017). Moreover, the interconnection between Big Data, analytics, the Internet of Things, information technology systems, and knowledge management practices is crucial in the data-driven environment, as is the role of artificial intelligence and cloud computing (Fletcher et al., 2020).

Fortunately, data-driven solutions and computational processing power are becoming more widely available to tourism SMEs through cloud computing, which democratizes access to platforms, infrastructure, and software (Samara et al., 2020). For example, small family-run hotels can store customer data in the cloud and make use of data-driven infrastructure and applications that would otherwise only be affordable for large hotel chains. Paradoxically, while these new technologies expand the pool of data that can support knowledge production and decision-making, they generate more and more insights that are increasingly difficult to interpret and understand by decision makers (Kaivo-Oja et al., 2015). This highlights that while data, knowledge management, and smart technologies are going to be key in the coming years, it is equally true that business managers are entering a new era of information saturation that they had better start learning how to govern.

Being able to accommodate all these new powerful techniques and tools in the tourism firm makes it essential to gather reliable data. If the organization does not obtain clean and reliable data, all analytics, artificial intelligence, and machine learning techniques are worth nothing. The firm would find itself in a scenario of the type "garbage goes in and garbage goes out". When data is unreliable, erroneous, or incomplete, this leads to bottlenecks in data processing and, ultimately, to failures in service with unfavorable consequences for the competitive performance of the organization. Yet this problem is not solved by removing empty spaces in data or fixing incorrect characters in datasets. This requires having a strong data governance practice in place, i.e., data must start to be treated as a core organizational asset of the firm as a preliminary step to gain a deep understanding of the relationships that exist between data, users, and organizational activities. Organizations must also develop a common vocabulary for data (Kiron, 2017) and appoint owners of various types of information (e.g., Chief Data Officers) who are responsible for creating reliable data and ensuring its accuracy. These data managers must take inventory of available data and determine how it should be collected. If the data does not exist, they will need to find alternative ways to collect it or access external data directly (Hume & West, 2020). The big question at this point is whether all tourism firms have the will and the resources to make those kinds of commitments.

Nobody doubts today that the role played by technology is fundamental; however, this is only part of the story. People are equally important, from leaders mobilizing the organization to embark on the Big Data and analytics journey, to frontline employees changing roles and responsibilities, to data scientists and IT engineers who collect and process a huge amount of information. Data-driven organizations need people with excellent analytical skills, the ability to manipulate and understand large data sets, and with the competences to interpret and apply the results. Without people who know what to do with data and how to leverage it, organizations cannot keep moving toward smartness.

Furthermore, people and technology need to be supported by management processes that guide the entire organization through the different stages that go from data collection to knowledge generation and dissemination. Without them, it simply won't be possible to make any noticeable progress. Many firms struggle to manage the information they have, even after they have collected, organized, and processed it. This is often due to a lack of robust processes to ensure the relevance, accuracy, and timeliness of information, which explains why so many data-driven initiatives fall short of expectations (Kiron, 2017).

One of the few antidotes that organizations can rely on to overcome these threats is leadership. The organization's leaders must be able to integrate data into decision-making with a blend of awareness, patience, expertise, and precise mobilization of resources and manpower. Leadership must focus not only on using data to improve the existing organization, but also on improving the customer experience; perform better operational processes; and design and implement new data-driven business models. Altogether, the harmonious combination of quality data and knowledge, people, technologies, and processes should lead tourism organizations to be more resilient. Resilience is a key feature of the smart organization that is not achieved by chance or accident, but through smart actions that stem from smart decisions made by smart leaders.

Discussion Questions

- What kind of difficulties do tourism firms routinely encounter when becoming data-driven organizations?
- What are the main barriers that firms in each of the tourism activities face to create value from data? How are the tourism activities different from each other?
- What kinds of public policies can be put in place to accelerate the transition from conventional tourism businesses to data-driven organizations?
- To what extent does the social, economic, and business context encourage or discourage the transition to a data-driven model? What regional differences exist?
- What business criteria should a tourism firm consider when deciding to implement a transformation strategy towards a data-driven organization?
- How should leaders of tourism firms prepare themselves to successfully tackle the transformation towards a data-driven organization?

Part 2 Smart Technologies

Part 2 of the book aims to help tourism firm owners and managers navigate the stormy waters of smart technologies that are changing the tourism ecosystem deeply and forever. The number, variety, and complexity of smart technologies has not stopped growing in recent years, as well as the increasingly powerful functionalities they offer to tourism firms. Not surprisingly, the world of smart technologies can be intimidating for many business leaders, as it happens to many specialists. However, all smart technologies have a common articulating factor: data. Therefore, this part of the book begins with a chapter that lays the conceptual foundations to understand data and that every owner and manager should know to enter the world of smart technologies. The following chapters will separate the wheat from the chaff as they place under the magnifying glass the critical technologies that are driving change in tourism firms: Big Data, data analytics, artificial intelligence, and the Internet of Things. At the end of this part, the reader will have a broad perspective on smart technologies and the opportunities that their applications offer, as well as the challenges and drawbacks that they pose for the management of tourism SMEs.

9 Understanding Data

Data are all kinds of values that reflect facts or events that occur in the context of the activity of the firm and that allow one to recreate or know a reality, either at the micro level of the organization (e.g, the firm, a department), or at the macro level (e.g., ecosystem, market). More specifically, the ISO/IEC-2382-2015 standard, focused on the information technology vocabulary, defines data as a "reinterpretable representation of information in a formalized manner suitable for communication, interpretation, or processing". Data is, therefore, a symbolic representation (numerical, alphabetic, algorithmic, spatial) of a quantitative or qualitative informational attribute or variable that represents an empirical fact, event, or entity, which can be reinterpreted. This means that the context surrounding the data is key for it to acquire its full meaning. Furthermore, data can come from different sources and take different forms. For example, data can be an image from a traffic camera, a message on social media, a voice in a conversation, the temperature of a room, the number of items sold by a store, or the exact time an order was delivered. In addition, data must be able to be communicated and processed, flowing from its source to a processing system. Eventually, all data has a life cycle within organizations that begins with collection, goes through storage and processing, and ends with the dissemination of the data converted into valuable information to users (Fig. 9.1).

The term data warehousing refers to back-end management, or those tasks, processes, and capabilities that support data management, including the tasks that need to be performed to prepare the data for analysis and which a user cannot access. Meanwhile, business intelligence refers to front-end (user-oriented) applications that provide data for analysis and decision-making support. The term analytics is concerned with advanced business intelligence methods that include quantitative data analysis (e.g., statistical analysis, predictive models, etc.) (Collier, 2012). As can be deduced from the data life cycle, each of the stages

DOI: 10.1079/9781800620261.0009

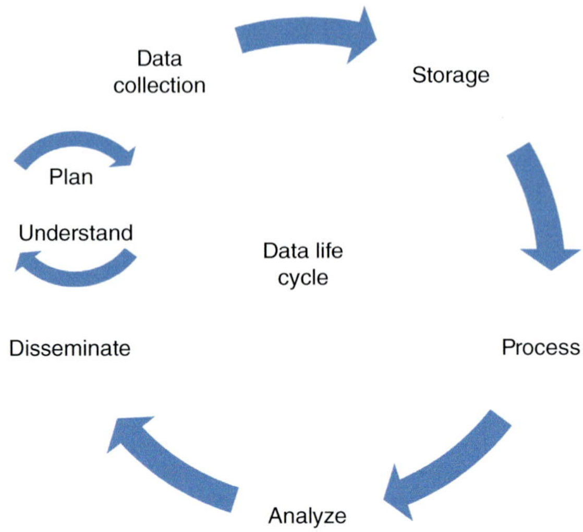

Fig. 9.1. The data life cycle. Source: own elaboration.

involved will require appropriate management practices, skills, competencies, and technologies to carry out efficient data management and meet the firm's performance goals.

9.1 Sources of Data

Tourism firms use different types of data that are differentiated by the way in which they are created and the media on which they reside. For example, social media have become one of the main platforms that contribute user-generated data to tourism firms due to the wide support they have received from tourists and society in general. Social media apps are convenient, easy to use, and consumers have them installed on their mobile devices. This greatly facilitates interactivity with tourists, and that many want to share their lives on the internet.

But social media apps are not the only data sources that allow tourism firms to access new data. As more and more people share their lives online and interact with smart devices and sensors of all kinds, large-scale structured and unstructured data is emerging, ushering in the era of Big Data. Moreover, due to the strong development of the Internet of Things (IoT), various sensors are being developed and used to track the movements of tourists and environmental conditions, providing vast amounts of spatio-temporal data (e.g., GPS data, mobile roaming data, Bluetooth data, etc.). In addition to this so-called unstructured data generated by devices, there is data from tourism operations, such as web searches, visits to web pages, online reservations, purchases, etc. (J. Li et al., 2018). These also represent large amounts of data of a transactional nature that have a high value for tourism firms to understand consumer behavior and improve their value offer to customers. Table 9.1 summarizes the main types of data to which tourism firms usually have access and which are generally divided into two categories (structured and unstructured data) and six subcategories (Lv et al., 2021).

Table 9.1. Types of data sources in tourism. Source: own elaboration based on Lv et al. (2021).

Type	Source
Structured data	Professional databases Business databases Government databases Sectoral databases
Unstructured data	User-generated content Device-generated content Web search and transaction data Social media

Structured data

Structured data is data that conforms to a standardized data model, has a well-defined structure and order, and is easily accessible to both humans and machines. At the dawn of the information age, structured data was the main source of information used by tourism firms, governments, and researchers – there wasn't much more. Nowadays, the main sources of structured data come from business databases, government databases, and industry databases.

Firms usually create a business database to record data related to their operations, transactions, and those events that are important to improve business management, including customer and user data, competitor data, or financial accounting data. Traditionally, CRM tools have been one of the most important sources of data to examine the customer profile in tourism firms, accompanied by manufacturing, supply chain, and accounting information systems.

Government databases are created and maintained by national and regional governments and usually contain data on the economy and the productive sectors of the economy. Examples of this type of databases are census data, data on national and international visitors to a destination, hotel occupancy, etc. In the case of tourism, some of the most used database include those related to the licenses of vehicles registered for tourist transport, the traffic flow of motorways, airports and all types of transport, employment in the hotel industry, etc. These data have traditionally been used to forecast tourist demand at the level of tourist destinations, as well as to know the flows of tourists, or the environmental impact of tourism. Of course, there are many other sources of government databases that can be used, such as the energy consumption of the tourism branches, the flows of cross-border travelers, employment, income per tourist, the number and occupancy of hotels, etc. However, a typical problem with these data sources is that it is government-owned information that is not always available or up to date.

Sectoral databases are those created and maintained by professional organizations belonging to the tourism sector. A good example is the tourism statistics database of the United Nations World Tourism Organization (UNWTO), which includes data such as the flow of tourists between markets of origin and destination, or the demand and capacity of tourist accommodation. Since this data is already structured, there is no need to carry out a data preparation process beforehand (e.g., data cleaning), so the data can be extracted and processed right away. Despite this great advantage, structured data does not usually adapt flexibly to the objectives pursued by tourism firms, so its effectiveness is generally very limited. This makes it common practice to combine structured data with unstructured data to take advantage of it.

Non-structured data

Rapid advances in internet technologies in recent years have led to the creation of unstructured data on a massive scale. Today's users can easily post their opinions about their last trip, share comments about what they think of this or that hotel through social media and digital platforms, and express their likes or dislikes about any product or service they have consumed. This is why internet giants like Facebook, Twitter, or Tripadvisor have become authentic hubs of user-generated content (UGC) in the form of reviews, videos, geolocated photos, etc.

All this UGC, together with the data generated through mobile devices and the data obtained through the digital footprint left by users on the web, has enormous value for firms that struggle to know their customers and users at an individual level. Let's take a closer look at each of these sources below.

User-generated content (UGC)

UGC typically includes two types of main data: 1) textual data related to the travel experience, such as those shared by tourists on social media and blogs; and 2) online photos and videos, such as those shared on Instagram and Flickr. Social networks, blogs, and online booking and review platforms (e.g., Tripadvisor, Booking.com, Airbnb, Ctrip, etc.) are powerful platforms through which tourists share a wide variety of information related to their tourist experiences, express their satisfaction or dissatisfaction with the tourist products and services consumed, or simply recommend a certain place or provider to other tourists. This type of data can be used by tourism firms as a primary source of information to analyze the movements and travel patterns of tourists. Nonetheless, they generally lack sufficiently detailed contextual information.

Review data is mainly used to assess tourists' satisfaction with the tourism products and services they consume (e.g., hotels, rural accommodation, restaurants, tourist attractions, etc.), as well as to inform changes that improve the experience of tourists. This data comes primarily from text data that tourists post on social media, as well as online reviews on travel blogs, accommodation booking sites, and travel websites. The volume of review data can range from several tens to hundreds of thousands depending on the location or attraction in question.

Blog data is used to analyze the sentiments of tourists and keep track of the places that are most popular. This data is also used to find out the recommendations made by visitors to a destination. Twitter and Sina Weibo are some of the main social blog platforms and blog data sources. By processing blog data, tourism firms can obtain valuable information about the temporal and spatial distribution of tourists, their feelings, and the places they recommend going, or what to do when traveling to a particular tourist destination.

Online photo and video data shared by users on social media and online platforms (e.g., YouTube, Flickr, Pinterest, Instagram, Tripadvisor) also contain very rich and varied information, such as information on the location, time of the visit, or even personal notes of the tourists. This data also provides information on travel recommendations to certain places, travel routes, and user preferences for certain accommodations or tourist attractions. In recent times, this type of unstructured data has increasingly attracted the attention of tourism firms, destinations, and researchers, as it offers a new perspective to study the behavior of tourists and the recommendations they make, which can have an impact on value offerings aimed at the market, the design of tourist routes, and marketing.

Some recent studies have used geotagged data in combination with other data sets to compare visitor movements. For example, after comparing Twitter geotagged data on tourist flows between countries with official tourism statistics, similarities have been found between both data sources. Geotagged data allows for detailed street-level accuracy, especially when compared to other traditionally used methods such as surveys or check-ins at accommodation. However, while this type of data can help tourism firms better understand the relationships between people and places, and provides insight into the context of visitors, it also raises issues related to the reliability of user-generated content, especially in the case of false reviews, or photos and videos that have been manipulated.

Device-generated data (DGD)

Vast amounts of data are also generated today from devices integrated with mobile signals (e.g., GPS positioners, Wi-Fi connections, etc.). Unlike traditional static data, such as that obtained through surveys and panels, device data is used to track the movements of tourists in space, thus allowing tourism firms to understand the behavior of tourists more accurately and efficiently in real time.

Mobile phone signaling data is increasingly being used to track the behavior of individual tourists. This type of data provides temporal and spatial information associated with certain data of mobile phone users that makes it possible to track the location of users and identify patterns of movement within and between tourist places (Xu et al., 2020). The signaling data usually includes the number of a station to which the mobile phone is connected, the time in which the connection takes place, and the identification number of the encrypted mobile phone. This can be useful to analyze the loyalty of tourists to a place, the temporal and spatial patterns of visitor movements, and the distances traveled by tourists. It can also be useful for marketing, as current technology allows mobile phones to be positioned, the origin of tourists to be tracked, and the temporal and spatial changes of visitors to be captured in real time both indoors and outdoors. However, there are several issues related to this data: it is difficult to access due to data privacy, and it often lacks accurate information on the location of users and their socioeconomic attributes, thus reducing its value for tourism firms. Additionally, on many occasions, this data does not distinguish between leisure tourists and business travelers, or between tourists and residents.

Data from GPS (satellite navigation system that provides geographic positioning and time information to a receiver) and GPS-enabled services is another important source of unstructured data. Tourism firms can use this data primarily to examine the spatio-temporal behavior of visitors in places and events. For example, GPS data from taxis has been used to track the movements of tourists in a tourist destination.

GPS data comes mainly from two sources: 1) GPS loggers carried by volunteers or installed in public devices used by tourists in a place (e.g., GPS data from public bikes, or scooters shared on city streets); and 2) GPS-enabled mobile apps installed on users' own devices. The former can be used in ad hoc analysis involving a small number of loggers or volunteers, like users who rent an electric scooter, bicycle, or motorcycle on the streets. The latter can provide information on the trajectory of tourists, travel routes, and their behavior in a destination, which is useful for modeling and predicting the behavior of tourists in tourist destinations (Z. Chen et al., 2021). In addition, this data is more flexible and cheaper to obtain than the first type, making it the preferred GPS data source. Unlike mobile phone positioning data, GPS data provides more precise information on tourist behavior in time and space, since it is generated continuously and without intervals (Shoval et al., 2014). It is also not affected by weather conditions. However, GPS data sample sizes are often smaller, especially if participant-based samples are used, which inevitably introduces bias. One solution may be to use GPS-enabled mobile apps.

Bluetooth and Wi-Fi data are used much less frequently. Bluetooth is a wireless communication technology embedded in smartphones, laptops, and mobile devices that can monitor a large number of users without prior notice to device owners. Mobile devices carried by people can be detected by sensors equipped with Bluetooth technology that can thus track the individual behavior of a large number of tourists. Due to the very short radio range of Bluetooth data reception, this type of data is used in very small-scale contexts, such as monitoring the movement of visitors in tourist activities or planned events (e.g., festivals, concerts, etc.). Bluetooth technology avoids the need for users to preregister (compared to mobile apps and other technologies), is low cost, and highly convenient. In addition, this data can be used in indoor environments where GPS or mobile phones may have problems due to weak signal.

Compared to Bluetooth, Wi-Fi data is more convenient and cost efficient as it is available on all modern smartphones. However, like Bluetooth data, Wi-Fi data has a very low coverage range compared to mobile data and allows users to be tracked without prior notice. Few tourism firms have used Wi-Fi data to date, although with the growth of Wi-Fi services in public spaces, expect Wi-Fi data to grow and become more accessible in the future. It should also not be forgotten that Bluetooth data and Wi-Fi data facilitate unannounced tracking of users, which may conflict with the principle of opt-in data collection.

Web search and transaction data

Unstructured data can also come from the traces left by users when browsing the web, in searches for information made through search engines, or purchases and reservations made online. Search engines like Google, Yahoo, or Baidu are some of the main sites tourists use to plan a trip or vacation. Search "footprints" are a valuable source of data on the tourism demand side and reflect the interest of tourists in places, products, and tourism providers. The data generated from tourists' online purchases and web browsing can be used to analyze things such as preferences and behavior patterns of tourists, as well as to forecast the arrival of tourists to a destination or a hotel (e.g., via Google Trends, Google Analytics traffic indicators, or the Baidu Index in China). It can also be used to predict economic indicators such as room demand or prices in a destination. Other transactional data, such as online room reservations or the purchase of travel products, registered by the transactional systems of tourism firms, can help owners and managers make decisions about what to offer customers or where to invest more. When tourists pay for their trips or reservations by credit card, the data is recorded and can be used later by firms to analyze the visitor's purchasing behavior and, where appropriate, design a personalized product.

9.2 Data Analytics

Analytics has been around since the 1950s and it is not a new idea. The origins of analytics were rudimentary, with access to a very small number of data sources, most of which came from internal business systems and were stored in a single repository, such as a data mart or data warehouse. The only types of analysis that could be done at that time were descriptive, and business intelligence consisted of a few reports on the most relevant business processes. In the early 2000s, companies like Google and Yahoo began to use Big Data to analyze the behavior of their users, and everything changed radically. The term Big Data became popular and data analytics and traditional business intelligence turned upside down.

Data analytics uses two sets of methods to answer different types of questions: descriptive analytics methods and predictive analytics methods (Andersen et al., 2018). Descriptive analytics analyzes data to answer the question: "what happened". This approach informs human decision makers about the past and present factors that explain the occurrence of a certain situation or event. Decision-making is therefore reactive, since it occurs once it is understood why things happened in the past. Techniques include descriptive statistics on combined or aggregated data to find relationships, patterns, and trends between them (e.g., cross-tabulation, correlation, and regression models). Typical examples of descriptive analytics are reports and dashboards for senior management based on key performance indicators (KPIs).

Predictive analytics is a data analytics approach geared towards answering the question: "what's going to happen". The analyst's attention in this case is more focused on the future than on the past, so predictive techniques become increasingly advanced and sophisticated (e.g., estimation of probabilities, predictive models that combine classical statistics and machine learning, etc.). Decision-making supported by predictive analytics is of an active type, since it tries to foresee what is going to happen and anticipate the firm's response.

An even more advanced level of analytics is prescriptive analytics, which is intended to guide the firm's decision-making and action plans. This approach involves a proactive mode of decision-making. The main differences between the active mode (predictive analytics) and the proactive mode (prescriptive analytics) lies in the organizational capabilities that the firm needs to implement the results of the analysis, rather than in the techniques used (Andersen et al., 2018). In the proactive mode, a combination of statistical models based on regressions (descriptive analytics) and probabilities (predictive analytics) are used to make rules-based, simulation, and optimization decisions. Current applications of prescriptive analytics include the analysis of dynamic and complex networks, often combining a variety of transactional, process, and sensor data streams.

Traditional tourism firms have so far focused almost exclusively on descriptive analytics. However, as the amount of data that firms can access has grown exponentially and become huge, it is possible for firms to start extracting more value from it, provided they are able to store and process data much faster. This has caused predictive and prescriptive analytics to gain significant momentum and firms realize that they must acquire new skills and capabilities to keep moving at a faster pace (Larson & Chang, 2016). Therefore, tourism firms must begin to familiarize themselves with increasingly advanced analytical techniques to move from reactive to active decision-making. This means that firms must take steps forward and evolve towards data science and learn to combine cutting-edge techniques such as machine learning. For many tourism firms, making this leap from descriptive analytics to advanced modes of predictive analytics will surely be difficult. Notwithstanding, owners and managers should be aware that increasing the knowledge and skills for data analysis has become a crucial driver of business competitiveness, without which data has no value.

9.3 Business Intelligence

Business intelligence includes the set of processes, technologies, and applications required to collect, analyze, and visualize business data not typically provided by regular reports and that can be used to support both operational and strategic decision-making (Larson & Chang, 2016; Marcello et al., 2018). Business intelligence empowers tourism firms by giving them an edge through information and knowledge management to improve business decision-making.

Modern business intelligence dates back to the 1990s when tools specialized in extracting, processing, and storing data in a central data warehouse began to emerge. These tools were used to organize, analyze, and visualize data in a descriptive way, such as through Online Analytical Processing (OLAP), which allowed sales, marketing, and dashboard reports to be generated for senior management by querying large amounts of multidimensional data (called OLAP cubes). However, the complexity involved in handling these tools made it difficult for inexperienced users with no prior knowledge of data analytics to perform basic business intelligence tasks on their own.

Business intelligence has continued to evolve ever since to provide tourism firms with historical, current, and predictive views of their key business variables and business processes, and increasingly to support decision-making activities that generally involve choosing between different alternatives. Today, firms use business intelligence to inform a wide variety of operational decisions, from product positioning and pricing, to generating insights into new markets, assessing the suitability of products and services for certain market segments, and measuring the impact of marketing and advertising strategies. In addition, with the exponential growth of the internet and the massive adoption by consumers of mobile devices, the IT industry has begun to address the difficulties related to the speed of data processing, which will result in tourism firms having tools capable of processing and analyzing large amounts of structured and unstructured data in real time, as well as cloud-based on-demand analytics capabilities. Ultimately, business intelligence becomes particularly powerful when it combines external data obtained from markets and customers, with internal data from the firm itself (e.g., financial data, customers data, bookings, etc.). This way, business intelligence has reached a point where it is no longer a complementary utility within a more general framework of business analytics applications, but rather an essential requirement for those firms willing to remain competitive.

9.4 Data Vocabulary

The world of data has become very complex in recent years and is expected to become even more so in the future. This, coupled with the fact that data is gaining strong momentum in tourism, means that new terms appear from time to time to refer to the innovative practices, tools, and capabilities that are shaping the Smart Revolution. Even terms whose meaning seemed conventionally accepted, sometimes acquire a new meaning in light of the latest technological and management advances.

In the face of this increasing sophistication in data management and its accompanying terminology, owners and managers need to understand the meaning of these key concepts, otherwise confusion sets in and they become unable to fully comprehend the true dimension of the phenomenon. For this reason, in this section we are going to review some key concepts with which every owner and manager of a tourism firm should be familiar, and which are relevant to framing and understanding the next chapters of the book.

Data Economy

As the volume of data has become really big, business organizations around the world have started to focus more and more on the "Data Economy". This circumstance is not only the direct result of the amount of data that firms are now able to collect, but also of the great variety of data captured, the general increase in the computing power available to extract value from data, and the substantial reduction in the price of computing capabilities, all of which have fueled the expansion of the so-called Data Economy. The Data Economy is defined as the set of goods and services whose value is based on the exploration and exploitation of existing databases with the aim of creating new value offerings that firms use to remain competitive. The Data Economy considers that data is at the center of all economic activity and that the development of Big Data, cloud services, and the IoT, are essential for the competitiveness of economies. In short, the Data Economy recognizes that data is a key intangible asset for value creation that, together with high-performance computing, is changing the way knowledge is created and shared. Hence, data is a catalyst for economic growth, innovation, and digitalization of SMEs in all economic sectors and society in general.

Data science

Data science is a broad term that encompasses the set of technologies and techniques that are used in advanced analytics on large amounts of data. As the name implies, it is a scientific approach of an interdisciplinary nature focused on extracting information from data in various ways to produce data products or models. Data science integrates next-generation data analytics fields such as statistics, data mining, and machine learning under a single scientific umbrella. As such, data science transcends the methods of traditional descriptive statistics, moving beyond identifying patterns in static and historical data to focus on combining static and dynamic data from very different sources that are then analyzed at high speed using techniques of descriptive, predictive, and prescriptive analytics. For example, the data about a customer's purchases may contain a pattern that tourism firms can learn in order to make relevant suggestions for future purchases by the same customer. These buying patterns can then be compared with other cross-sectional buying patterns from customers and correlated with seemingly unrelated data (e.g., geographic location, weather conditions, socioeconomic context, etc.) to gain a more comprehensive understanding of customer buying behavior under a multitude of conditions (Unhelkar, 2017). Figure 9.2 illustrates the different disciplines that are integrated within modern data science, which combines broad and multidisciplinary skills that include mathematics, statistics, computer science, and knowledge at the sector or industry level.

As Big Data evolves, the need to analyze fast-moving, non-static data will increase. Advances in data science are making it possible to combine large amounts of static data with real-time event processing, dynamic application of business rules, and the incorporation of cognitive and machine learning algorithms, thus transcending traditional statistical analysis techniques. Therefore, data science has become a key toolkit for developing Big Data initiatives and

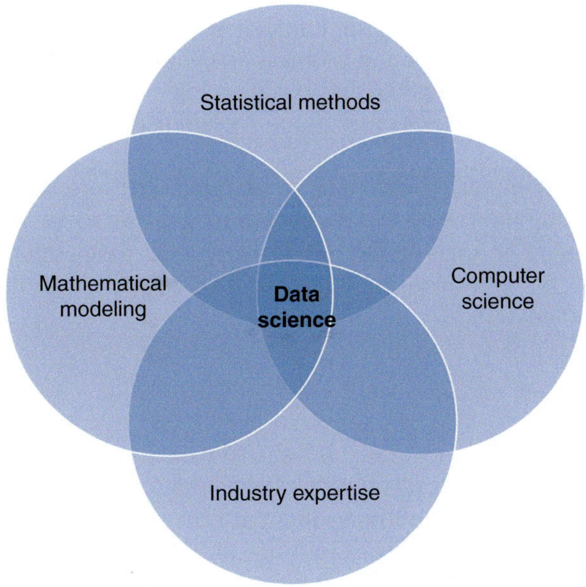

Fig. 9.2. Main disciplines involved in data science. Source: own elaboration based on Del Vecchio, Mele et al. (2018a).

supporting the role of the knowledge worker in business firms. However, as is often the case with other disruptive technologies, the challenge does not lie so much in data science or Big Data technologies, but rather in the process of organizational transformation that each firm should properly implement as a result.

Data scientist

In the context of the Data Economy and the growth of Big Data, it is crucial for business owners and managers to understand what kind of skills are needed to succeed. As more tourism firms get involved with data, new career profiles are created, among which the data scientist stands out. Data scientists are an evolution of the traditional data analyst role that already existed in many organizations. What sets data scientists apart is that they combine computing and hard science skills that go beyond the skills of a data analyst. According to IBM (2022), a data scientist is someone with the following skills and abilities:

- Knows and applies statistics, mathematical modeling, and the scientific method.
- Uses a wide variety of tools and techniques to evaluate and prepare data, ranging from SQL to data mining and integration methods.
- Uses predictive analytics and artificial intelligence, including machine learning and deep learning models, to extract insights from data.
- Writes code for applications that automate processing and calculations with data.
- Is a storyteller who conveys the meaning and importance of results to decision makers and stakeholders.
- Explains to others how the results can be used to solve business problems.

Therefore, data scientists are competent beyond the use of classic software tools and data management and business analytics. They know how to choose well the problems to investigate using business logic, build and execute code to create models and prototypes of solutions, including data visualization tools, and use the results obtained to develop products or services based on valuable information that was previously hidden. As these "unicorns" are scarce in the labor market, it is usually more effective and easier for tourism firms to create working groups made up of a combination of professionals (both internal and external) who together bring these skills and can add value to the organization (Del Vecchio, Di Minin, et al., 2018).

Data mining

Data mining is the analysis of large data sets stored in databases, warehouses, or other repositories of information to extract hidden information and uncover potentially useful patterns for use by stakeholders. As such, data mining is an interdisciplinary field that brings together machine learning, statistics, databases, and visualization techniques (Castaldi et al., 2018). The objective of data mining is to efficiently build predictive or descriptive models based on a large amount of data that, besides explaining them, can be generalized to new data. A typical data mining process includes these steps (F. Chen et al., 2015):

1. **Data preparation:** It usually encompasses three steps: integrating data from various data sources and cleaning the noise contained in the data, extracting some parts of the data (target data) and loading it into the data mining system, and preprocessing the data to facilitate data mining.
2. **Data mining:** It consists of applying algorithms to target data in order to find patterns and evaluating the knowledge discovered from those patterns.
3. **Data presentation:** It is about visualizing the data and delivering the extracted knowledge (insights) to the users.

One of the most compelling applications of data mining is through artificial intelligence. With artificial intelligence, data scientists seek to mimic the efficiency of the human brain in processing information and capturing critical aspects of data in a way that enables future use. In this way, data mining offers the automatic discovery of previously unknown patterns and the prediction of trends and behaviors. These new technologies complement conventional decision support tools, giving business analysts and marketers new opportunities to analyze the business more quickly and efficiently. For tourism owners and managers, data mining opens up new opportunities to create products and services capable of responding to rapidly changing market conditions and remaining competitive.

Machine learning

Machine learning is a branch of artificial intelligence in which the machine or system can learn from data, find patterns, and automatically make decisions based on this learning. Machine learning stems from computational learning theory and provides an informational system with the ability to learn automatically by detecting patterns, without having to be explicitly programmed, and then make predictions (Noor & Haneef, 2020). Machine learning techniques are now being used in complex fields as varied as biology, medicine, social media, astronomy... and tourism, to find hidden insights within data.

Discussion Questions

- What limitations or barriers do tourism SMEs have to generating large amounts of data and exploiting it to its full potential?
- How do tourism SMEs store and process their data? Are there differences between small and large tourism firms?
- What stage of the data life cycle is currently the most challenging for the tourism firm?
- What types of data sources are most used by tourism firms? What mix of structured and unstructured data is optimal?
- What are tourism firms doing to integrate data and knowledge management capabilities in the organization?
- What mode of decision-making currently prevails in tourism firms (i.e., reactive, active, proactive)? What reasons explain it?
- What is the potential of machine learning for the creation of innovative tourism products and services?

10 Big Data

The progress made by internet technologies and the widespread adoption of mobile devices have made it possible for users today to generate vast amounts of data, also known as Big Data. Ordinary people in their daily activities generate an avalanche of information, which spans different types of sources, from social networks (e.g., Facebook, Twitter, Instagram, and YouTube), review platforms (e.g., Tripadvisor), reservation systems (e.g., Booking.com, Hotels.com), blogs, discussion forums, collaborative maps, and many others (Peterlin et al., 2021). This massive user-generated content (UGC) data, along with other passive sources of data generated from devices (DGD), is one of the trends that is having the greatest impact in today's business context.

Big Data has become a core organizational asset and a crucial competitive factor in any tourism business strategy. Consequently, tourism organizations around the world are seeking innovative new techniques to help them maximize the potential of Big Data and address the new challenges created. Historically, IT departments managed the systems in which business transaction data (e.g., orders, sales, shipping, inventory) was generated, stored, and processed. But things have changed a lot in the last few years. The amount and variety of data being generated, how data is now processed, and what can be done with it, are being made possible by new advanced data analytics techniques tailored to large amounts of data (Larson & Chang, 2016). These changes in the treatment of data are affecting practically all industries at an ever-accelerating pace of innovation. For example, some of the most recent developments focus on analyzing the mood of tourists and determining their state of mind (i.e., when a new product or service is introduced or when tourists have a problem). This gives an idea of the depth of the changes that are taking place and how firms are trying to find new answers to who, when, and where.

Business owners and managers must be aware that the earthquake caused by Big Data is not a spontaneous event. In fact, it is part of a transition that is

DOI: 10.1079/9781800620261.0010

taking place on a global scale in economies, which involves the metamorphosis of value creation from tangible assets to intangible assets that have special economic properties (Mihet & Philippon, 2018). Parallel to the increasing availability of large amounts of data, in recent years there has been a remarkable advance at a theoretical level and in the practical applications of what can be done with the data. These advances, mainly focused on data science, are allowing the development of high-performance algorithms that help maximize the value of data for tourism firms. This has proven that Big Data is no longer considered a passing fad and has become a key tool for detecting patterns, understanding consumer behavior and satisfaction in a more accurate and deeper way, and predicting key variables, such as the arrival of tourists, hotel occupancy, or the profitability of firms (X. Li & Law, 2020).

Ironically, even though tourism firms have more data than ever before, only a small fraction of the data is actually serving a purpose. This is not surprising because the challenges facing any organization willing to extract value from data are considerable, such as how to handle large volumes of data and integrate dozens, if not hundreds, of different sources, and provide consistency to the various formats in which the data is stored. New technologies, such as cognitive computing, promise to address this challenge. These technologies are specifically designed to integrate and analyze large data sets and extract meaning from different types of data, thus representing a big leap forward from classical computing, since they mimic some aspects of human thought when it comes to evaluating information but without the biases introduced by human cognition (Castaldi et al., 2018).

In summary, Big Data promises exciting new opportunities, including unlocking new insights that can accelerate the creation of new products and services, boost customer relationships, improve operational processes, and even embrace innovative business models. Furthermore, Big Data is the perfect companion to traditional "small data". However, Big Data and all its potential will come to nothing if organizations are not capable of integrating it, analyzing it, and understanding it. And this is a highly complex process that goes beyond Big Data, as it transcends the mere capture and processing of information, and requires organizations to equip themselves with specialized techniques and knowledge with which to analyze the results and extract value from them.

10.1 Concept of Big Data

It is not easy to define what Big Data is, or to draw a clear dividing line between what is "big" and "small" data. This difficulty is reflected in the number and variety of conceptualizations that can be found about Big Data in the literature.

Perhaps one of the most common and widely accepted ways of defining Big Data is through the "feature-oriented perspective", which characterizes Big Data according to three Vs: volume, variety, and velocity. Unlike other types of data, Big Data cannot be stored on an ordinary PC or portable hard drive, as it typically exceeds 100 terabytes or even petabytes of information (hence the volume). Big Data emanates from a wide variety of structured and unstructured sources, and can have many formats (e.g., texts, sounds, images, videos). In addition, the data is usually spatially and temporally referenced (variety). Likewise, the speed of data creation and analysis occurs in a very short time, which allows decision-making to be adapted to very short time windows that may lead to the initiation of corrective and/or adjustment actions practically in real time.

But these are not the only characteristics that help us define Big Data. The conceptualization based on the three Vs has been extended several times to accommodate four more Vs, as shown in Fig. 10.1 (Mariani et al., 2018).

- **Veracity,** which refers to the reliability, validity, and completeness of data.
- **Value,** which highlights the role of Big Data as a tool used to create value for both consumers and firms.
- **Variability,** because Big Data often consists of unstructured records whose meaning can change depending on the moment and the context.
- **Visualization,** or the need for the insights obtained from Big Data to be displayed in a visually attractive and understandable way for users so that they can do useful things with them.

In addition to the "feature-oriented perspective", Big Data can also be defined according to a "process-oriented perspective". This approach is based on highlighting the processes that are inherent to Big Data, such as the collection, storage, processing, and analysis of Big Data, and the technologies that support them. Using this approach, Big Data is the set of data that is difficult to collect, manage, analyze, and visualize in a limited time with the available technologies of today (Chen & Zhang, 2014). This entails the need for organizations to implement processes aimed at managing Big Data and technologies that overcome the limitations of traditional technologies.

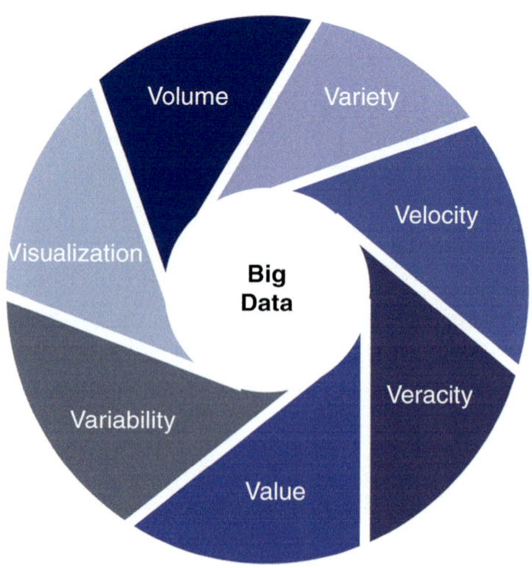

Fig. 10.1. Big Data dimensions. Source: own elaboration based on Sebei et al. (2018).

Other characteristics that distinguish Big Data, apart from those mentioned in the "feature-oriented" and "process-oriented" perspectives, are the following:

- Big Data often collects and explores entire populations rather than samples, thus challenging conventional statistical tools and inferential methods.
- Big Data provides a high level of granularity in the data, which allows the analyst to focus on very fine aspects or very specific qualities of the information available, which can be reflected in finer and more accurate decision-making.
- Big Data can be used flexibly when analyzing different data collections and extracting meaning from them.

Owners and managers should keep in mind that when talking about Big Data, we are not only referring to large public or internal transactional structured datasets (e.g., sales, customers, inventories, data from the population register, vehicle registration, etc.), but also peripheral and non-transactional unstructured data generated by sensors, smartphones, and radio-frequency identification (RFID) chips, which is used to track the dynamics of visitors in a territory, learn about the online behavior of consumers and their interactions in social media, and predict decisions and trends faster and more accurately (Morabito, 2015; Xu et al., 2020).

Big Data is also synonymous with data analytics and disruptive technologies (Unhelkar, 2017). Analytics has been around for some time now and has gone through various stages, from the invention of spreadsheets that allowed for simple calculations to today's sophisticated analytics tools. Data analytics relies heavily on statistical techniques, including both descriptive and predictive modeling, to understand not only "what has happened" but "what will happen". From a statistical point of view, Big Data analytics allows the identification of patterns, makes predictions, and provides advice for better decisions making (i.e., which products to deliver to customers, or which services to bundle). For their part, Big Data technologies allow complex analytics to be applied to large and highly dispersed data sets, providing very fine granularity and, in essence, leading to accurate decision-making. A technological framework widely used by firms that work with Big Data is Hadoop. The Hadoop ecosystem enables programmatic management of all kinds of data science-related tasks (using open source languages such as Java, Python, and R), ranging from data storage in a distributed database architecture to data manipulation, automation and business analytics.

10.2 Big Data Technologies

Big Data would not exist without a strong pace of technological innovation. In recent years, there has been a real explosion in the availability of data as a result of the number of devices connected to the internet, which has increased exponentially. This phenomenon has attracted the attention of the technology industry and the interest of computer and data scientists, who work together to develop new methods that go beyond traditional data storage and management tools and are able to extract value from Big Data (Mariani et al., 2018).

As has happened in other industries, tourism has widely recognized the need to address these new technologies and use Big Data to implement a customer-centric approach that improves consumer experience and satisfaction. Big Data-based approaches solve many of the problems associated with working with representative samples of data, as they can encompass almost the entire population under scrutiny. In this way, Big Data becomes a powerful tool to address new and innovative research questions that can ultimately drive the creation of new value for tourism firms and their customers.

Compared to conventional relational database management systems, which are considered standard for structured data management, Hadoop and NoSQL are technological solutions focused on handling Big Data. Both are complementary and compatible with each other, but there are also substantial differences between them. The Hadoop framework is often used when data sizes are really big. Its origin was a published article on the design and implementation of the Google File System (Ghemawat et al., 2003), a scalable distributed file system for data-intensive applications, supporting a virtually unlimited number of computers in a network (nodes) to process petabytes of data simultaneously. Hadoop is currently a collection of open software utilities built for both storage and distributed parallel processing of data sets. It is available through the Apache distribution or from providers such as Cloudera, MapR, and HortonWorks. Hadoop is made up of four main components (Fig. 10.2).

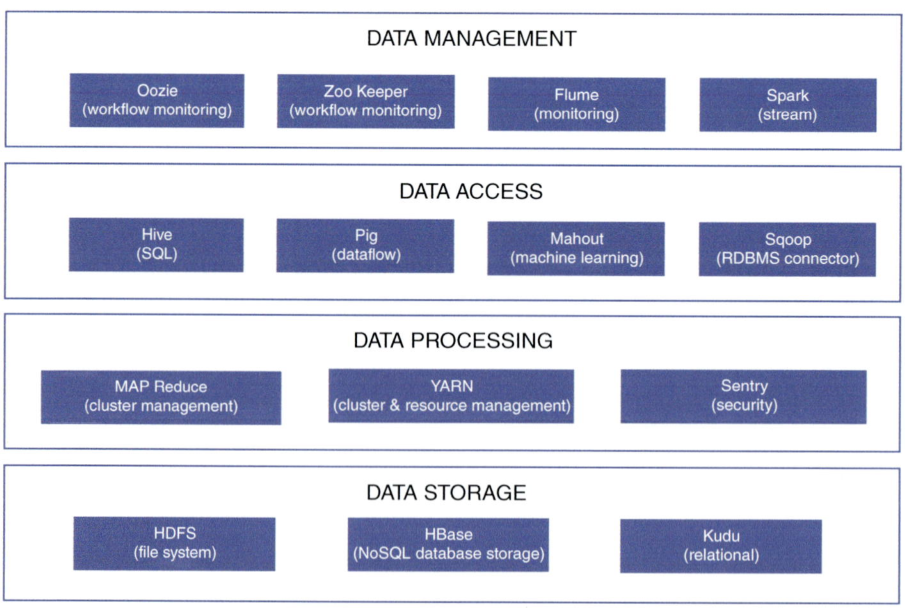

Fig. 10.2. The Hadoop ecosystem. Source: own elaboration based on GeeksforGeeks (2021) and Cloudera (2022).

- **HDFS (Hadoop File System),** is the storage unit for large sets of structured and unstructured data and, as such, is the main component of the Hadoop ecosystem. With HDFS, data can be stored on thousands of machines (called nodes), and metadata can be maintained in the form of log files for fast and efficient access.
- **MapReduce,** is the programming-based processing unit of the Hadoop ecosystem, which makes it possible to write applications that transform large data sets into manageable sets using parallel and distributed algorithms. Hadoop splits files into large blocks and then sends the packaged code to cluster nodes to process the data in parallel. With MapReduce the process is done on the slave nodes and the final result is then sent to a master node.
- **YARN (Yet Another Resource Negotiator),** is the resource management unit of the ecosystem and the one that manages the resources in the clusters of nodes, programming and allocating resources for the Hadoop system and making sure the machines are not overloaded. Hadoop YARN operates as an operating system for Hadoop built on top of HDFS.
- **Hadoop Common (also called Core),** consists of utilities and libraries to start Hadoop and support the rest of the modules.

Other tools and software packages that can be installed on top of or alongside Hadoop include Spark, Hive, Impala, Pig, Hive, HBase, Kudu, Kafka, ZooKeeper, Flume, Sqoop, Oozie, and Storm, all of which can work collectively to provide services such as data absorption, analysis, storage, and maintenance.

One of the main benefits of using Hadoop is that the firm can use business machines as data nodes, so the system is highly scalable and allows any data set to be processed faster and more efficiently than with a conventional computer architecture. This can lead to significant savings for firms since they do not have to invest thousands of dollars in very expensive data nodes. In addition, the Hadoop framework is written primarily in the Java language, with some native C code, giving users the flexibility to program it either with Java or any other non-Java programming language through Hadoop Streaming.

Depending on the specific needs of the firm, it may be convenient to use Hadoop and NoSQL separately, or together in mixed architectures, thus taking advantage of the characteristics of each one to find the most efficient answers. NoSQL databases (non-relational databases) are a more flexible and scalable solution than conventional relational databases as they are designed to manage and retrieve data on a massive scale in formats other than tables (as relational databases do). Since NoSQL databases are distributed databases (where data is stored on multiple servers), new data can be added to them without having to be defined upfront in the database schema, thus allowing rapid processing of big volumes of data of all kinds. In this way, NoSQL databases cope with the scalability and performance problems of conventional relational databases in situations of concurrence of thousands of users and millions of daily queries. As data continues to grow, simply add more hardware to keep up, without slowing down performance. Some of the most popular NoSQL platforms are MongoDB, Elasticsearch®, and Redis®.

Ultimately, NoSQL databases provide a different data management system than relational systems and the Hadoop framework, and, although both can operate autonomously, they are compatible. The integration between NoSQL and Hadoop systems is almost native, making it easy to integrate with Hadoop. Sometimes it can be very helpful to connect to Hadoop to do analysis from NoSQL, where the information is stored. This dynamic NoSQL database schema with Hadoop software is valuable for agile development when rapid and continuous iteration is required. Big Data and Agile can go hand in hand: Big Data allows the firm to be agile, and the agility provided through Hadoop-NoSQL gives the firm the capabilities to formulate successful Big Data strategies. As Big Data continues to grow, it is realistic to think that the combination of NoSQL database and Hadoop software will become a powerful framework that will allow firms to reach their full potential with data.

10.3 Impacts and Opportunities

In the era of Big Data, great opportunities are continually being created for virtually all functions of tourism firms. Through Big Data, the firm can accumulate competitive benefits, which can be conclusive if the firm also learns to use them on an efficient scale and to extract knowledge from data with the appropriate techniques. Next, we examine some of the major impacts and opportunities that Big Data can bring to tourism organizations in the development of innovative strategies.

Managerial practices

Big Data is having a considerable impact on the management practices and business models of tourism firms (Centobelli & Ndou, 2019). The exploitation of Big Data by tourism firms represents an opportunity with enormous potential to create new value for both organizations and consumers, as well as to improve the planning and management of tourism firms. The implementation of increasingly powerful capacities to extract and use different types of information and knowledge that remains hidden in sites such as social networks, mobile applications, etc., can be used by tourism firms to improve their daily operations. Owners and managers of tourism firms, supported by their teams and data specialists, need to decide on the data sources that are appropriate (e.g., transactional data, social networks, reservations, etc.); collect and store data on a large scale using the latest data manipulation and storing techniques (e.g., traditional data warehouses, data warehouses, data lakes); clean and validate data considering privacy and security in data access; extract knowledge from Big Data, emphasizing speed and its multi- and omni-channel applications; and use the knowledge generated to disseminate it throughout the organization and decision makers.

The analysis of user-generated data (UGC), such as reviews, tagged videos, and photos has become an inexhaustible source of insights to improve tourism products and services to the needs of customers, uncover unknown patterns in tourism demand, and, in general, extract valuable information that would otherwise remain hidden. By exploiting Big Data, tourism firms can improve their marketing strategies and analyze the performance of their products and services. For example, firms can apply sentiment analysis techniques on the content generated by users themselves, or apply text analytics techniques to multiple data sources (e.g., customer reviews) to examine the quality of the information. In addition, UGC typically incorporates geotagged information related to users' travel routes, duration of visits, places of interest, sources of tourists, etc., which can be a good starting point to characterize the geographical preferences of tourists. This information can also be key to planning tourism marketing actions.

It is key that owners and managers do not limit their efforts exclusively to the collection, experimentation, and analysis of Big Data. The real challenge to Big Data and data analytics success is using every insight gained from data to make decisions that lead to greater business benefits. Surely most owners and managers will not initially make their decisions based solely on data but will use whatever knowledge they have to support their decisions. Even in more mature phases, it is almost certain that they will want to continue complementing Big Data with data obtained from "traditional" or "small data" sources (e.g., traditional customer surveys) to enrich the 360-degree views of their customers. Therefore, it is foreseeable that tourism firms will continue to conduct surveys of their customers as they have done in all these years to find out customer perceptions, while progressively integrating the "small" data with Big Data more focused on understanding the actual behavior of tourists.

Organization

Organizations do not change easily, and the value of Big Data and analytics may not be evident to everyone. Therefore, firms must continue to empower employees, customers, and suppliers to change their daily habits and behaviors and guide them through the data-driven transition without giving up. Tourism organizations also need to establish who will own and sponsor Big Data and analytics initiatives. They must devise a system of incentives aligned with business objectives that stimulate and reinforce behaviors based on data. Only in this way can it be ensured that Big Data ends up integrated into the firm's operating and decision-making processes.

Tourism firms must set up their approach to Big Data deployment from the outset, which means they need to allocate data collection and ownership tasks among different business functions. All this should be done based on a well-structured plan whose mission is to generate new knowledge that is of value to the business. Of course, this plan should also include other key aspects, such as those related to integration with the technological infrastructure, privacy policy, and access rights. In short, a Big Data implementation plan will have to lay the foundations for the governance framework for Big Data and analytics within the organization.

Business owners and managers must be aware that data alone cannot create value without powerful data processing capabilities. Data science is not a panacea to harness the full power of Big Data. Even more important than the resources dedicated to data are the capabilities to manage these resources. For this reason, organizations must embed Big Data management skills in their employees and take this into account when carrying out recruitment and selection processes. Likewise, organizations must implement human resources practices aimed at democratizing access to and use of data, as well as exploitation and experimentation with Big Data, seeking to achieve a balance between leadership, talent management, and culture that enhances opportunities for value creation.

The organizational model that firms may adopt to make the transition to Big Data will largely depend on its business objectives and operating model. In a very simplified way, there are four models from which organizations can choose to optimize Big Data in the organization (Morabito, 2015), which vary according to the degree of centralization in decision-making.

- **Independent division within the organization:** Each division of the organization has different sets of data that are managed independently of the rest and based on which they make their own decisions. In this model, the level of supervision and monitoring by a central unit that controls or directs the Big Data is very low or does not exist.
- **Divisions within the organization with central support:** This is an organizational model in which divisions make their own decisions about data, but where divisions cooperate with each other in developing Big Data initiatives. For this, they have the support of a central unit that, in addition to coordinating the initiatives, also supervises the results obtained.
- **Center of Excellence:** In this model there is an autonomous center that supervises the implementation of Big Data within the organization, so that the divisions within the organization follow the initiatives set by the center, which is also responsible for coordinating them.
- **Completely centralized model:** In this model there is a corporate center that is responsible for prioritizing Big Data initiatives, allocating resources, coordinating the implementation of initiatives, and supervising the results obtained.

In practice, it is difficult to opt for one model or another universally and for all types of organizations in all industries, so each organization must evaluate its organizational starting point, the resources and talent it has, how far it is willing to go, and if it prefers to focus more on exploitation or exploration activities. After this, the organization must choose the degree of centralization or decentralization that it considers most appropriate.

Employees

The role of employees in the adoption of Big Data and data analytics is crucial in the context of tourism firms. Tourism firms have more and more access to Big

Data generated by users through digital platforms, which makes them highly dependent on having qualified employees who know how to manage, interpret, and use data from their customers and users effectively and efficiently. Employees' abilities to interpret data and extract meaning from it based on varying contexts becomes crucial to firms. This means that firms need to hire, or at least have access to, data scientists who can help them deeply understand consumers in the markets in which they operate.

Employees are the main drivers of organizational knowledge creation. From this perspective, a firm is an organization that integrates the knowledge that resides within and through individuals. Therefore, the greater the organization's ability to democratize data and to expand data applications, the greater the probability of increasing the potential value created from Big Data (Shamim et al., 2020). At an employee level, this is achieved by reinforcing the interactions between the employees themselves within and between the departments of the firm, so that Big Data can be integrated into the different functions and decision-making areas of the organization. These emerging interactions between individuals, coupled with increasingly data-intensive business processes, make employees better equipped to respond quickly to changing consumer needs.

In addition to this ability to exploit Big Data, the ability to experiment with data is also crucial for tourism firms. Experimentation with data is key to creating value from Big Data, as it encourages trial and error and encourages an innovative attitude among the organization's employees. The most successful firms in the use of Big Data and analytics are those whose employees are continuously involved in the process of generating and experimenting with ideas, which allows them to constantly absorb new information, adjust their strategies as new opportunities arise, and update their capabilities to generate competitive advantages.

Last but not least, the employees of tourism firms must have skills related to the management of Big Data, that is, skills that enable them to activate the knowledge generated from the data. Owners and manager should note that while many firms are capable of collecting a significant amount of data, not all are capable of responding in a timely manner to the opportunities that arise from the exploitation of that data. Therefore, without the activation of knowledge, data cannot be transformed into real value for the business. The bottom line is that to harness value creation through Big Data, tourism firms need to devise strategies to develop Big Data management skills among employees and stakeholders. For example, organizations through human resources practices can enhance Big Data democratization, experimentation, and execution capabilities, leading to value creation and ambidexterity in the organization. Similarly, organizations can also emphasize Big Data capabilities through their recruitment and selection processes.

Real time decision-making

Real-time business intelligence is the ability to deliver information about business processes as they occur. This capacity is linked to the ability to perform automated analysis of Big Data. Real-time data generation and business intelligence are promising new sources for creating valuable insights that don't yet exist in the business about on-site consumer behavior through ubiquitous mobile apps and data mining. When applied to data mining and predictive learning, advances in artificial intelligence also offer great potential to enhance the intelligence capabilities of tourism firms, helping them understand the hyper-competitive markets in which they operate. Examples of these new capabilities include those offered by mobile customer relationship management applications (m-CRM), which can automatically detect customers' business opportunities and immediately communicate with their smartphones to respond to them; or real-time travel patterns extracted from users' mobile devices, which can be used to build predictive transportation and urban mobility models, or recommendation systems. In these cases, the ability to act on data in real time improves the match between customer needs and the products and services offered by the tourism firm.

Analytical performance

Big Data technologies serve the new analytical paradigm consisting of explaining the causes of tourism phenomena from exploited data and not from the traditional deductive method based on establishing hypotheses that are later refuted by empirical evidence (Xu et al., 2020). Given the large scale of Big Data, it is now possible to reduce or eliminate the traditional limitations of conventional statistics due to the small size of the samples, thus overcoming the sample bias (J. Li et al., 2018). This brings to the table great advantages and opportunities to understand small- and large-scale tourism phenomena, making it possible to identify and visualize new behaviors that have remained hidden until now in view of conventional statistical techniques (i.e., the unequal distribution of the environmental impacts of tourism).

Second, the different data sources used by Big Data allow the same old phenomena to be explored from different perspectives simultaneously, thus increasing the degree of complexity of the tourism reality under analysis. For example, data obtained through GPS, smartphones, search engines, and social media in a place could be used to monitor the flow of tourists and address problems related to carrying capacity and overtourism. In other cases, geo tagged tourist sentiment data from social media could make it possible to explore human behavior in relation to a particular tourist product, place, or asset, or even attempt to explain tourist satisfaction. Structured data from public administration datasets could in turn be used to monitor the impact of tourism activities on the rest of the economy, or the global trends of tourism demand.

The multidimensional nature and granularity provided by modern data mining and Big Data analytics make it possible to exploit mixed methodologies to study tourism phenomena, both at a macro level (to understand the tourism system as a whole) and a micro level (to provide contextual information about more fragmented subsystems within the tourism ecosystem). In these cases, since the data sources are usually in the hands of different stakeholders, it will necessary to establish the collaboration of all of them to meet the final goals. Sometimes, gaining government support will also be key to guiding concerted efforts from tourism stakeholders and beyond and aligning Big Data-focused initiatives as much as possible (e.g., open data initiatives).

Fourth, since Big Data systems constantly capture data, usually through automatic data collection methods, there are historical series that allow longitudinal studies to be carried out that help tourism firms to understand disruptive changes and trends. Finally, Big Data opens new paths to understand the behavior of the individual within broader and more complex socio-cultural contexts. In other words, Big Data offers the possibility of making more precise projections on the interactions that occur at the micro level between individuals and at the group level, which would allow establishing connections, for example, between these interactions and the preference for certain products/services, and predict bookings or sales.

All the above could provide tourism firms with knowledge of enormous value when it comes to improving their products/services and supporting their business decisions. The opportunities seem endless. Furthermore, by incorporating Big Data into the organization's culture, firms could increase their sales as they capture new customers, test new segments, and retain existing ones; firms could experiment with new innovative products/services and customize them according to the tastes and preferences of consumers; they could identify current trends and set their pricing strategies in real time to be more profitable; and the list could be much longer (Del Vecchio, Mele, et al., 2018a; Peterlin et al., 2021; Stylos et al., 2021).

However, all these advantages and benefits come at a cost. The development of a culture based on Big Data will most likely require tourism firms to reengineer their product/service delivery processes, making them more efficient in terms of operating costs and adapting them to the level of service expected by customers. In short, a culture based on Big Data implies the transformation of the core elements of the firm's business model, reconfiguring them from top to bottom and making them depend not on their tangible assets, but on the data and the organization's capacity to extract value and knowledge from them. There is no doubt that the above would have a significant impact on the design of business marketing and on the effectiveness of the marketing actions of tourism firms.

10.4 Big Data Applications

Big Data applications are many and varied and are contributing to the transformation of tourism firms through a new range of technologies and technological innovations that have the potential to change the business of tourism forever. With Big Data technologies and data analytics, today it is possible to collect a large amount of data generated by tourists throughout the travel process and then extract it in the form of actionable insights for business. Therefore, business owners and managers need to start thinking about how to make use of Big Data applications so that the firm becomes "smarter" and its benefits have an impact at the level of the consumer, the organization, and the tourism sector in general.

Consumer applications

With Big Data tourism firms can analyze various types of tourist behavior, including behavior in the spatial, temporal, and spatio-temporal dimensions, as well as that associated with decision-making (Lv et al., 2021). Due to the popularity achieved by mobile devices and the development of technologies that allow these devices to be monitored, the Big Data thus generated allows the spatial and temporal behavior of tourists to be explored with considerable precision. For example, UGC and DGD can be used to analyze tourist behaviors in a time interval. For its part, the Big Data generated from online information searches and the information exchanged by users after purchases can provide valuable information to understand the consumption behavior of tourists and their decision-making process.

The UGC data from the opinions and attitudes of tourists in relation to their travel experience is a highly valuable source of Big Data that can be used to explore the attitudes of tourists – mainly those related to satisfaction/dissatisfaction and preferences. Unlike traditional data (e.g., surveys), which unveils "stated preferences", Big Data uses UGC data for "revealed preferences" from a much larger and therefore more representative sample of the public. Through the Big Data from online platforms and social media, tourism firms can analyze consumer satisfaction/dissatisfaction and preferences (e.g., quality of hotel service, location, mobility, etc.) thus avoiding biases caused by subjective sampling and predefined attributes.

Organizational applications

At an organizational level, Big Data applications are mainly focused on improving marketing management and contributing to better decision-making. Through Big Data firms are beginning to transform their marketing strategies, customer service, and even the way they promote products/services. Big Data can provide valuable insights for tourism firms to improve their marketing strategies and operations, and act on customer product/service personalization and recommendation. For example, the exploitation of Big Data obtained from customer reviews (extracted from platforms such as Tripadvisor) could be used to segment hotel customers; likewise, the analysis of large volumes of travel photos published by users on social media could serve to discover the attributes preferred by tourists in relation to a product/service or a place.

The use of Big Data and data analytics tools allows organizations to obtain unprecedented amounts of information that can be analyzed immediately and shorten the time cycle for decision-making. The implementation of Big Data in the organization makes it easier for the firm's employees to analyze large amounts of data in order to create more value for the firm, for example, by identifying the areas of the organization in which costs can be reduced, automating repetitive tasks, etc. Big Data can also be of great help to business leaders, who can make knowledge-based decisions such as what is the best way to meet customer needs through new products and services.

Industry applications

From the tourism sector point of view, the use of Big Data and advanced data analytics techniques promises to improve the predictions of various indicators of tourism demand. For example, Big Data can be used to predict the volume of tourist arrivals in cities and countries, using image and web search data from Google Trends. Other applications can focus on building predictive models of hotel occupancy (weekly, monthly) in certain destinations using Big Data time series obtained from web traffic data. Structured Big Data can also be combined with unstructured Big Data to increase predictions accuracy, for example, by combining data from government sources with web browser traffic data to predict tourists arrivals (Lv et al., 2021).

10.5 Challenges

The use of Big Data represents a breakthrough in the management of tourism firms. The incorporation of tourism firms into the era of Big Data implies far-reaching challenges that drive organizations to find a holistic approach to data, analytics, and information technology, to enable timely and accurate decision-making and sustained competitive advantage. The biggest challenge of Big Data is not only technical but also cultural and managerial (Kung et al., 2015; McAfee et al., 2012), meaning that organizations entering the era of Big Data not only need to change traditional IT architectures, but must fundamentally develop a data-centric culture that enables them to efficiently and continuously extract value from data. There is no doubt that most Big Data is of great value to the firm and, if used correctly, can become a key organizational asset that helps the organization achieve competitive advantage. However, most organizations make limited use of Big Data because they lack the necessary tools and/or do not understand the value of data. Not to mention that the level of general knowledge in organizations about the value of Big Data for decision-making is low (Olszak & Zurada, 2019).

To reap the optimal benefits of Big Data, organizations must transform and incubate a Big Data mindset, finding the best ways to generate insights and extract value from data. This requires firms to develop new strategies that guide them in a new direction. Ultimately, the firm is faced with a complex situation that must be addressed with a strategic vision that optimally combines these interdependent factors and reinforce them to generate unparalleled innovative solutions.

Be that as it may, the pace of adoption of Big Data (and by extension of data analytics) in tourism firms is slow and to a certain extent disappointing. This situation is highly influenced by the lack of vision on the part of owners and managers when deciding to integrate the organization's data with the firm's operational needs, but also by the lack of analytical skills of employees and a culture that does not guarantee that departments will share data instead of limiting its use (Stylos et al., 2021). No less important is the fact that any Big Data strategy is, by definition, expensive in terms of time and resources, which makes it difficult for many firms that are not able to find the pace that best suits them to adopt Big Data. Figure 10.3 shows the typical challenges tourism firms face when adopting Big Data.

Despite everything said above, owners and managers should note that Big Data is not yet implemented at a sufficient level in tourism firms to create formidable value, so there is not yet enough knowledge on the challenges and barriers that tourism firms actually face. Notwithstanding, some of the main challenges that tourism firms will most likely have to address with Big Data are discussed below.

Technical challenges

The tourism firm faces multiple technical challenges when dealing with Big Data and data analytics. Most of these technical challenges arise from the need to manage the large volume of data, its heterogeneity and noise, and to ensure the quality, reliability, and accuracy of the data. There is no doubt that tourism is a challenging ecosystem to address the processes of collecting and managing data sources and, in general, to put into practice the four Vs: volume, variety, velocity, and veracity.

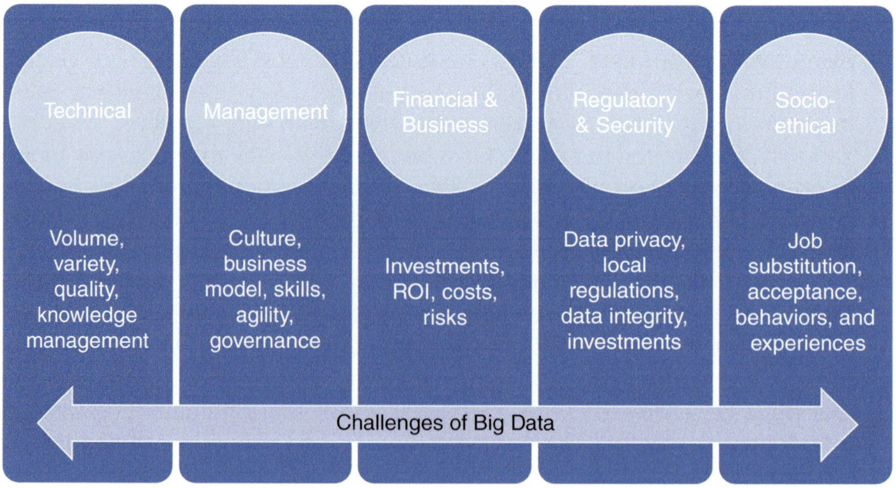

Fig. 10.3. Challenges of Big Data for tourism firms. Source: own elaboration.

In this challenging context, data silos play a critical role. Instead of data being collected in one place, it is often found siloed across different systems within the organization, therefore compromising data collection and processing, as well as affecting its quality and accuracy. It is still quite common for organizations not to share or enable data integration due to privacy regulations (Stylos et al., 2021). Most of the time it is simply because departments consider the data to be part of their business's own "competitive advantage" and prefer to have them safely stored away.

In practice, the underlying problem in many organizations is that data is not seen as a core business asset; nor is data analytics seen as one of its most profitable capabilities, but rather as a "competitive advantage" unique to a single department or business unit. Therefore, to develop innovative products/services from Big Data, organizations need to first uncover their hidden data reserves in departments and integrate the data stored in multiple internal systems with all kinds of external information. This highlights the importance of establishing knowledge management infrastructures in tourism firms that allow organizations and their key stakeholders to pool valuable knowledge resources that can be used at all levels of the organization.

Another major source of technical challenges for the tourism firm is found in the lack of a data standard. Data standards are important because they can speed up and make more efficient the processes of collecting, processing, and exploiting Big Data. Data standards prevent data from being subject to the arbitrariness of people or the departments in the organization that decide what, how, and why some data must be collected and managed to the detriment of others. The tourism sector should think about developing and adopting its own data standard as soon as possible if it really wants to give Big Data and data analytics a definitive boost and mitigate many of the risks that are currently associated with Big Data management.

In addition, there is the critical challenge of data access. Not all Big Data is easily accessible and shared as open data. While it is relatively easy to access data from social media and online platforms, data from mobile phones or sensing devices is subject to strict personal data privacy regulations. This makes data difficult to access or inaccessible to most actors in the tourism ecosystem, even though this data is often the key to analyzing tourist behavior and making decisions about business operations.

Closely related to data access is also the challenge of data quality, especially UGC. Consumers may provide false or low-quality reviews on platforms, social media, and online blogs, which do not correspond to real and objective situations (Lv et al., 2021). The situation is even worse if you think that there are organizations that create fake data to attract customers (i.e., by hiring spammers, who send huge numbers of fake reviews in favor of a firm or a product/service). This is making confidence in the quality of Big Data a growing concern for stakeholders in the tourism ecosystem.

The development of new analytical solutions to handle the large volume and variety of data is another endless source of technical challenges for tourism firms. As research into analytical methods advances, tourism firms' interest in these techniques has grown to unprecedented levels. Today, business owners and managers are increasingly concerned with accessing this type of technology and data scientists with applying it to everyday business practices. Meanwhile, tourism SMEs that cannot afford to hire data scientists seek new ways to collaborate with other firms, inside and outside of tourism, to pool shared resources and carry out data analytics tasks.

Many of the new analytical techniques are aimed at avoiding the algorithmic bias that firms face when using Big Data and data analytics. This bias occurs when the data used as inputs are collected, selected, or used based on certain human attitudes or values (Samara et al., 2020). As a result, the supposed neutrality of Big Data and analytics models is often lost, reducing the "explanatory" power of Big Data (e.g., machine learning-based models). Additionally, the fact that most of the algorithms currently used work in "black box mode", that is, they have internal weights and relationships between variables that are unclear and not transparent at all, means the loss of control over the process of data processing and the ability to make corrections or adjustments. Another factor that affects the quality of the algorithms has to do with the population bias. Generally, the samples subject to Big Data correspond to the part of the population that is more likely to use mobile devices, such as young people and the population with a higher educational level that knows how to use technologies. Therefore, the findings obtained from Big Data cannot always be extrapolated to the rest of the population (Xu et al., 2020), which negatively affects the explanatory power of the algorithms.

Management challenges

The most important challenges posed by Big Data are not of a technological nature but are found in how to create a data-centric culture and how to build business models that integrate business functions with those of IT and data management. The former is one of the biggest barriers to the effective implementation of Big Data in the tourism firm. It manifests in contexts marked by a lack of managerial leadership, the absence of an integrative view of data with IT and business processes, the resistance of employees, or a lack of resources to tackle the transformation towards a data-centric organization. This situation contrasts with the growing dependence that tourism firms have on Big Data and the provision of services that are complex and of a multidimensional nature.

When an organization is new to Big Data, it typically does not have staff with adequate Big Data management skills, its data management infrastructure is poor, and it does not have the processes required to acquire new types of data (Russom, 2013). Firms need to start increasing their efforts to train their employees and hire external staff who can help, usually consultants. Getting started is always tricky, but once serious Big Data efforts begin and the organization starts to move along the learning curve, there needs to be as smooth a flow of communication as possible between the business leaders and data and IT specialists. The choice of relevant data, the methods of data collection, processing, and analysis, and the means for interpreting and disseminating knowledge in the organization will be most effective when there is a fruitful dialogue between those who know about business operations and those who know about data and IT. A Big Data strategy that includes a data governance framework will surely contribute to managing Big Data projects more effectively and efficiently.

As regards the challenges of business models, firms that are capitalizing on the use of Big Data and analytics have realized that the key is to integrate Big Data as deeply as possible into their business models. This is the fastest and most efficient way to ensure that Big Data and analytics consolidate in the organization and that improvements are made to the competitiveness and performance of the firm. It is certainly important to strengthen data management infrastructures and skills as Big Data becomes more relevant, otherwise the firm will end up delaying the use of Big Data and not creating value for the business (Russom, 2013). However, focusing entirely on adding more ICT resources and capabilities alone will not help the organization start extracting new insights from its databases. Big Data just doesn't work like that. Firms that accept that data is a core asset of their business model are clear about their objectives and integrate state-of-the-art information and analytics as a critical component of all important decisions (Morabito, 2015). In this same sense, business leaders must answer questions such as: How far is the firm willing to go? How is Big Data going to improve business performance? What should the firm focus on?

Financial and business challenges

The challenges in this category are related to the costs that must be incurred and the investments that tourism firms must make to address Big Data and analytics. Some of the most relevant are the following:

- Hesitation from many owners and managers about the Return on Investment (ROI), since they fear that investments in Big Data and analytics are very high, and it is difficult to obtain an acceptable return on them.
- Concern about the high costs that firms must incur, both in terms of new hardware and software costs, as well as the "silent" (hidden) costs that are inherent in far-reaching cultural and organizational transformation processes whose results are uncertain.
- The need to develop a new business model if the firm really wants to take advantage of the full potential of Big Data and analytics. These new models must either focus entirely on the exploitation of data or look for hybrid models that combine the exploitation of the opportunities offered by data and intelligence with the firm's traditional sources of income.

In general, owners and managers are afraid to invest in Big Data and analytics initiatives due to the perceived high risks of failure and the belief that the organization do not have the resources, capabilities, and skills needed to succeed in the transformation towards a data-centric model. However, tourism firms very often forget that the true cost associated with the implementation and adoption of a new paradigm, such as Big Data and the data-driven model, is not to acquire new hardware and software to continue doing basically the same things, but to achieve what is known as a "disintermediation effect", that is, a reduction in the (transaction) costs that are generated from the adoption of the new paradigm together with the competitive advantages that arise when replacing the old business rules with new ones.

Regulatory and security challenges

Advances in smart technologies are bringing great benefits to businesses, users, and consumers in general, but they also bring significant threats to privacy and data security that may be more damaging than expected. The fact that different regulations concur in the same data, together with the scope of the different national legal frameworks, make Big Data management even more complex. Therefore, firms carrying out smart transformation must have a good understanding of the regulatory frameworks that are applicable and the implications that there are for the technical, strategic, and governance dimensions of Big Data (Kemp, 2014). Although these concerns do not yet reach the importance of other challenges, it is realistic to say that as the adoption of Big Data and data analytics increases, they will gain in relevance.

Data privacy issues are currently a source of great concern to firms of all kinds, consumers, users, governments, and the academic community around the world. So, what can tourism firms do to solve Big Data privacy issues?

Traditional access control and data storage mechanisms to ensure privacy have proven insufficient with the growing demand for more stringent privacy requirements in the era of Big Data. Data privacy issues affect not just a few in the organization but encompass all those involved in the data life cycle, including those who collect the data, those who perform data mining, and those who use the information to generate knowledge and make decisions. Therefore, it is essential for firms to have much more precise and granular data storage and access control mechanisms that guarantee all aspects of privacy. The fact that each country has its own data privacy regulations makes it difficult to have universal data protection standards, sometimes leading to unethical behavior in which the privacy of consumers is violated, and the data is resold to intermediaries (resulting in higher data cost for firms). Moreover, as firms increasingly operate their data in cloud environments, there is the additional problem of having to guarantee the privacy of data in the cloud, which makes everything even more complex.

Preserving information privacy is a major concern, but providing security is another big issue that tourism firms urgently need to consider carefully. Data security has always been a problem for business, however, in recent times, with the exponential growth of data volume and the growing threat of cybercriminals, the problem has taken on a new dimension. Firms around the world face significant information security issues that are both specific to each organization and to the channels through which information flows massively (e.g., social media, mobile apps, etc.). Additionally, the amount of sensitive information that needs to be protected in tourism is constantly increasing, creating ongoing security challenges (Larson & Chang, 2016; M. Singh et al., 2018).

The conventional security measures used until now to protect static and small-scale data are no longer suitable for Big Data. This poses a huge challenge for firms that don't have enough time and resources to deal with security issues. One of the biggest security risks is the privacy leak, which has caused severe trouble to many firms. Data integrity issues are another major concern. The veracity and value of Big Data can be compromised if the firm is not equipped with a robust process capable of ensuring data integrity and disaster recovery. Furthermore, Big Data poses significant technical challenges associated with virtual machine backup, multi-site replication, as well as data governance.

Those firms that take data security seriously work to secure the entry points to information systems, so they can detect cybercriminal attacks and alert users before they even happen. These firms use advanced information encryption techniques to protect access to their data and frameworks that include identity management, firewalls, and encryption. However, despite all the efforts made by firms and the fact that there is a thriving multi-million dollar industry focused on developing solutions that protect the information in the hands of organizations and mitigate the damage caused by cybercrime, the reality is that current technologies cannot fully address all security issues. Therefore, a coordinated effort between firms, the technology industry, and governments is necessary to take administrative and legislative measures that regulate the privacy and security of all parties interested in the use of smart technologies. The introduction of the General Data Protection Regulation (GDPR) in Europe in 2016 is an example of how the regulatory framework can determine how Big Data is collected, processed, and made available.

Socio-ethical challenges

Among the most relevant socio-ethical challenges of Big Data are the substitution of work by technology and the growing intervention of data and automatic processing capacity through robotics and artificial intelligence in the development of society (Samara et al., 2020). In the case of work, the challenge is not only in the effects that machines can have in replacing humans with robots and intelligent machines, but in how we humans can use them to increase our skills and capabilities. Despite the great potential that the latter presents, the fear of job loss due to the threat of machines is one of the main social concerns. For this reason, this is a field that continues to spark heated debates and in which researchers strive to find an optimal solution that allows human–machine and human–robot peaceful coexistence for the benefit of humanity.

Another concern that is relevant from a socio-ethical perspective is the acceptance by modern societies of the growing role that Big Data, data analytics, and artificial intelligence applications are starting to play in the workflows and tasks of our daily life. According to some authors, the very acceptance of robotics (Murphy et al., 2017) can make users feel more and more isolated, which would affect the uses and customs of consumers and their behavior patterns.

In the case of tourism, it is worth asking: to what extent is the use of these smart technologies aimed at improving the experience of consumers who enjoy their vacations and not at replacing humans with robots? The answer is that there is evidence that demonstrates both intentions, although it is still early to take sides with one or the other. Only time will tell how we managed to use these tools that changed our lives forever.

Discussion Questions

- What levers of change are accelerating (or delaying) the implementation of Big Data in tourism firms? What organizational changes are necessary to optimize the performance of Big Data in tourism firms?
- What is the set of skills and capabilities (business, analytical, and technical) required to handle Big Data successfully? Do tourism firms have the necessary skills to handle Big Data? What role does education and training play in building these new capabilities?
- How do the employees of tourism firms perceive the Big Data phenomenon? Are they preparing at the speed and with the skills required to tackle the challenge?
- What are the critical success factors to increase the performance of Big Data in tourism firms?
- What kind of policies and regulations could be adopted to deal effectively with data security, privacy, and governance issues? And to improve democratic access to data by stakeholders in the tourism ecosystem?
- How is Big Data impacting the existing business models of tourism firms?
- What are the complementary investments that tourism firms must make to exploit the potential of Big Data? What is the cost of not adopting Big Data for a modern tourism firm?

11 Data Analytics

Data analytics is the process of discovering, interpreting, and communicating insights that are meaningful to the firm from patterns observed in massive amounts of data. In other words, analytics uncovers information that is hidden in the data and that the firm might never discover if it were not for analytics methods and technologies. When analytics is in the context of a business vision and serves the specific goals of a firm it is also known as business analytics.

The data itself has little meaning and adds no value to the firm. Firms could try to collect all the data in the world and dedicate unlimited resources to store it, but unless they do something with it, treat it, organize it, and process it, all the effort will be wasted, and they will never recover their investment. Fortunately, advances in analytics technologies are creating new opportunities for tourism firms to capitalize on their data and act on the insights it provides. Modern analytics techniques are increasingly advanced and sophisticated and allow firms to explain not only what has happened through descriptive techniques, but also to carry out predictive analysis that anticipate what could happen, self-learn and self-adapt to changes in the environment to uncover hidden patterns in the data, and answer key business questions. These tools are becoming more intuitive and easier to use, even without being an expert or having advanced knowledge of mathematical methods and statistics. They also incorporate powerful visualization capabilities that allow millions of data to be interpreted in an instant. They are also mobile and allow business owners and managers to see data signals anytime, anywhere, before their competitors.

These advances represent an enormous potential available to tourism firms to improve business management. Firms can now "talk" to their data and trust it to make more informed decisions that help increase sales, reduce costs, and deliver speed and simplicity to customers. Understanding what data analytics is about and implementing an analytics framework integrated with Big Data is therefore a critical step in any smart transformation strategy.

DOI: 10.1079/9781800620261.0011

11.1 The Data Analytics Workflow

Data analytics is about capturing and processing historical data that affects the business, analyzing that data to identify patterns, trends, and causes, and making decisions based on the insights generated. Therefore, it is a top-down approach made up of many data-intensive components that must work together to generate the answers the firm is looking for.

Although data analytics tools and technologies process data and create knowledge through reports, graphs, and other visualization tools, they cannot solve the problems of data warehousing, data searching, data sharing, or real-time analytics. They also need the right infrastructure in place to incorporate that data. Furthermore, the practical application of the insights generated on the behavior of tourists to something that is useful for the business it is especially challenging for any firm considering that data analytics and business intelligence applications are still rare in tourism firms (Mariani et al., 2018). These capabilities will need to be enhanced as firms increase their capabilities to collect real-time tourist data (i.e., through mobile devices, social media, platforms, websites, vendors, etc.) and seek to track travel patterns in a more effective way.

The integration of data analytics and Big Data management provides the processes, infrastructures, methods, and technologies required for effective and efficient data management in modern smart organizations. Both are increasingly critical capabilities for tourism firms around the world as they manage increased volumes of data and address smart transformation. Typical components that are part of the data analytics workflow are shown in Fig. 11.1.

- **Data collection/generation:** As the data that feeds data analytics comes from very diverse sources (e.g., spreadsheets, databases, mobile applications, social media, Internet of Things (IoT) devices, etc.), it is necessary to start by grouping and storing them in a centralized repository so that people can access them later. There are multiple options for storing data: in some cases, data is stored in data warehouses or data lakes, in other cases, cloud databases are used.

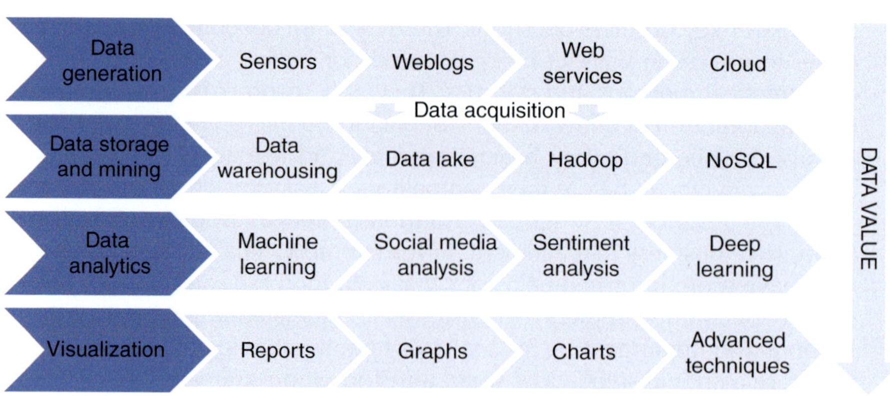

Fig. 11.1. The data analytics workflow. Source: own elaboration.

- **Data storage and mining:** Once the data has been collected and stored, the firm must classify, order, and process it. At this stage, it is very useful to use machine learning algorithms because they help to quickly recognize patterns and execute repetitive actions at high speed (e.g., setting metadata), allowing data scientists to focus their attention on exploiting the information instead of wasting their time on repetitive manual tasks that machines can do more efficiently.
- **Data analytics:** It can be descriptive or predictive. The first tries to answer the question "what is happening" and "why is it happening" and helps to better understand the story behind the data. The latter consists of building predictive models based on trends and descriptive analysis results, thus informing future decisions on operational, organizational, and business issues.
- **Visualization:** These tools and techniques help users interpret analytics results, thus making this task easier for the human brain (i.e., by combining types of graphs, using colors, mixing reports with text and numbers, etc.). Visualization skills are intended to enable anyone, whether a data scientist or an owner or manager, to quickly and easily understand the meaning of the new insights generated.

A smart data analytics toolbox should be filled with cloud analytics tools, which allow organizations to consolidate data from different areas and departments (e.g., sales, marketing, human resources, finance) and thus obtain a unified view of the firm's key figures. Taken together, the visualization, prediction, and scenario modeling tools provide all sorts of unique insights across the organization.

11.2 Methods and Technologies

A robust technological architecture that supports data analytics in the tourism firms should include several levels:

1. The level of the data source systems, from which data is extracted, transformed, and loaded (ETL) in the data warehouse or data lake.
2. The data integration-presentation level, where data is combined, manipulated, cleaned, and validated.
3. The analysis level, where multidimensional queries and optimized analytics are performed.
4. The visualization level, where data is presented to users using data mining techniques, statistical analysis, and other analytics technologies.

The most used methods and technologies at each level are discussed below.

Data source level

One of the biggest challenges facing organizations seeking to move forward with advanced data analytics is how to automatically collect massive amounts of data in non-standard formats and where to store them for further processing. This issue becomes even more critical when the organization must combine different and heterogeneous data sources (both internal and external), as data integration and quality issues arise. Nearly 70% of the time organizations spend on data analytics initiatives has to do with identifying, cleaning, and integrating data. "Good" data is often difficult to locate because it is scattered across many applications and systems, and must be reformatted to make it easier to analyze and update (Chessell et al., 2014). In short, data collection for analytics is often a heavy workload for the teams in charge to provide the right quality data within the organization.

To address these challenges, many organizations implement data lake solutions. The idea of a data lake is quite simple: it is about putting the data in a data repository in its original format, thus eliminating the tasks and costs of transformation before loading it. Once the data is loaded into the data lake, anyone in the organization can access it whenever they need to for research and reporting. The data lake's ability to store and process vast amounts of highly varied data at low cost has made it the perfect place to perform ETL operations and meet the needs for greater agility pursued by smart organizations (Fang, 2015).

Data lakes are different from data warehouses (Table 11.1). These are meant to consolidate data from many different sources and create a single version of the truth. A data warehouse is a highly structured system with a data model that must be carefully designed before loading the data, because if you want to change it later, it is complicated and time consuming. A data warehouse typically works in batches and is ready to support thousands of simultaneous accesses from users who create reports or perform analysis.

On the other hand, data lakes are centralized repositories that can store structured and unstructured data at any scale in raw format, without the need to transform the data beforehand. They are closely related to the Apache Hadoop ecosystem of open source solutions, which is a popular framework

Table 11.1. Differences between data lakes and data warehouses. Source: adapted from AWS (2022).

Characteristic	Data warehouse	Data lake
Data	Relational from transactional systems, operational databases, and business applications	Non-relational from websites, mobile apps, social media, IoT devices, and corporate apps
Schema	Designed in advance of data storage ("schema on write")	Written at the time of analysis ("schema on read")
Performance/cost	Faster query results with higher storage cost	Faster query results with low storage costs
Data quality	Curated data	Data that may or may not be curated (raw data)
Users	Business analysts	Business analysts, data scientists, data developers
Analytics	Batch reporting, business intelligence, and visualizations	Machine learning, predictive analysis, data discovery, and profiling

because it provides a cost-effective and technologically sound way to meet the challenges of Big Data analytics in business firms. Data lakes support different types of analytics, from simple dashboards and visualizations to Big Data processing and machine learning at a lower cost. However, data lakes also have some drawbacks. One of the most important is its inability to determine the quality of the data, and the risks associated with data security and access control, as data is often loaded into the data lake without content monitoring. In terms of performance, data lakes and data warehouses offer similar metrics.

As we can see, the data lake and the data warehouse serve different needs and use cases, so the tourism firm may need both types of storage solutions as it moves through its smart transformation journey. That said, can the data warehouse and data lake work together? Yes, they can. This can be achieved by converting the firm's data storage system into a hybrid framework consisting of a data warehouse and a data lake. This way the firm extends the capabilities of the data warehouse by allowing data from the data lake to feed the queries that users perform against the data warehouse. With such an architecture, data sourcing and analytics become transparent to end users, as the data system itself determines which data is used to answer each query.

Integration and presentation level

Probably the second most difficult question that data and analytics-centric organizations must answer (after how to extract and integrate data for processing) is how to extract "useful" insights from data to support better decision-making. Data mining allows firms to understand the patterns that are hidden in large volumes of data (i.e., patterns in users' web browsing, patterns in past purchase transactions, etc.). This knowledge is highly valuable for firms because it can help them plan and launch new marketing campaigns quickly, and to know if the firm's products and services are attractive enough for customers or if the firm needs new suppliers or partners.

Data mining can be applied to many types of data, such as database data (e.g., relational databases, NoSQL databases), data in data warehouses, data streams, spatio-temporal data, time series, IoT data, etc. (F. Chen et al., 2015). One of the richest data sources that tourism firms can apply data mining techniques to is e-commerce, as it injects many data records into data warehouses (e.g., customer data, product data, billing data, etc.) that can be exploited to extract valuable knowledge for the firm. Some of the main data mining techniques available to the tourism firm are discussed below.

- **Classification:** It is the process of finding a set of functions (or attributes) that describe classes of data, to accurately predict the class of an object whose class is unknown. Classification is important in decision-making processes. It can be used, for example, to identify customer segments with low, medium, or high credit risk.
- **Clustering:** It is an unsupervised learning method that analyzes data objects without even considering a previously known class model. Clustering

uses algorithms that divide the data into meaningful groups so that the patterns in one group are similar and the patterns in other groups are different. For example, a search engine may group billions of web pages into different clusters, such as news, reviews, videos, etc.

- **Association analysis:** It is the discovery of (association) rules that show that certain attributes appear normally in a data set. This type of analysis generates a general and qualitative knowledge that can be very useful for decision-making processes, such as when analyzing data on commercial transactions, shopping baskets, etc.
- **Time series analysis:** It encompasses a set of methods designed to analyze a set of temporal data objects (usually large volumes of data with high dimensionality and regularly updated) and extract relevant statistical significance and other features.
- **Outlier analysis:** It seeks to find patterns in data that are different from the rest of the data set. These patterns normally contain information that is valuable in explaining the abnormal behavior of a system based on the recorded data of its operation.

These data mining techniques are often used in combination to try to extract more information and relevant knowledge from a specific data set. For example, association analysis is usually combined with clustering analysis techniques to provide insights into the combinations of products that were purchased by customers, thus encouraging customers to buy other related products that they may have forgotten or overlooked.

Other common data mining use cases include:

- Monitoring and analysis of user behavior during web browsing to find similarities and browsing patterns that can be useful in designing more user-friendly websites that respond to user needs.
- Analysis of the preferred content consulted by users to identify and group similar profiles to whom potentially interesting content should subsequently be offered through recommendation systems.
- Analysis of online transactions to improve the accuracy of demand forecasts for a given product or service and optimize product recommendations to users.

There are also good examples of data mining use cases in destinations management, such as analyzing the needs of tourists to predict the performance of a certain product or service; the automation of operational decision-making processes; or the search for association rules between some phenomena linked to tourism and its spatial dimension (i.e., the relationship between the concentration of tourists in certain areas of the city and the consumption of certain products and services).

Analysis level

There are three main types of data analytics techniques that tourism firms can use to drive decision-making: descriptive, predictive, and prescriptive analytics. These methods become especially powerful when used together (Fig. 11.2).

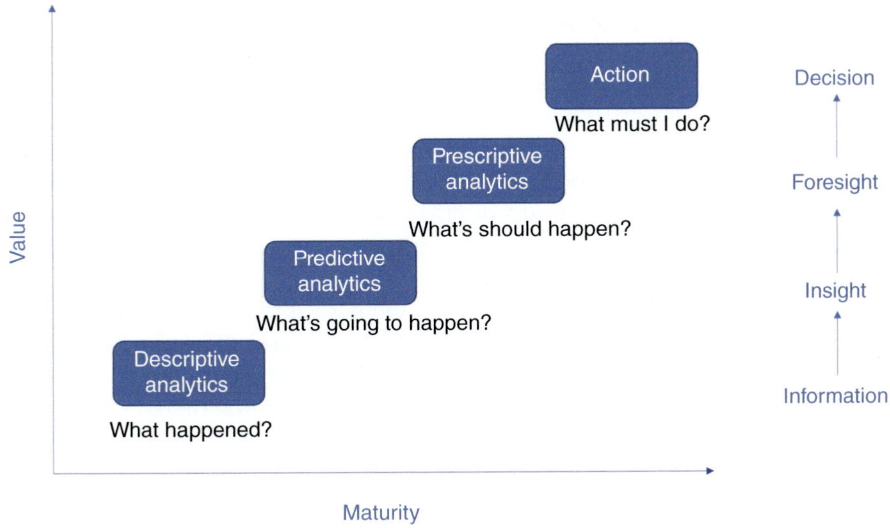

Fig. 11.2. Data analytics value creation scale. Source: own elaboration.

Descriptive analytics focuses on the analysis of historical data to understand what has happened in the past. It uses basic techniques based on data aggregation and data mining, therefore any findings are usually easy to understand for a business audience. Descriptive analytics typically displays outcomes using visual data representations such as graphs, charts, reports, and dashboards that can be easily aggregated to provide a clear picture of the operations of the firm (e.g., revenue reports, inventory reports, sales reports, etc.). Another applicable use case can be found in social media with tools such as Google Analytics and Facebook Audience Insights, which provide data groupings based on counts of events such as clicks and likes. Descriptive analytics is often the basis for delving into more complex analyses.

Predictive analytics is a more advanced analytics technique that provides information about what could happen in the future. It uses statistical models and machine learning methods to determine the probability of future outcomes based on historical data. Through predictive analytics new opportunities arise for decision-making through controlled experiments that can help inform key operational business processes (i.e., organizations can test different hypotheses and then analyze the results). This way business owners and managers can differentiate between causality and correlation and, therefore, improve the consistency of the results and the rigor of the business opportunities detected. For example, organizations are using predictive analytics to find behavior patterns in tourist movements; predict whether a product or service is going to be successful in the market; or optimize marketing campaigns by detecting cross-selling opportunities. One growing field within predictive analytics is deep learning, which mimics the decision-making process of the human brain to make even more sophisticated predictions.

A weakness of descriptive and predictive analytics methods is that they are often based on official records produced by public statistical organizations, whose reliability has sometimes been questioned due to the lack of harmonization of collection methods, the updating of the data, and the statistical procedures used. On other occasions, the data is based on surveys carried out in the "traditional" way, which means that many types of tourists go unnoticed. One potential solution to improve the quality of data for analytics is to harness the digital footprint left by millions of people online across technology platforms and connected devices. However, this does not ensure that the analytics will take into account all the tourists who travel to a certain place or consume the firm's products and services.

Prescriptive analytics shows firms what should happen in the future or which option is the best among several alternatives. These techniques use a variety of statistical and mathematical methods that are applied to a variety of descriptive and predictive data. Prescriptive algorithms detect possible decision patterns that could have different consequences for the business. Through prescriptive analytics owners and managers can know in advance the possible repercussions of certain decisions and decide what is the best course of action before putting them into practice, thus reducing the costs of making mistakes. Prescriptive analytics can guide the organization in the management and allocation of resources more efficiently and effectively, since it uses optimization engines and statistical methods that allow millions of possible alternatives to be analyzed fast and the best decision to be recommended.

Last but not least, modern data analytics techniques require knowledge of some computer programming language, and database management systems to get the most out of it. However, the workforce of tourism firms does not always have these skills and capabilities. Similarly, firms require information systems that are well designed, organized, and managed. At times, tourism firms, especially smaller ones, will need to consider establishing collaborations with other firms and stakeholders, both within the tourism ecosystem and externally, to overcome these difficulties and have access to the personal, technological, and managerial resources necessary to make Big Data analytics a reality.

Machine learning

A particularly relevant set of methods and algorithms that is often confused with data mining is machine learning. This is actually not that surprising, because both data mining and machine learning use algorithms to discover patterns in data, analyze correlations, and extract insights from unstructured data. However, although there are many people who assimilate them, the outcomes they provide are different.

While data mining aims to extract patterns from large amounts of data, machine learning is an artificial intelligence that allows a system to learn from data rather than through programming. In other words, data mining relies on human intervention and determines an outcome based on the collected data, whereas machine learning trains an algorithm to perform complex tasks without relying on human intervention, so that as it ingests more training data it can learn and generate more accurate models based on that data. After training, when the model is fed an input, the machine learning algorithm will output a forecast based on the data used to train the model. Table 11.2 shows a summary of some of the most common predictive algorithms used in data analytics.

Machine learning techniques are essentially used to make predictions in supervised, unsupervised, and reinforcement learning environments (Noor & Haneef, 2020). Supervised learning is performed on training data; i.e., the machine uses some input data that has been previously labeled or classified for learning. Labeled data contains associated information in the form of a label, tag, or class, which gives them a significant meaning. For example, an image of a woman with a tag that identifies it as an "image of a woman".

Unsupervised learning is performed with training data that does not have associated classes, tags, or labels to describe it; therefore, it needs to be grouped based on similarity. For example, an image of a woman that does not have a title that allows the machine to understand that it is an "image of a woman". Unlabeled data can be photos, videos, or audio without a title or explanation. Social networks are usually analyzed using unsupervised learning techniques as social data is unstructured.

Reinforcement learning differs from the other types of learning because the system is not trained on data, but instead learns by trial and error as it interacts with the changing environment. Every time the system performs an action, it gets feedback in the form of "punishment" or "reward", so a sequence of decisions that are correct will reinforce the process, since it better solves the problem.

Table 11.2. Most common predictive algorithms. Source: own elaboration.

Algorithm	Rationale
Linear regression	Supervised learning technique that helps model linear relationships between a dependent variable and one or more independent variables
k-nearest neighbors regression	Supervised learning technique that searches the observations closest to the one to be predicted and estimates the output value as an average of the most similar data points
Logistic regression	Method that estimates the probability of a binary qualitative variable as a function of one or more quantitative independent variables (or predictors)
Bayesian networks	Probabilistic graphical model that allows the representation of the dependency relationships between a set of variables
Decision trees	Predictive model of decisions and their possible consequences in the form of a tree
Artificial neural networks	Predictive model in the form of a graph of nodes that emulates the behavior of neurons in the brain

Deep learning

Over the last decade, the increased availability of computing hardware that works in parallel, advances in machine learning algorithms, and the availability of labeled data have led to the renaissance of artificial neural networks. Deep learning is a machine learning method that consists of complex neural networks made up of three or more layers that iteratively learn from data. Over time, "neurons" iteration makes systems based on deep learning "smarter" and capable of uncovering hidden information, relationships, and trends, thus uncovering new opportunities, from consumer preferences to optimization of key marketing processes. Deep learning automates much of the mining process, removing some of the human intervention that is often required. It also allows the use of large data sets, which makes it work better with unstructured data. A deep learning model typically requires more data to improve its accuracy, compared to a regular machine learning model.

Since 2012, the world has witnessed an unprecedented interest in the use of deep learning techniques to solve an increasingly wide range of complex problems related to vision and spoken and written language. Today, practical applications of deep learning can be found in social media (e.g., object recognition in photos) and on smart devices (e.g., Alexa or Siri). Google has more than 1000 deep learning projects currently in development across all of its major products, including Google Search, Android, Gmail, translation, Maps, YouTube, and self-driving cars. Another example is DeepMind, which is a deep learning program that learns to play dozens of Atari games as well as (or better than) humans without being given specific instructions on how to do it. This program, trained differently, became AlphaGo, which defeated the GO board game champion in 2016.

Deep learning is a field at the forefront of research initiatives and funding programs to improve artificial intelligence. It has become virtually a prerequisite for all entrepreneurs building sophisticated software applications, such as those that incorporate natural language processing and vision (Parloff, 2016). There seems to be no end to these developments, as today almost anyone has access to free software and researchers are focusing their efforts on new and more powerful algorithms that lead to cumulative learning, including the not-so-distant possibility that smart computers can write new programs themselves.

Sentiment analysis

Sentiment analysis is a subfield of natural language processing (NLP) that allows text to be automatically classified by valence. With the help of sentiment analysis techniques it is possible to examine texts that contain explicit consumer opinions about firms, products/services, brands, events, or any other topic that interests people (Bazzaz Abkenar et al., 2021; Kauffmann et al., 2019). The importance of sentiment analysis stems from its ability to understand the polarity of sentiment in large volumes of text (i.e., classify reviews as positive or negative). For example, sentiment analysis can help determine the attributes of a product or service that should be improved, how customer services can be improved, how to design better marketing campaigns, etc. Sentiment analysis can also be used to rank customers or users based on their opinions, or to improve recommendation systems using positive and negative customer feedback.

The three most common methods used today for sentiment analysis are:

- **Lexicon-based methods:** use a set of predefined word lists, corpus, and dictionaries (e.g., SentiWordNet, Affective Lexicon, SenticNet) to extract subjectivity and polarity from opinions and comments.
- **Learning-based methods:** use machine learning algorithms (both supervised and unsupervised) to classify texts into positive or negative classes.
- **Hybrids:** combine lexical-based and learning-based approaches.

Sentiment analysis can be applied at these three different levels (Z. Chen et al., 2021):

- **At the document level,** the goal is to determine the general opinion about a particular entity, such as a product, hotel, tourist destination, etc.
- **At the sentence level,** the polarity of opinions is analyzed further by dividing a document into sentences that can contain different opinions and are classified into positive or negative classes.
- **At an aspect level,** which seeks to recognize expressions of sentiment and identify different opinions about an entity when there is a variety of attributes. For example, what are the most negative terms that dissatisfied business customers use during their stay in a hotel.

Sentiment analysis shares many of the same problems as NPL. In some domains or cultural contexts a word or expression may have an opposite meaning; there are sarcastic phrases that may express a sentiment that is the opposite of that indicated by their grammatical structure; there may be sentences that do not use sentiment words but contain sentiment information, etc. (Jandail & Ratan, 2014). Moreover, there is a bottleneck in sentiment analysis due to the scarcity of dictionaries and tools in different languages. Most of the resources are only available in English and there is not enough corpus for languages like Chinese, Spanish, French, and others (Bazzaz Abkenar et al., 2021).

Another major challenge is fake reviews, which render platforms and consumers unable to tell the difference between a real review and a fake one. When a customer wants to buy a product/service online, it is very common to spend time reviewing the ratings made by previous customers. The more positive reviews the product/service has, the more likely it is that the customer will decide to make a purchase. This confronts tourism firms with the difficult task of striking a balance between promoting online reviews while providing useful and truthful information that drives consumer decision-making. Unfortunately, although there are commercial algorithms that are capable of detecting these fake reviews in increasingly precise and sophisticated ways (i.e., systems that analyze both reviews and reviewers), it is very difficult to detect with absolute certainty whether a review is fake or not. Sometimes this situation is exploited by organizations and individuals to use spammers (or even bots) who submit huge number of fake reviews that increase the popularity of their product or discredit their competitors' products. These deceptive practices are known as "review spam".

Visualization level

Data visualization is the graphic representation of information and data. Data visualization tools provide a way to understand historical series, trends, outliers, and patterns in data by using visual elements such as charts, tables, and maps. They are essential when working with Big Data analytics and to support data-driven decisions. Good visualization always tells a story, removes noise from data, and highlights the information that is most useful, such as trends and outliers. However, getting a good visualization isn't as easy as just creating a chart and embellishing it. Effective data visualization is a delicate balancing act between form and function (Tableau, 2022).

Some of the most frequent types of data visualization techniques used today are the following:

- **Univariate analysis:** A single feature of the data is used to analyze almost all of its properties (e.g., distribution plots, box-and-whisker plots, violin plots, etc.).
- **Bivariate analysis:** Data is presented graphically comparing two variables (e.g., line plots, bar plot, scatter plot, etc.).
- **Multivariate analysis:** Data sets with more than three variables are represented graphically (e.g., scatterplot matrix, hyperbox, trellis displays, etc.). It is the most challenging visualization method compared to the others.

11.3 Social Media Analytics

Social media analytics (SMA) is about extracting useful insights from social media data by collecting, cleaning, analyzing, summarizing, and visualizing data, with the aim of helping organizations and individuals to improve their decisions in different spheres of life. Traditionally, data on consumer behaviors and interests was collected through questionnaires (still widely used), but with the rise and popularity of social media it is now possible to collect vast amounts of data on consumer behavior. In this vein, social networks (e.g., Facebook), microblogs (e.g., Twitter), and media-sharing platforms (e.g., YouTube, Instagram) are platforms where online users are not only consumers, but also producers of content and social data that can be exploited using Big Data techniques (Sebei et al., 2018). The most popular social media provide application programming interfaces (APIs) (e.g., Facebook API, Twitter API) through which data analysts can send queries and extract large amounts of relevant data that can be later converted into useful knowledge for firms. When data is not accessible via APIs, web crawlers can be used to crawl social platforms and extract the desired data.

Since social media data is unstructured and produced in large quantities, traditional analytics techniques are no longer capable of processing it at the scale of volume, speed, and variety that is generated. This raises the need to develop new technologies that help improve the capacity and performance of traditional analytical techniques and achieve reliable insights. The combination of Big Data technologies with SMA techniques, known as Social Big Data, have enormous potential to extract value that is useful for the business (Sebei et al., 2018). Social media platforms themselves are a clear example of Big Data sources, since they are characterized by an exponential growth of large volumes of heterogeneous data (e.g., text, photos, videos, audio), thus making the application of SMA something natural.

Tourism activities and tourism firms have been strongly influenced by social media in the last decade as more and more tourism consumers join every day. SMA in tourism uses ICT to collect, process, analyze, and visualize data, and transform it into useful knowledge that serves to improve both tourism products and services and the tourism experience (Viñan-Ludeña, 2019). The insights extracted from social media are essential today for tourism firms that seek to identify new business opportunities, find hidden risks, optimize their performance, and increase their competitive edge. For this reason, tourism social media are an ideal field of application of SMA with great potential ahead.

Social media data

Social media data can be categorized into three different types: user-generated content (UGC), device-generated content (DGC), and transactional data. Let's review each of them in more detail below.

User-generated content (UGC)

UGC mainly includes two types of online data: text data and photo data. Text data includes a wide variety of tourism-related information such as reviews, recommendations, and experiences in which tourists talk about their satisfaction with products and services on social media platforms, travel blogs, and booking sites (e.g., Tripadvisor, Facebook, Booking, Expedia, Ctrip, Twitter, etc.). Photo data published by tourists on social platforms contains very rich information, such as location, time, or comments, which can be used to analyze the behavior of tourists on a trip, describe their profile, and know the routes most traveled and the duration and places visited on their trips.

Device-generated content (DGC)

DGC includes GPS data, mobile network data, roaming data, Bluetooth data, and Wi-Fi data. GPS data is global and quite accurate compared to other technologies, as well as relatively easy to obtain and inexpensive. With GPS data tourism firms can analyze the spatial and temporal behavior of users who use mobile devices with integrated GPS. For example, firms can identify the trajectory of tourists, model tourist routes to predict the behavior and plans of tourists in places, segment users, and offer itinerary recommendations considering the information collected from weather stations.

Mobile network roaming data with its great coverage and rich information can be useful to provide new perspectives for tourism research. Although the exploitation of roaming data has just begun, many destinations use this data for the management of tourist flows and the spatial distribution of visitors, loyalty, and marketing actions, optimization of travel times etc. However, acquiring this data is difficult because it can lead to privacy issues by monitoring users, thus limiting its use.

Bluetooth data is generated on mobile devices carried by tourists that can be detected by Bluetooth sensors. Bluetooth technology can reveal the individual behavior of a large number of tourists and provide valuable information about their location and the route traveled. However, due to the limited reception range of Bluetooth data, the number of applications is limited to very specific areas or planned tourism activities (Z. Chen et al., 2021). Wi-Fi technology surpasses Bluetooth technology by allowing greater spatial coverage through wireless access points that facilitate monitoring of tourist behavior. Just like Bluetooth data, Wi-Fi data introduces privacy issues with user data.

Transactional data

Transactional data comes from two sources: website search data and online transaction data. Firms can use the traces that users leave each time they perform an online search through a search engine to learn about the online behavior of tourists, understand what tourists pay attention to, and know what the online purchase decision-making process is like. This data can also be valuable in predicting the traffic to a certain tourist place, or the number of customers to a hotel (e.g., using Google Trends, Baidu Index).

Data from other tourism-related transactions, such as online reservations, ticketing, and online purchase of travel products, are important sources for analytics. For example, a hotel website registers data such as reservations and online purchases from its guests, has data on room occupancy, and knows data on consumables such as electricity, water, and waste, which can be valuable in helping owners and managers to improve hotel management. When a tourist makes a purchase of a travel product/service with a credit card, the data is also recorded and stored, and can be valuable for analyzing the tourist's purchasing behavior and designing a personalized experience.

SMA methods

SMA examines vast amounts of data available on social media platforms through advanced analytics techniques. These techniques reveal hidden patterns, discover unspecified correlations, and generate valuable insights not possible through other traditional analytical techniques. Data sets from social platforms are growing so fast and are so complex that getting meaningful insights from this huge amount of data, opinions, and sentiments is a challenge for any organization. As a result, owners and managers must recognize the need for advanced technological tools and infrastructures in addition to analytical techniques (Mirzaalian & Halpenny, 2019). The SMA workflow uses a combination of techniques that range from web crawling, NLP with machine learning, and statistical methods.

Web crawling technology is used to extract implicit information from travel-related textual data at the collection stage. Methods used in data preparation and analysis depend on the purpose and scope of the analysis, and include data cleansing, tokenization, word stemming, and part-of-speech tagging (POST). Some multifunction tools can handle these data processing tasks, including WEKA (Waikato Environment for Knowledge Analysis), LingPipe, and TextBlob. Pattern discovery is also important in text data mining, including tasks such as sentiment analysis, text classification and summaries, statistical analysis, dependency modeling, and latent Dirichlet assignment (LDA) (Z. Chen et al., 2021). Unlike text analytics, which is broad in scope and seeks to analyze textual data to extract machine-readable facts, sentiment analysis analyzes subjective opinions and feelings about a product/service using different analytic methods like NLP, computational linguistics, and analysis of UGC texts. Together, these methods allow organizations to monitor trending topics, mine opinions, and analyze georeferenced information. Specific uses in tourism include analyzing the differences between satisfied and dissatisfied customers, understanding preferred attributes of hotels, or identifying top hotel customer concerns.

When it comes to analyzing social media photos, the original online photo data is pre-processed using data cleansing and text extraction techniques to generate metadata. These metadata can then be grouped using clustering techniques by traveler origin, travel time, type of tourist attractions, etc. With this information, the travel trajectory can be studied considering the sequence of points of interest and the time intervals between them to establish the travel route (spatial dimension) and the travel plans (temporary dimension). Next, techniques such as the Markov chain (a probabilistic method used to analyze sequences and determine if some sequence occur more frequently than others) can help predict the next destination of travelers and generate more detailed information about a likely travel route based on the visitor's location (J. Li et al., 2018).

Owners and managers should keep in mind that choosing an analytics tool is an important responsibility that generally depends on the ease of use of the tool, the type of social media and data it supports, the speed of the process, the accuracy and frequency with which the tool is used, and how often it is updated by the manufacturer or supplier (Bazzaz Abkenar et al., 2021). In general, firms use the commercial tools available, but also the analytical tools provided by the social networks themselves, such as Google Analytics, Facebook Audience Insights, Instagram Insights, etc.

Last but not least, firms must interpret the results obtained from the analysis phase. For this, there are tools that help display two or three dimensions of social data, such as time, location, and other attributes. Geo-visualization is an especially useful tool to show the underlying network structure of data and content published on social media. However, as DGC data lacks information that can be linked to the user, the geographic visualization of social media content primarily focuses on UGC data (Z. Chen et al., 2021).

Applications

SMA can be useful in almost every tourism business management function where decisions are made that are influenced in some way by social trends (Noor & Haneef, 2020). When a firm analyzes data from social media to see if one of its products or services is gaining or losing popularity over time, monitors the health of its brand, or assesses the reasons why there are angry customers, it's not just processing relevant information about the degree of satisfaction or dissatisfaction of people, but gathering first-hand knowledge about the trends and challenges present in a specific community, that of its customers and users.

Through SMA, firms can get timely feedback on what they are doing or plan to do and gain instant insight into the needs and interests of their customer base. Furthermore, the firm can compare its products/services with those of its competitors, identify its competitive advantages, predict the future of certain products, and establish effective strategies focused on its target community. This makes it easier for owners and managers to focus more on implementing new ideas, products and services that improve the ROI, and save time and effort in decisions when there is not much information about the target audience and the competitors.

11.4 Big Data, Artificial Intelligence, and Analytics

The accelerated development of Big Data and analytics has meant that conventional algorithms and data processing software are no longer capable of handling large and complex amounts of data. As a result, many organizations have chosen to combine the use of Big Data, data analytics and artificial intelligence (AI) to make the most of the potential offered by Big Data and make tourism firms smarter and more innovative. The combined use of these technologies generates important synergies, since AI requires Big Data to learn and improve decision-making processes, and Big Data leverages AI to extract better insights from data analytics. In other words, by combining massive amounts of data with the ability of machines to develop human-like intelligence, it becomes possible to harness the power of predictive analytics to extract high-value information for the business and set up the basis for the adoption of the data-driven organizational model. AI can help in all phases of the Big Data and data analytics life cycle, including identifying potential connections between data sets, automatically exploring data without human intervention, and accelerating modeling of data. From the consumer's perspective, Big Data analytics and AI (BDAI) can bring closer the ultimate personalization of the tourist experience and enrich it before, during and after the trip. Meanwhile, from the perspective of the tourism firm, BDAI promises to increase revenues and economic benefits.

BDAI allows firms to better forecast demand, using sophisticated algorithms to process structured and unstructured data. Other BDAI-related techniques, such as neural networks, classification trees, and regression, can be used to predict customer loyalty through the identification of customer attraction characteristics (Samara et al., 2020). BDAI can also help tourism firms create products and services at lower cost and with higher quality, mainly through automation, robotics, and digital augmented processes. Using BDAI, it is possible to speed up or even replace repetitive tasks, as well as reduce the number of errors made in the provision of services (e.g., room assignment, tourist itinerary planning, etc.).

Another valuable utility of BDAI is the ability to promote products and services at the right price to the right audience with the right messages (Samara et al., 2020). Thanks to BDAI systems it is possible for firms to set an optimal price for a given customer, automatically identifying purchase intentions, and improving customer satisfaction. This increases the customer's intention to increase loyalty to the firm or to buy the firm's products/services and, by extension, for the firm to increase its revenue. Many hotels and airlines around the world have already shown interest in using BDAI for strategic revenue management and capturing customer preferences.

On the other hand, BDAI methods are behind the possibility of targeting accurate advertising and promotional messages to customers, offering personalized information based on context and location, thus meeting customer requirements. This makes it possible for firms to design individualized marketing campaigns and increase ad "click-through rates" using BDAI. In other cases, BDAI supports customer recommendation systems that provide customization capabilities to web-based applications.

11.5 Implementation Challenges

Firms that evolve in the adoption of data analytics are better prepared to turn the challenges of smart transformation into opportunities. The more data-driven a firm is, the better financial and operational results it typically has, so analytics has competitive value that is directly related to firm performance (Zeng & Glaister, 2018). However, tourism firms face considerable challenges when it comes to implementing analytics-driven management and creating value from it, making it necessary for owners and managers to understand the nature of these challenges, why they arise, and how organizations can address them. A tool that can contribute to that understanding is the maturity process of implementing data analytics in business organizations. Data analytics maturity generally depends on the degree of analytical skills organizations have and to what extent they have been transformed using analytics. Some authors describe maturity in three levels of capabilities with respect to the analytics implementation process: aspirational, experienced, and transformed (LaValle et al., 2011):

- At the aspirational level are organizations that focus on data analytics to improve efficiency or reduce costs (i.e., by automating existing processes). These types of organizations generally have few of the core components necessary to incorporate or act on analytical insights, such as people, processes, and tools.
- Once the organization gains some experience with analytics after having successfully implemented several projects in the aspirational phase, it will seek to go beyond mere operational efficiency and cost reduction. These organizations strive to develop better ways to collect data and generate insights from analytics with an eye towards driving effective improvements at all levels of the organization.
- At the transformed level, organizations have gained substantial experience in applying analytics across a wide range of functions and decision-making. Possibly they have already automated an important part of their operations through the effective use of analytical insights and are no longer so concerned with efficiency and cost reduction. They prefer to use analytics to optimize processes, manage their material and human resources, and create differential advantages that increase customer profitability and ROI.

Interestingly, and contrary to popular opinion, the main challenges firms face when starting with analytics are not related to getting the right data or technologies, but to cultural and management barriers, as there is usually a lack of understanding of how to use analytics to improve business (LaValle et al., 2011). Business owners and managers as leaders will need to interpret and communicate the complex knowledge that results from the application of data analytics, so that the organization can absorb its meaning as quickly as possible and contribute to successful decision-making.

Leadership and direction

Owners and managers are likely to find it difficult to move from making decisions based on intuition and personal experience to making them based on data, especially when that data often contradicts their own intuition or conventional wisdom in business management. For this reason, it is essential that the firm has a specific strategic direction before making an "opportunistic" use of data analytics, otherwise the efforts will end up stalling and resources will be wasted. Not to mention that an image of mistrust will be created about the true value that analytics can have for the organization.

It is very important that owner and managers are convinced of the value of analytics and its potential to improve management. Only in this way can they become its main promoters. It's all too easy to lose focus when outcomes are uncertain and seem distant; therefore, unless the leader has been through a similar process before and understands the pitfalls, he or she may not be prepared to give the firm the boost it needs and keep it going for the long haul. Business leaders should learn all they can about how to lead and manage data analytics initiatives and how they can help the organization maximize value creation through analytics. Implementing an agile approach to developing data analytics projects can be a very good way for owners and managers to help the firm achieve the benefits of analytics more quickly and effectively and, in the process, get the stakeholders to learn and be part of the firm's objectives.

The need for agility

Firms apply data analytics to solve business problems in highly varied and complex contexts that are changing at an ever-increasing rate. This often leads to firms engaging in so many different IT and analytics initiatives that management becomes chaotic. Each project has a different sponsor, goals, and funding sources, which poses the challenge of how to get all these initiatives up and running together. Moreover, when firms try to integrate analytical tools and technologies, they find themselves in an environment plagued by contradictions and inconsistencies that affect data structures, technologies, and processes.

Having reached this point, how can the firm address all these challenges?

In a challenging scenario like this, the firm requires some guiding principles that assist the execution of analytics initiatives in a decentralized manner, but without giving up a governance model at the organizational level that avoids lack of coordination and entering into total chaos (Kemp, 2014). When the firm gets equipped with a governance model that is consistent and well designed, it ensures that there will be a balance between short-term and long-term analytics initiatives, and that it will have enough discipline to maintain the course. If the organization adds effective leadership to the above and a standardized architecture of systems and analytical processes consistent with the structure of the organization, then the firm as a whole will be on the right track to be more agile and flexible. From that moment on, the firm will have an easier time integrating the new data sources, technologies, and analytical processes that are necessary to mature in smart transformation. Notwithstanding, business owners and managers should never give up on developing an agile culture in the firm, though it will require significant time and organizational resources.

Implementation

Data analytics is a huge effort that leaves little time and few resources to capitalize on, so analytics initiatives often stall before they deliver results. All too often organizations succumb to the temptation to first gather all the available data before starting to perform analytics on it, without even thinking about whether it is reliable and what kind of value they will be able to extract from it. Instead, organizations should start at the end, first making explicit what kind of insights they expect to generate from analytics and what questions they want to answer, before moving on to identify the data they will need. By putting the expected insights and key questions first, organizations can then focus on tackling specific problems and using specific data and analytics without spending time guessing. Once the insights are gained and the questions answered, the organization will be closer to recognizing the gaps that exist in its capabilities, resources, and business processes.

As the firm gains experience with analytics in particular business units or management functions, it will start seeking to broaden the scope of the results and make more advanced use of these capabilities. Sometimes the creation of a centralized analytics unit (i.e., center of excellence, center of competence, or similar) is a mechanism that can bring together increasingly advanced analytics capabilities within the organization and lead the implementation of an analytics governance framework by setting priorities and standards.

So, where can the organization begin?

If the organization is at an aspirational level, owners and managers might be better off trying to gather the best people and enough resources to start one (or a few) analytics initiatives that can deliver effective, fast, and visible results. These initiatives should focus on some of the main business challenges that everyone recognizes as key to the organization and that can be addressed through analytics. Next, the available data with which to respond to the challenge will have to be collected, as well as choosing an appropriate analytical technique. Once the results have been obtained, they must be communicated to the rest of the organization, highlighting the value they generate and the potential for improvement for the firm if progress continues along this line.

If the firm is at an experienced level, it will have been adding new capabilities and achieving some success with analytics from the aspirational level (e.g., in finance, marketing, customer service, etc.), so now it's time to focus on the key problems of the organization and extend the use of analytical decision-making to additional functions. To achieve this, it is highly recommended that the different departments of the firm establish collaborations that promote the creation of value for the business without compromising the work and needs of each one. At this stage of the analytics implementation, the firm could consider preparing an agenda or roadmap that identifies the practices and tools necessary to achieve an effective development of analytics in the organization. This agenda should have the adherence of those in charge of business strategies and those who manage data and analytical tools, otherwise each department or unit may end up exploring new developments independently, making it more difficult for the organization to realize the business benefits promised by analytics.

If the organization is at a transformed level, it will be in a position to start reaping the benefits of everything learned so far and apply all the lessons learned to other fields and functions with minimal changes. The firm may need to decide at this stage whether to continue to extend the value of analytics to all levels of the organization or, instead, delve into specific challenges, such as the development of scenarios and simulations, the application of analytics to business processes, the creation of specific dashboards and scorecards, or the use of more advanced analytics techniques. In either case, the organization should be committed to innovative improvements in the way analytics are used and continue to demonstrate new ways to help improve business performance.

11.6 Benefits

Data analytics provides great benefits that can be relevant to all levels of the organization. When data from all departments is consolidated and integrated into a single information repository and everyone is synchronized in the analytics process, the following benefits may begin to unlock:

- **Data-driven decisions:** With data analytics, firms can start making smarter decisions, which basically means decisions that are backed by data. For example, the firm can explain certain phenomena and trends that have the most impact on the business and thus design better marketing campaigns, create new products and services, make changes to the budgets, etc.
- **Accessible knowledge:** Data analytics allows the firm to collect vast amounts of data that is difficult to interpret and prepare visualizations that are easy to understand. The insights so produced will be much more accessible to members of the organization and, when used by more employees, the innovative potential of the organization can be significantly improved.
- **Scenario modeling:** Through predictive analytics, firms can start building models to identify trends and patterns that help shed more light on their future. In the past, these abilities were reserved for a few specialists who had the knowledge and tools to exploit them. However, with modern data analytics tools, users can generate models without prior knowledge of algorithms or data science, increasing the ability to pose business scenarios, build models, and experiment with them.
- **Augmented capabilities:** When machine learning and AI are combined with data analytics tools and capabilities, the potential to create value from data soars even further. This augmented analytics is distinguished by its ability to self-learn and adapt to large amounts of data and by automating processes and decision-making without human intervention.

Discussion Questions

- What is the maturity level that tourism firms have reached in relation to the implementation of data analytics? Is this level above or below other service industries?
- What analytical method is most used by tourism firms today? Why?
- What is/are the data source(s) most currently used by tourism firms to feed analytics? What other data sources have potential for the tourism firm?
- Which stage of the data analytics workflow is the most critical from the tourism firm point of view? Why?
- What opportunities does machine learning offer to improve decision-making in tourism firms?
- What role does owners and managers' leadership play in extracting value from data through analytics?
- What kind of data visualization techniques are most valuable to tourism firms?

12 Artificial Intelligence

The term artificial intelligence (AI) was first introduced in the 1950s as a branch of computer science concerned with the automation of intelligent behavior (McCarthy, 2007). Since then, AI and associated techniques have developed rapidly to the point that today there is still no fully accepted definition of what AI is. Most frequently, AI is associated with the ability of a machine to sense its environment, learn from experience, make decisions based on input and goals, and ultimately perform human-like tasks. With the swift progress made by Big Data technologies and the ever-increasing capacity of computer storage and lightning-fast speed of data processing machines, AI is being revitalized. Considering those advances AI could also be defined as "the ability (of a computer system) to correctly interpret external data, learn from such data, and use those learnings to achieve specific goals and tasks through flexible adaptation" (Kaplan & Haenlein, 2019). These "intelligent" systems use all kinds of algorithms and machine learning techniques to learn from data and make decisions aimed at achieving specific goals.

After many years of promising expectations, AI is gaining momentum in business organizations all around the world and applications extend to a wide-ranging variety of fields and contexts: from image recognition, video surveillance, telemedicine, robots, autonomous vehicles, and social media websites, to voice recognition, machine translations, online games, autonomous planning and programming, spam filters, weather forecasts, etc. (Ivanov & Webster, 2019). There is no doubt that AI has enormous potential, and that computers and robots are slowly approaching ever closer levels of human intelligence. Does this mean that AI is becoming an increasingly serious competitor for human jobs? The debate is served. Perhaps the day is not far off when human supremacy over machines is questioned. Be that as it may, everything indicates that AI is going to bring great changes to our daily lives and society in general.

DOI: 10.1079/9781800620261.0012

The impact of AI on tourism firms and employment is expected to be massive, resulting in highly interconnected organizations whose decision-making will be based on analytics and autonomous exploitation of Big Data to cope with increasingly intense global competition.

12.1 Types of Artificial Intelligence

Contrary to popular belief, AI does not make us humans smarter. Certainly it is not going to replace the owners and managers of the firms in making business decisions in the short term. However, AI may make business leaders a little less prone to bias, or at least start to question it, so they can better allocate resources where they're needed and perhaps make organizations more agile (Andersen et al., 2018).

There are three main types of AI according to the literature on the subject:

- **Narrow artificial intelligence** (also known as weak AI) is that AI with limited learning and adaptation abilities and that can only be used in routine and/or repetitive tasks. Narrow AI is applied in very specific fields where AI is better and works faster than humans (e.g., to identify tumors, play chess games, play a videogame). This type of AI represents the current stage of AI development.
- **Artificial general intelligence** (known as strong AI) is more powerful than weak AI and is focused on more complex tasks that involve the processing and learning of Big Data. Therefore, it is an AI very close to human intelligence and can even surpass it in some specific fields (Lv et al., 2021).
- **Artificial superintelligence** is the self-aware AI, i.e., the AI that performs better than humans in all areas. For now, this AI is more science fiction than reality, but it seems that it is the way we are heading.

Another very popular classification of AI is the one proposed by IBM, which differentiates four types of AI according to the degree of human intervention and the type of systems in which the AI runs:

- **Assisted intelligence** is AI based on systems preprogrammed by humans that do not have the ability to learn from their interactions with the environment in which they operate. This type of AI can perform routine and repetitive tasks at high speed and is therefore good at helping humans perform certain tasks in a more agile and efficient manner.
- **Automated intelligence** is AI intended to automate manual or cognitive tasks, whether routine or non-routine. This type of AI is good when it comes to automating existing tasks that don't involve creating new ways of doing things.
- **Augmented intelligence** aims to help people make better decisions. It is an AI oriented to constantly learn from humans and the context that surrounds them.
- **Autonomous intelligence** is AI focused on automating decision-making processes, for which it adapts to the different contexts and situations that

arise. This AI works autonomously (without human intervention), which means that it is the type of AI most similar to the human being.

AI is one of the fields of science that arouses the most admiration today, but it is also one of the most feared, especially when we refer to artificial superintelligence. Hollywood productions are tapping into this burgeoning El Dorado as a great source of inspiration and are largely to blame for society's not very favorable image of AI. However, in real life, machines are still in the early days of achieving this kind of superintelligence, and narrow artificial intelligence is as far as software engineers have come. This is the kind of AI this book will be referring to.

12.2 General Applications

Perhaps the biggest challenge society and business face when it comes to AI is how to reap the benefits that AI technologies promise and how to seize the opportunities in fields as diverse as the creation of new products and services or the improvement of productivity, while avoiding the dangers of job loss and wealth redistribution. Some of the most characteristic applications of AI have to do with improving decision-making, reinventing business models, and reshaping the customer experience. According to Gartner's latest surveys of technology trends, AI ranks at the top of global trends in strategic technologies. This situation contrasts, however, with the fact that only 59% of business organizations worldwide are collecting information to develop their AI strategies.

The new wave of AI technologies is helping to improve the ability of businesses to make predictions using data, with AI being a key driver that is substantially lowering the cost of making predictions and bringing them closer to smaller organizations. In general, both firms and the AI industry seem to agree that AI offers significant opportunities to improve data analytics and decision-making, as well as process automation, such as communication with customers, accounting processes, supply chain, reservations, etc. A recent investigation by Davenport and Ronanki (2018) on 152 projects spanning a wide range of functions and business processes based on AI systems, classified AI applications into three categories:

- **Process automation:** It is the automation of back office and financial processes using robotic process automation (RPA) tools. This is the most common and cheapest type of AI application in firms today (i.e., data transfer from email systems and call centers to registration systems, reconciliation of failures in charging for services, reading of legal conditions and contracts). RPA is a more advanced automation technique than previous tools, since "robots" act like humans, processing vast amounts of information from multiple IT systems.
- **Cognitive insights:** Detect patterns in large volumes of data and interpret them using machine learning algorithms (i.e., anticipate what a customer is most likely to buy, identify when fraud is about to occur, automate personalized targeting of online ads). It is the second most common type of AI applications among firms.

- **Cognitive engagement:** These are applications in which employees or customers get involved through chatbots, intelligent agents, and machine learning, tackling issues from password requests to tech support questions to product and service recommendation systems. In general, these types of AI technologies are more used by firms to interact with employees than with customers.

12.3 Applications in Tourism

In the past few years there has been a rapid growth in the adoption of AI technologies by tourism firms. AI technologies have been progressively incorporated into tourism firms on the basis of enabling technologies such as Big Data, analytics, Internet of Things (IoT), cloud computing, voice and facial recognition, service robots, etc. (M. Li et al., 2021). One factor that has favored a higher pace of AI adoption has been the public health emergency caused by COVID-19. The COVID-19 pandemic has made AI technologies go mainstream to prevent, or at least reduce, social interactions in service sectors, which in the process has affected management decision-making and partially replaced some human tasks. From the customer's point of view, AI applications have reduced the level of person-to-person (and face-to-face) interaction and enabled new technology-enabled service encounters. These have impacted customer experiences and behaviors to the point of reshaping service interactions in many cases. In addition, owners and managers have felt compelled to gain insight into what AI-enabled service encounters look like and how they may affect customer service performance. Some of the main applications that AI technologies are having in tourism firms are presented below, among which recommendation systems and customer sentiment analysis stand out due to the frequency of cases found, followed by prediction solutions, facial and image recognition solutions, and service personalization (Pinheiro et al., 2021).

Big Data and AI

Big Data technologies are among those that have fostered most the development of AI in recent times and contributed to its current boom. Big Data analytics done by humans is time-consuming and using AI techniques can help make sense of Big Data faster and more effectively. The benefits of adopting Big Data strategies with AI in tourism firms range from greater efficiency in processes and improved productivity levels to improved customer experience, which can also be more personalized (Samara et al., 2020). Despite these well-known benefits of the combined use of AI and Big Data, there remains a pressing need to better understand the synergies between the two, given the increasingly important role they are playing for tourism business competitiveness and resilience (Duan et al., 2019).

Sentiment analysis in online reviews

The increasing use of social media is making consumers increasingly willing to share their experiences and opinions about the tourism products and services they consume. The volume of online reviews that appear on social media and specialized blogs on different aspects of tourism activities has not stopped growing and consumers are increasingly using them to inform themselves and make decisions about their next trip. This information is of great value to tourism firms, which by knowing the comments (positive and negative) of tourists, can address issues relevant to the business and make decisions on specific aspects of their value offering and service delivery.

Personalized recommendation systems

The huge amount of information that is continuously generated about places, attractions, activities, accommodations, restaurants, and the ratings provided by tourists have made travel planning a highly demanding and time-consuming task. Consumers use recommendation systems as decision support systems, as well as a mechanism to overcome information overload. These systems are increasingly used through online platforms with the aim of actively recommending relevant information to customers, based on an increasingly precise knowledge of behavior patterns and particular interests of consumers.

Demand forecasting

Forecasting visitors' demand is important for business owners and managers to understand what is driving consumption of their products and services, how they can respond to increased (decreased) demand, and how they can anticipate certain future events. With this knowledge in their pocket, tourism firms can establish much more effective management strategies that help them maintain a quality offer and adjusted pricing policies, allocating their capacity and the working hours of their staff more efficiently.

Cancellations forecasting

In the hotel industry, for example, reservation cancellations affect management decisions and prevent firms from making accurate demand forecasts and effective revenue management. To mitigate these undesirable effects, many hoteliers resort to overbooking strategies and cancellation policies that are restrictive, which in turn can have a negative impact on revenues and on the firms' own reputation. When reservations that are likely to be canceled are identified early enough, hotels can take action in advance to avoid cancellation by offering extra services, discounts, or other benefits that customers like.

Monitor price changes

It is important that owners and managers are aware of the factors that affect the market in which they operate and closely monitor price changes as they occur. By monitoring prices, they can anticipate changes that may unexpectedly affect their business. This is especially important because tourism products and services cannot be stored and must be consumed on the go. The issue of prices also has a direct relationship with the online reputation of firms, since part of the experience perceived by customers depends on the relationship between the value and the perceived quality of tourist services.

Image recognition

Image recognition has become an innovative tool to identify and classify people and scenarios in many industries. It has been around for some time since mobile apps began helping visitors to national parks in South Africa identify animals, and as a way to captivate visitors while there was a lack of professional rangers. Today, these types of AI-based tools are used for facial recognition and to analyze emotions or measure customer satisfaction when tourists consume a product or service. Other use cases of AI in tourism include identifying influential attributes of a place, evaluating how online reviews affect firm performance, evaluating employee satisfaction, and market segmentation. The use of chatbots for recommendations and the application of facial recognition systems in hotels, attractions, or airports is also becoming very common.

12.4 Service Encounters and AI

A service encounter is the type of social exchange that has traditionally taken place between service actors in person-to-person mode. This double-sided view has so far emphasized the role of customers and employees, ignoring non-human factors involved in service interactions, such as service facilities, atmosphere, or the environment. However, with the introduction of AI technologies, service encounters have started to change drastically. These changes have been further accelerated by the COVID-19 pandemic, which has seen many tourism business owners and managers implement AI for the benefits it offers in the new post-COVID-19 business environment. The question to be elucidated is how service encounters are going to change after the implementation of AI in frontline services.

With Big Data entering the scene, AI has rapidly penetrated service processes in a quest for tourism firms to remain competitive. More and more tasks are being automated and becoming "smarter" as customers play a more prominent role in services due to AI and co-creation of service value. Ultimately, human–AI interactions have been embedded in service encounters, the quality of which depends not only on customers and employees, but on smart technologies. Today, AI applications facilitate, guide, replace, and even enhance human–robot interactions by providing information that is relevant to the customer. This is driving tourism firms to deepen their knowledge of the specific attributes of AI that can favor these interactions and improve the experience of customers and users. Based on the new features of service encounters and AI, four modes of AI-based service encounters can be defined (M. Li et al., 2021).

- **Supplemented with AI:** These encounters take place when AI and human employees serve customers separately and independently. Among the main capabilities that AI can provide is that of offering guidance, i.e., the AI integrated into a mobile application can guide a hotel guest when ordering a meal, although it will be the human employees who will serve the food on the table. When AI is combined with Big Data and machine learning technologies in mobile applications, service providers can gain more flexibility to interact with customers and, in the process, provide smart recommendations based on the customer's search and browsing patterns or the purchase history. Smart recommendations are based on built-in AI that automatically matches customer needs with services based on customer information. AI-enhanced service encounters on social media platforms take place through the exchange of ratings and recommendations. Virtual reality is another relevant application of AI that can be used to complement employee service encounters, creating a simulated world that customers can immerse themselves in and enhance the service experience. Smart virtual reality service encounters can bridge the gap between customers' service expectations and actual service performance, as well as add novel experiences for tourists.
- **Generated by AI:** AI technologies are so powerful that they can independently serve customers at service points without the presence of employees. AI can replace routine human work and decision-making and directly engage with customers in a fast and personalized way. In the end, AI service encounters reduce unnecessary human interaction and provide customers with efficient and convenient services (i.e., when queuing), creating novel experiences, and achieving greater flexibility and customer satisfaction. Technologies in this category include self-service devices, smart home systems, and chatbots. As far as employees are concerned, AI-generated service encounters lack the human touch.
- **Mediated by AI:** These are technology-mediated remote service encounters that expand the capabilities of human employees, such as when a robot performs production or distribution tasks (i.e., robots that deliver food and drinks in a hotel). In this mode, the AI is used to overcome the barriers of time and space of services and reduce costs for customers and service providers. For example, customers can book accomodation through online services offered by social media. Compared to traditional face-to-face service encounters, AI-mediated services can lead to decreased customer satisfaction. However, combining the right AI attributes with employee personality can offset this adverse effect.

- **Enabled by AI:** These are service encounters where AI technologies and employees jointly provide services to customers. These services are based on cooperation and interaction between AI, employees, and customers. AI can register information about customer preferences and access Big Data over the internet, which can better feed a customer relationship management system (CRM) and deliver more efficient and personalized customer services. This modality differs from the previous ones in that the robots have a higher level of autonomy, intelligence, as well as a greater storage and data retrieval capacity. Special attention must be paid to the design of the robots (i.e., its operability, way of making notifications), the characteristics of the customers (i.e., cultural, socio-economic level, personality) and service policies (i.e., complaint management, levels of service) as these are key factors for customer acceptance and satisfaction with the service.

Owners and managers must reflect on the AI technologies that best fit their business processes and services. For example, there are hotels that have implemented contactless smart services to allow customers to check in quickly, locate their rooms by following service robots, have an informal conversation with chatbots, or get information without interacting face-to-face with employees. It is important that when evaluating AI technologies, owners and managers assess those factors related to customer segments, internal business processes, and the market in which they operate in order to transform threats into opportunities.

12.5 Challenges of Artificial Intelligence

The rise of AI has sparked many controversial opinions in recent years. On one side are those, like former IBM CEO Ginni Rometty, who argue that AI technologies will help us become better and elevate the kind of things the human condition can achieve. On the other hand, relevant figures such as Stephen Hawking or Bill Gates have stressed that the development of complete AI could mean the end of the human race and that this should concern us all. Surely what these very different views are telling us is that it is essential that we humans continue to dig deeper into how people can co-exist with AI and how we can minimize the negative impact of these technologies.

Both AI developers and large corporations in the tech industry continually make claims about the great benefits and substantial impacts that AI is going to have for consumers and firms. Even some surveys, such as Davenport and Ronanki (2018), have estimated that three-quarters of firms believe that AI will substantially transform their business in about 3 years. But, what are the main challenges facing business organizations in relation to AI? One of the main challenges is decision-making. It is widely believed that AI can help employees in organizations to make better decisions, increase their analytical skills, and, in general, increase their creative capacity and respond to changes in the environment. However, with the advent of AI, what appears on the horizon is a new human–machine symbiosis that leaves questions such as: How can humans and AI complement each other in organizational decision-making? And, how can firms optimize the collaboration between humans and AI and work effectively in the human–machine interface? The answers to these questions are not simple, even more so when most AI systems do not reveal their functioning or how the algorithms work. This lack of transparency prevents AI users from knowing why decisions are made in a certain way. Solving this problem requires both researchers and industry to make AI more understandable and explain how the algorithms work. Most likely we will have to wait a few years to answer all these questions with clear evidence. Furthermore, most claims about the great impact that AI is going to have are not sufficiently supported by empirical evidence or rigorous academic research, so it is still difficult to know precisely how, why, and to what extent AI systems are going to affect tourism firms.

We could add the problem of how to measure the benefits (and impacts) that AI has on decision-making, encompassing the social, economic, and business perspectives (Duan et al., 2019). As with most technologies in the smart ecosystem, both researchers and the technology industry have focused on selling the great benefits of the application of AI in the different contexts of tourism, such as travel, destinations, and hospitality. However, much less effort has been devoted to measuring the impact that AI has on the firm and delving into the transformation of organizations. This is all the more relevant as it is clear that organizations committing to the use of AI face more than the usual hurdles when adopting any unproven and largely unknown technology.

To harness the full potential of AI, tourism organizations will need to reorganize themselves differently from within. They will also need to implement new management techniques and interact with new scientific methods that allow them to respond more quickly and effectively to their environment (Andersen et al., 2018). Before embarking on AI initiatives, tourism firms need to have a good understanding of what type of technologies do what tasks, and the strengths and limitations of each of them, as well as the barriers to implementation and the critical success factors.

The advancement of AI technologies continues at breakneck speed, with faster and more advanced systems appearing every day that can tackle more complex tasks that require cognitive skills, such as sensing emotions, making unspoken judgments, and executing processes without human supervision. In short, the changes brought about by AI are likely to be as disruptive as those of the Industrial Revolution, if not more. The speed at which changes will occur will depend on the rate at which AI technologies can automate and replace non-repetitive mental tasks currently performed by humans, and smart software (capable of developing new programs on its own) becomes available.

12.6 The Future Role of Humans

In the new context drawn by smart technologies and AI, business owners and managers are starting to ask themselves: What will be the role of humans when computers and robots can perform practically all our tasks the same or better and much cheaper? What will we humans do then (Makridakis, 2017)? There are four scenarios that can give us some answers to these questions:

- **The optimists:** The optimists led by Ray Kurzweil (Kurzweil, 2005; Kurzweil et al., 1990) predict a utopian future dominated by genetics, nanotechnology and robotics (GNR) that revolutionizes everything and in which humans are capable of taking advantage of speed, memory, and the ability to share knowledge through their brains connected to the cloud. Genetics would make it possible for humans to change their genes to prevent disease and delay aging; nanotechnology, through 3D printers, would allow virtually any physical product to be created from cheap materials and information, leading to unlimited wealth creation; and, ultimately, robots would do all the real work, leaving it up to humans to decide how to spend their time doing activities of their choosing.
- **The pessimists:** They argue that the most powerful technologies of the 21st century (robotics, genetic engineering, and nanotechnology) are a real threat to humans to the point of becoming an endangered species (Joy, 2000). For them, the optimists underestimate the true magnitude of the challenge and the risks that thinking machines and intelligent robots entail. As social problems become more complex and machines more intelligent, people will tend to let machines make important decisions for humans, in the belief that decisions made by machines will get better results than those made by humans. Machines will eventually take control of all important decisions for people, who will be reduced to second-class status, some say even the equivalent of computer pets. What will then happen to society and our daily lives when non-conscious but highly intelligent algorithms know us better than we know ourselves (Harari, 2016)?
- **The pragmatists:** They believe that humans should learn to exploit the power offered by computers to increase their own capabilities and always be one step ahead of AI, or at least not be at a disadvantage (Markoff, 2016). In case of danger, pragmatists suggest that all thinking machines can be disabled to render them inoperable and propose that AI technologies always be controllable through "OpenAI" and effective regulation.
- **The doubters:** They do not believe that AI will even be possible and that it will ever pose a real threat to humanity. For them, human intelligence cannot be replicated or captured in formal rules and AI is nothing more than a fad brought by the computer industry (Dreyfus, 1972). Moreover, computers will never be able to be creative, as this would require breaking the rules and become anti-algorithmic (Jankel, 2015). It is true that some of the criticisms of the doubters have been valid for much of the last century, but today in the face of new developments in AI it is much more difficult to maintain them. For example, they criticized Herbert Simon's prediction that a computer would be able to beat the chess champion in a few years, and Deep Blue became the world chess champion in 1997. Moreover, we are not too far from machines being able to do all the work that humans can do, as there are autonomous vehicles, robot-nurses that take care of

the elderly, and Google Search, which knows what we are looking for better than we do.

The scenarios outlined above provide arguments of all kinds for and against the adoption of AI technologies. It is perhaps too early to opt for any of these scenarios considering that we are still at a very early stage in the development of AI. Furthermore, uncertainty surrounds everything that is going to happen in the coming years with AI and how tourism firms will be affected. Müller and Bostrom (2016) posed the following question to hundreds of AI experts in a series of conferences: "[...] assume that human scientific activity continues without major negative interruption. By what year would you see a (10%/50%/90%) probability for such HLMI (high-level machine intelligence) to exist?" The median answer for the 10% chance was 2022, the 50% chance was 2040, and the 90% chance was 2075. Ultimately, most AI experts thought that AI was not that far away and more believed that it will have a positive effect on humanity.

Discussion Questions

- What implications will AI have for the owners and managers of tourism firms?
- What do you think the role of humans will be when computers and robots can do their job as well or better, and much cheaper?
- What are the main applications that AI is having in the tourism firm? What will be the next applications?
- Which actors of the tourism ecosystem are the most benefited (affected) by the implementation of AI?
- What negative effects can the application of AI-based technologies have on the experience of tourism consumers? How can these effects be avoided?
- What internal and external factors does the implementation of AI in tourism firms depend on? What role does leadership play in them?
- What are the consequences of the adoption of AI-based technologies for the people and internal resources of the organization?

13 Internet of Things

The accelerated changes in internet applications and technology are reflected in the development of the so-called Internet of Things (IoT), which describes a global network of connected and integrated devices with ubiquitous computing capabilities. Similar to the profound transformation that the internet has undergone, led first by infrastructure (e.g., fiber optics, routers, Ethernet cables, etc.) and then by business models, the same is happening today with the IoT. Firms and the technology industry are devoting a huge amount of resources to the design and implementation of infrastructure (e.g., sensors, gateways, base stations), but in a few years the focus will radically change towards business opportunities.

The key idea behind the IoT is to interconnect physical "things" through digital communication networks (e.g., via Wi-Fi, Bluetooth, Ethernet) to allow the "things" to exchange information with each other and help the user to make smart decisions for themselves. This ability of the IoT to convert each physical object into a potential digital artifact, a process known as "leak" or "liquification" of information resources, means the transformation of a physical object into a series of information resources that characterize it, on which it is possible to act (Ng & Wakenshaw, 2017). In an IoT context, each physical object is associated with a set of digital data (current and historical) that contains information about its physical properties, the origin, ownership, and the environment, which can lead us to imagine a world in which each physical object can be part of the internet.

IoT ecosystems are going to be central to Big Data collection. In fact, it is expected that the main source of Big Data in the coming years will be IoT systems, followed by data from humans and robots and AI applications (Kaivo-Oja et al., 2015). Through the IoT, firms will be able to track, communicate with "things", and monitor them to exploit their status. But even more important is

DOI: 10.1079/9781800620261.0013

that due to the programmability of "things" it will be possible for firms to add new instructions and modify their behavior throughout the life cycle of things (Fig. 13.1). Ultimately, this ability for everyday "things" to think and interact will be the driver of the IoT, unleashing limitless opportunities for tourism firms, most positive and some negative. Therefore, the IoT represents a new paradigm that has the potential to revolutionize the way we live and work thanks to the vast amount of data and services that flow from the interactions between a large number of heterogeneous connected devices (Palattella et al., 2016).

13.1 Concept of IoT

One of the first to use the term IoT was MIT technologist Kevin Ashton, who explained how the IoT could be created by adding radio frequency identification and other sensors to everyday objects (Ashton, 2009). Over time, the term IoT has evolved to refer to a "network of entities that are connected through any form of sensor, enabling these entities, which we term as internet-connected constituents, to be located, identified, and even operated upon" (Ng & Wakenshaw, 2017). Today, the IoT is considered one of the technologies that is going to lead the way to the Fourth Industrial Revolution and that interconnected physical devices will automate skills and tasks, significantly impacting the way consumers, businesses, and governments interact (Shim et al., 2019).

The IoT encompasses a network of physical objects including anything from a smart watch to a critical infrastructure control system. Each object in turn consists of some type of sensor (e.g., temperature, light, movement, etc.) or actuator (e.g., screen, sound, light, motor). Other components of the IoT ecosystem are the embedded technologies to communicate with and sense the

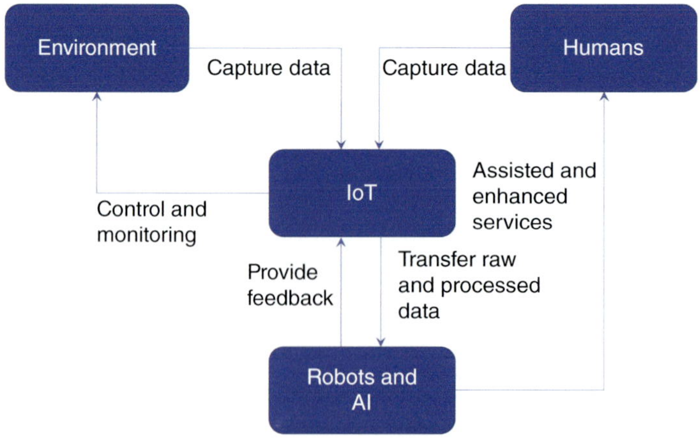

Fig. 13.1. IoT operations framework. Source: own elaboration based on Kaivo-Oja et al. (2015).

surrounding environment and interact with its internal states, such as computational resources to process the data detected locally or remotely, and a short-range (Bluetooth, ZigBee, radio-frequency identification (RFID)) or long-range (internet) means of communication to transmit the data.

IoT devices detect or read the environment in which they are embedded without human intervention. The data so detected is then processed and analyzed to acquire information and knowledge, which is distributed to those who will decide whether to take any action. Indeed, the essence of the IoT is to make any information available to anyone at any time. This relegates the traditional separation between physical and digital objects to the past (Fleisch et al., 2015), and merges both worlds to create smart ubiquitous services that offer endless opportunities for business and society.

But, one may wonder, why do we need to connect all "things"? The answer seems evident: the IoT is a big step forward towards automating processes that improve centralized control and management (Pramanik et al., 2018). The automation provided by the IoT can reduce operational costs compared to manual tasks, by automatically controlling and managing many independent devices that are connected and communicate with each other. The number and variety of communication technologies is currently so big that the IoT encompasses a great diversity of application domains and almost infinite possibilities for communication and interaction. This is being made possible, among other things, by the reduction in broadband internet costs, advances in wireless network technologies, access to the cloud, the reduction in prices of sensors with detection and communication capabilities, and the arrival of 5G cellular systems. Some forecasts, such as those of the Swedish technology company Ericsson, suggest that there will be around 29 billion connected devices by 2022, of which around 18 billion will be related to the IoT. This will generate trillions of dollars of value for businesses and national economies. The enterprise IoT platform market alone is estimated to reach, according to consulting firm Gartner, US$7.6 billion in 2024 with a compound annual growth rate of 31%, driven by growth in on-premises and cloud solutions in all industries.

For many specialists, the IoT represents the next step in the computer revolution, which will be based on the incorporation of information and communication technologies in home machines, in the workplace, and in a wide range of industrial processes (De Saulles, 2016). The massive interconnection of devices and the incorporation of network sensors in everyday objects will provide tourism firms with unimaginable amounts of data. Moreover, AI and machine learning applications to analyze and make sense of this data will lead to a new wave of innovation in business processes. Together, the IoT, Big Data, and AI will eventually drive a radical reshaping of industries and business models by providing new insights into how business ecosystems and customer relationships work, unlocking new revenue streams through unprecedented use cases. Ultimately, new firms will emerge that are able to harness and capitalize on IoT data, while established firms will need to adapt the way they operate or face extinction, just as they did when steam, electricity, or the internal combustion engine came along and the old ways of working became obsolete.

IoT vs IoE

The IoT is built on the data created by things. This is a significant difference from how the internet has developed up to now, mainly based on data created by people. Building on this idea, some big companies in the technology industry (including CISCO and Qualcomm) have adapted their view of the IoT and used the term "Internet of Everything" (IoE) instead. Unlike the IoT, which is made up of "things", the IoE encompasses four pillars: people, data, processes, and things (Fig. 13.2). In the IoE, devices are permanently connected to the internet, including machine-to-machine (M2M), person-to-machine (P2M), and person-to-person (P2P) systems. This makes it possible for business and industrial processes to be put at the service of improving people's lives.

For IoE advocates, the big issue is not so much "things", but the connections between things, people, processes, and data, which is what really creates value. They believe that the growth of many organizations in the Internet Age occurs in waves that are "S" shaped, such that with each successive wave new features and greater network connectivity are added, eventually leading to qualitative and quantitative leaps in the number and variety of opportunities (and of risks) and firms reaching the IoE.

The IoE offers the potential to extract and analyze real-time data from the millions of sensors connected to IoT platforms that can then be used to optimize human-centric automated processes. For example, the IoE can be used to achieve public policy objectives, environmental sustainability, economic, and social objectives, etc. The combination of the IoE with Big Data processing in urban spaces is another example that shows great potential. As cities become "smart cities" they will increase their use of the information and knowledge they generate and will be able to address highly specific problems and challenges that citizens have (Miraz et al., 2015).

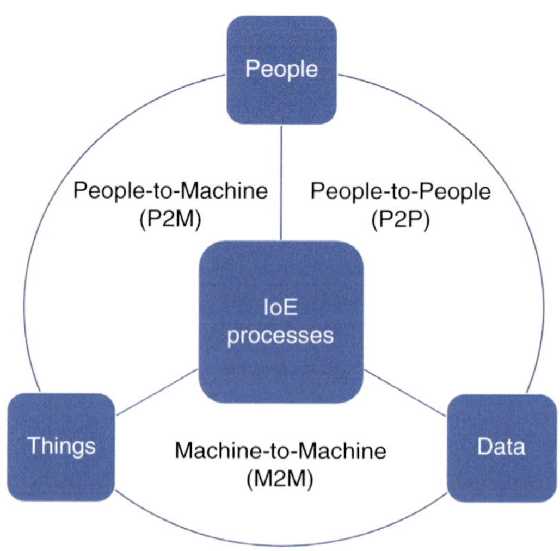

Fig. 13.2. Internet of Everything. Source: own elaboration based on Miraz et al. (2015).

It must be noted that the IoT is somewhat different from the Internet of People (IoP). The latter uses a PC, tablet, or smartphone and runs apps that connect to a remote server to do something (i.e., browse the web, send an email, shop online, watch video). Unlike IoP, IoT is based on devices that can sense events on their own and, using analytics and business intelligence, respond faster and better than humans. The adjustments and adaptations of those responses are made without human intervention.

13.2 How is the IoT Inside?

To fully understand what the IoT is like on the inside, it is first important to understand how its assemblage works, i.e., how connected objects/devices work together and how the ability to do things at a systemic level emerges that no object could do on itself (Hoffman & Novak, 2015). The IoT is made up of a wide variety of heterogeneous sensors and components, different communication protocols, and processing technologies that resemble the modules of an architecture and interact with each other in three different ways:

- Between one component and another (i.e., within the assemblage).
- Between one component and the assemblage (i.e., a part with the whole).
- Between one assemblage and another (i.e., between assemblages).

Assemblages are not fixed but are constantly changing as new components or modules can be added or removed (Ng & Wakenshaw, 2017), thus forming a network-like structure that is designed to connect whatever "thing" may be identified as an object, capture its data, and process and distribute the information anytime, anywhere (Ma et al., 2013).

As the IoT is a very broad concept and its components are so varied and heterogeneous, there is no standard and uniform architecture that defines it. This often causes interoperability issues when it comes to integrating all the different types of technologies. To address this problem, the IoT architecture can best be seen as an intelligent multi-layered system that enables the identification, positioning, tracking, monitoring, and management of devices, and the data generated by them. More specifically, architecture based on four layers is the most common in the IoT, as shown in Fig. 13.3 (Attaran, 2017; Ma et al., 2013):

- **Sensing layer** (also known as layer of things) is the layer that connects the physical and cyber worlds. This layer contains the sensor-like devices (things) that collect the information, such as RFID readers, infrared sensors, GPS systems, monitoring cameras, etc. and actuator type things. Through the sensing layer, the IoT assemblage achieves the ubiquitous perception that distinguishes it from other networks, as well as the real-time monitoring and management of the behavioral properties of the objects that comprise it.
- **Network layer** (also known as communication layer, or gateway and network layer) is the layer where things connect (i.e., the medium through which the data collected by devices is transmitted to higher layers for further analysis and processing). This layer consists of an IoT gateway made

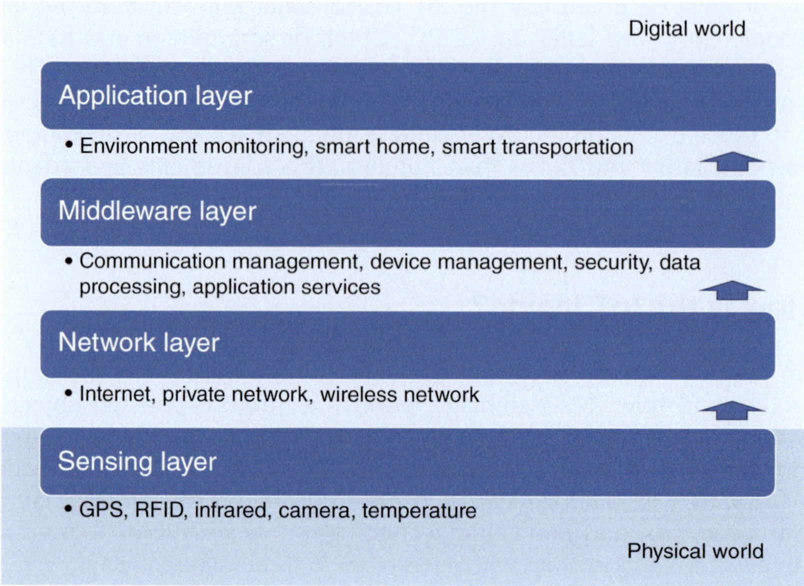

Fig. 13.3. Four-layer IoT architecture. Source: own elaboration based on Ma et al. (2013) and Fleisch (2015).

up of an embedded operating system, signal processors, microcontrollers, and gateway networks like LAN (Local Area Network), WAN (Wide Area Network), Wi-Fi, etc. The gateways route the data that comes from the sensor and the network layers and moves it to the next layer, which is the service management layer. This layer has a large capacity to store data from sensors, RFID tags, etc. and delivers consistently reliable performance across public, private, and hybrid networks.

- **Service management layer** (middleware layer, or data processing layer) is a crucial part of the IoT architecture, as it is responsible for carrying out data and text mining, information processing and analysis, information security, device management, and providing service interfaces for user applications. This data management process is necessary to extract information from the huge amount of raw data collected by sensor devices and to generate a valuable outcome.
- **Application layer** is an interface between the end user and the general IoT architecture that builds on top of the three lower layers to provide end-user IoT applications in various domains. These applications can be as simple as turning off an electrical switch or as complex as monitoring critical infrastructure.

One thing to keep in mind is that IoT layers cannot be created or function independently of one another. Therefore, an IoT solution is not the mere addition of layers, but an integration process that extends to the physical level, and where the whole is greater than the sum of its parts. Furthermore, a well-structured IoT architecture is the foundation that will help firms create innovative services. For this, it is important that the IoT architecture is designed considering the existing technological infrastructure, its scalability, as well as low energy consumption and cost-effective start-up, maintenance, and updating (Ghule & Sakhare, 2017).

Most IoT platforms are designed to support a wide variety of applications and services, but they are different in performance, security, customization, ease of use, and management (Lee, 2019). This is why choosing an IoT platform is not easy and tourism firms need to consider things like scaling and cloud access options, disaster prevention and recovery plans, security and data backup measures, communication protocols and standards, and the type of technical support offered. In addition, firms must decide whether to implement private IoT infrastructure, third-party IoT infrastructure, or a hybrid (private and third-party) IoT. Furthermore, owners and managers will need to assess and decide on the IoT infrastructure (i.e., total cost of the infrastructure related to hardware, software, labor, and maintenance), and on the main qualitative elements, such as the intangible benefits and risks, the level of customization required, and the organizational skills required.

13.3 IoT Data and Processes

Data management in the IoT involves data collection, processing, storage, and analysis processes. In general, IoT data shares these five distinct characteristics (Ma et al., 2013):

- **Heterogeneity:** IoT data is of different types and has very different properties that are used to describe the state of "things". IoT data ranges from integers to characters, to semi-structured and unstructured data (e.g., audio, video, images).
- **Inaccuracy:** Data inaccuracy is one of the most limiting factors affecting the adoption of IoT. For example, in the case of RFID systems, there are studies that show that only between 60% and 70% of the data can be used, since most of it is unreliable or unreliable. This situation is repeated with other data sources and technologies, which creates difficulties for the direct use of data.
- **Massive data in real time:** The connection of millions of devices and the communication between them through networks generate a huge volume of heterogeneous data that flows uninterruptedly, at high speed, and in real time. Consequently, processing IoT data requires scalable and strong storage, classification, and value creation capabilities.
- **Implicit semantics:** IoT data is data extracted from digital representations of objects with weak and low-level semantics, since it has little or nothing to do with human perception. For firms to build applications up to the level of smart services, data semantics must be abstracted from the mass of low-level data using an event-driven perspective.

In accordance with the characteristics described above, the use of IoT data requires some key processes to be carried out as shown in Fig. 13.4 (Ma et al., 2013).

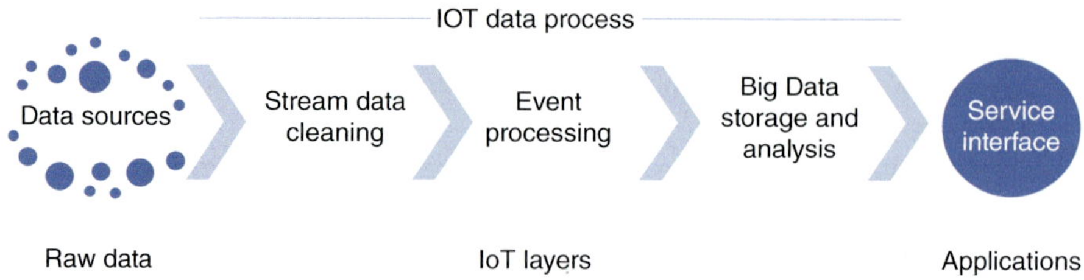

Fig. 13.4. IoT data processes. Source: own elaboration based on Ma et al. (2013).

- **Data cleansing:** It focuses on overcoming the lack of reliability that plagues IoT data, which is often incomplete, erroneous, and duplicate. Data cleansing is an essential step for the correct analysis and interpretation of IoT data. Cleansing data is a real challenge for most organizations given the enormous volume of data generated in IoT, the variety of sources and errors that exist, and the need to process it in real time. Once the IoT data is clean, it can be put to use.
- **Event processing:** Clean but still unprocessed data only provides very simple information that cannot be used to support business processes and management decision-making. Firms need to transform this raw data into a high-level business logic that is understandable by users. This is achieved by applying event processing techniques, such as CEP (complex event processing) semantics, which is very similar to the tasks performed by the human brain when trying to recognize patterns in the information it receives. CEP abstracts information by aggregating low-level events into high-level events that make sense to a user and can be responded to.
- **Data analysis and storage:** Once the data has a business logic, the next step is to analyze, store, compress, and exchange it. Data must be able to be shared between different IoT devices and systems; however, this is not always possible due to platforms that are heterogeneous. In this case, it becomes necessary to have a multi-platform information exchange language (similar to XML) that covers different application domains.

13.4 Creating Value from IoT

IoT ecosystems represent an opportunity for business owners and managers to go beyond traditional ways of doing business and create new sources of value. From an internal point of view, owners and managers should see the IoT as an opportunity to reach new levels of organizational efficiency and effectiveness and improve innovation processes, thus generating long-term value for the firm (Lu et al., 2018). By enabling "things" with the ability to be identified, monitored, and encrypted, it is possible to increase the speed, accuracy, convenience, and quality of transactions and processing of products and services.

But, which firms can benefit most from the IoT? Surely it will be those more advanced firms in the data value chain, which have already captured data that requires analysis. Actually, most Big Data analytics vendors already offer solutions to generate new revenue streams from the IoT, such as IBM Watson IoT platform, Google DeepMind, AWS, and Microsoft Azure.

Tourism firms should understand the potential of the IoT for business process management, technology strategy and management and, ultimately, for the design of long-term value creation strategies. They should learn that the main mechanism by which IoT technologies create value is through the generation of refined real-world information, on which they can base the optimization of technological and business processes (Del Giudice, 2016). However, more evidence is needed on how the IoT is changing the way firms manage their core business processes and the impact on knowledge flows, innovation, and competitiveness. A recent survey of 500 business executives who had led IoT initiatives in their organizations asked them to identify which functions were top priorities for IoT (Insights Team, 2017 as cited in Nord et al., 2019). Although the findings revealed that the IoT is affecting many parts of organizations, three areas were identified as priorities:

1. The first priority area is the customer experience. This is a key area of business management in which other authors agree that the IoT can help in several ways (Gillet, 2018).
   - Monitoring and improving customer experience through the firm's value offerings.
   - Personalizing the firm's interaction with each customer.
   - Improving and learning over time through automatic updates of products and services.

2. The second priority is measuring and monitoring the performance of the firm in real time, so that IoT can support financial decision-making and provide a more holistic view of the firm that complements the data generated by accounting and financial information systems.

3. The third priority area is tracking and managing firm assets, such as computing devices, tools, vehicles, machines, etc. By using IoT sensors attached to assets, firms can actively track information about their location and condition without the need for human intervention.

The IoT offers a host of hidden opportunities to create value from the huge streams of data it handles. Unlike conventional data sources, IoT devices generate huge amounts of data in real time, allowing owners and managers to quickly act on business processes and service delivery based on the expectations and conditions of the market. Firms can leverage the Big Data resulting from the IoT to perform valuable analytics on consumer behavior, efficiency, and effectiveness of business processes, and the performance of products and services, which in turn can support strategic business planning.

Production systems incorporating IoT technologies can reveal operational inefficiencies in the organization of human work that can be remedied by leveraging the refined insights that the IoT generates (Sestino et al., 2020). The IoT can be used to monitor production, avoid downtime, and rapidly adapt production methods, all of which can make business processes more effective and efficient, speed up cycles for creating new products and services, and reduce costs.

In the field of services, IoT devices such as "Voice Assistant" can be integrated into traditional services (i.e., hotel room service) to improve their planning and delivery quality. Furthermore, smart products and services imbued with IoT can provide insights into how they are used by consumers, allowing firms to quickly adapt to real market needs and increase productivity. Owners and managers can also use IoT data from past events to identify trends and simulate future scenarios using predictive analytics machine learning techniques and process automation. In addition, when large amounts of data about employee activities are available, firms can use that information to better organize human work, simplifying employee effort and enabling customer self-service, which can increase trust and result in streamlined operations and staff empowerment.

IoT technologies are a great opportunity to create stronger bonds with customers and improve consumption experiences. As going online has become the norm, people have become accustomed to consuming personalized services at high speed. Features built into IoT devices, such as voice commands, can satisfy these desires and encourage consumers to think of their devices (e.g., smart watches, smartphones, etc.) as assistants during pre- and post-purchase activities rather than just tools. Promotional strategies, advertisements, and more customer-friendly direct marketing campaigns based on geomarketing and geolocation can be activated thanks to data extracted from IoT devices and applications. In fact, IoT data increases dramatically when organizations consider all the touchpoints they have (e.g., websites, screens, interactive totems, voice assistants, etc.). With all this information at their fingertips, firms can provide value offerings that are based on highly personalized business strategies. These opportunities are further expanded when firms consider the use of AI technologies.

Additionally, the intellectual capital of firms in the form of knowledge, information, and ideas is a crucial element to create value from the IoT. This is because IoT technologies are changing the way firms acquire and manage knowledge inside and outside organizations. Firms should exploit the IoT through knowledge management systems that incorporate advanced ICT and explore external sources of knowledge, which in turn would translate into better and innovative performance. This will drive firms to adopt an open approach to fostering knowledge flows and improving their internal knowledge management capabilities, both of which are prerequisites for increasing the firm's capacity for innovation (Santoro et al., 2018). When adopting an open innovation approach, firms will need to establish collaborations with other players in the tourism ecosystem to acquire new knowledge and create new value from it. Those firms that first recognize this challenge and approach the transformation process quickly and effectively will be able to gain a competitive edge.

13.5 IoT Technologies

IoT technologies offer the potential to change the tourism marketplace by helping firms develop compelling business strategies. At the organization level, the value created by IoT systems and applications is critical to their adoption and the opportunity to create a new connected world with great potential for firms, consumers, and governments. But new problems and challenges inevitably arise.

One is that IoT technologies have different and often competing standards, which leads to a fragmented market for IoT products and services. The need for a market standard is key because it would lead to network effects (i.e., increasing value to users as more members join), but also because end users, be they home consumers or firms, will continue to be reluctant to invest in systems that they fear will be compromised, become obsolete, or simply are not able to communicate with other IoT devices. It is expected that this trend will normalize in the medium term and that some of these standards (and industry players) will prevail over the rest and begin to dominate the IoT market. Next in this section, the technologies that are playing a more relevant role in the development of the IoT are described (Fig. 13.5).

IoT hardware

At the core of IoT technologies is the link between the physical and virtual worlds. Critical to this is having the right hardware, which can be sensors, microprocessors, RFID chips, and antennas, as well as full-fledged devices like fitness trackers, thermostats, lighting, smart meters, door locks, etc. that are embedded in consumer environments, smart cities, or smart industrial environments (M2M). The main hardware components that form the technological base of IoT systems include (De Saulles, 2016):

- **Sensors:** As a result of the massive adoption of smartphones, sensors that a few years ago were very expensive and only available to a few have drastically

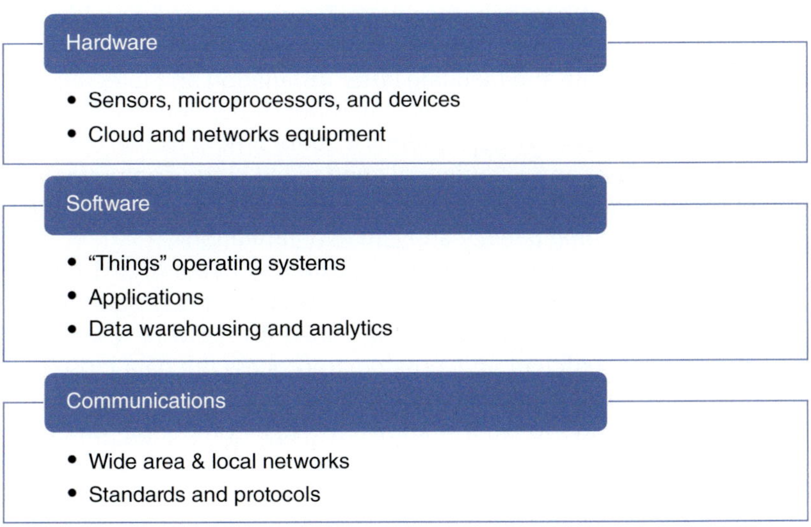

Fig. 13.5. IoT technologies. Source: own elaboration based on De Saulles (2016).

reduced their cost and their use has spread, thus facilitating the development of IoT systems. Sensors are not only hardware sensors or sensor networks, but also sensor software. All of them form a system capable of capturing conditions of interest in the real world, such as the presence of people. Most likely, sensors are going to be the main source of data in the future thanks to their high speed, lower power consumption, and high resolution. Common sensors include accelerometers, GPS chips, and cameras.

- **Smartphones:** Smartphones can behave as sensors and controllers for IoT devices through mobile applications (e.g., smart watches and smart home applications).
- **Processing capacity and bandwidth:** The deployment of 3G and 4G cellular networks in most of the developed world, as well as cheap broadband connections for businesses and homes with Wi-Fi capabilities, allow wireless connectivity capable of supporting IoT developments cheaply and ubiquitously.
- **Big Data infrastructure:** Back-office infrastructures put into service by major players in the technology industry have been increasing in recent years. As a result, firms in all industries have powerful facilities at their disposal to store and process the vast amounts of data generated by IoT devices at low cost.
- **IPv6:** The old IPv4 protocol, first implemented in 1983, had a limitation of 4.3 billion unique addresses. In a world in which some analysts (like IDC) predict more than 41 billion connected IoT devices by 2025, capable of generating 80 zettabytes of data, it was essential to develop a new protocol. The new IPv6 protocol provides 3.4×10^{38} addresses, an almost unimaginably large figure, and more than enough to cover IoT needs for decades to come.

The integration of IoT with the cloud (CoT) is another vast source of opportunities. For example, it can help manage IoT resources and provide IoT services more cost effectively and efficiently. In addition, it can simplify the flow and processing of IoT data and facilitate faster installation and integration at a lower cost (Atlam et al., 2018). However, CoT is not straightforward and presents significant challenges for IoT applications that are time sensitive or where internet connectivity is poor and a traditional centralized cloud computing architecture is not possible. In cases like this, a more advanced cloud computing paradigm is needed to overcome latency and capacity limitations.

One of these technologies was suggested by Cisco in 2014 and is known as Fog Computing. Transmitting massive amounts of raw data over a network to a centralized location places a heavy load on network resources. To make this process more efficient, it is possible to process the data close to its source and send only the data that has value to the remote data center. Fog Computing (similar in many ways to what is known as edge computing) processes data as close to the source as possible, so intelligence is transferred from the cloud to the edge. Time-sensitive data (e.g., for connected vehicles, video streaming, gaming, and virtual/augmented reality) is processed at its origin in the edge computing architecture, or sent to an intermediary server located close to the source. Meanwhile, less time-sensitive data is sent to the cloud for long-term storage or historical analysis (Attaran, 2017). With Fog Computing, IoT data processing and storage capabilities can be provided locally on IoT devices instead of being sent to the cloud, so the "fog" offers a higher quality and faster response service. Fog Computing is all in all a good technical solution that enables IoT to provide efficient and secure services for many IoT users.

IoT software

While the hardware is often the most visible part of an IoT system, especially when it comes to our homes or the devices people wear, the software running on the hardware is arguably even more important from a business perspective. Software within the context of IoT is not just limited to the software behind the "things" operating system but includes the applications that manage communications with other "things" and with hubs, the middleware that connects with other applications, and the software that supports data capture and analysis in the back-end (De Saulles, 2016).

Much of the IoT software focuses on processing the data generated by the IoT (i.e., making sense of the data and improving business processes and decision-making). This often requires complex back-end systems to store and analyze the results. The software used for managing conventional databases is often not appropriate for the much larger data sets generated by the IoT. Database software solutions built around frameworks like Hadoop and NoSQL have proven capable of handling massive volumes of data at very high speeds. Other software vendors offer data warehousing and analytics solutions based on these technologies, including the Microsoft Azure platform, GE Predix platform, and Amazon Web Services (AWS). Once data from an IoT system has been captured and processed within one of these data platforms, analysis and visualization of the information is crucial if meaningful decisions are to be made. Vendors offering enhanced data visualization and reporting services include some of the established business intelligence players like SAP, SAS, Tableau, Splunk, or Qlik.

Another key factor for the development of IoT systems is how to make them smarter. This involves uniting IoT with AI. AI can transform the IoT into an entity that works intelligently and behaves with decision-making capacity based on data and past events. By incorporating reasoning and decision-making capabilities into the IoT, the system can automate the management and control of IoT functions autonomously (e.g., self-configuration, self-protection, self-organization, self-healing, self-sufficiency, self-learning, self-adjusting), thus reducing the operating cost (Pramanik et al., 2018). What's more, the IoT system can automatically train, learn, and fix future problems to a certain extent.

Furthermore, a wide variety of communication technologies covering many domains of application and communication requirements have gradually emerged. Some technologies are domain-specific applications, such as Bluetooth Low Energy in personal area networks or Zigbee in home automation systems. Others, such as Wi-Fi, Low Power Wide Area Networks (LPWA), and cellular communications (such as 3GPP – 4G Machine Type Communications, or MTC) have a much broader reach. These niche solutions are often very dynamic, constantly evolving into new technologies or application domains. The advent of 5G communications means a potentially disruptive factor for IoT technologies. With the expected increase in data speed, lower end-to-end latency, and improved coverage over 4G, 5G is expected to meet the most demanding IoT applications in terms of communication requirements and number of connected devices. Furthermore, 5G may become a unified interconnection framework, facilitating seamless connectivity of "things" with the internet by integrating heterogeneous access technologies (Palattella et al., 2016).

IoT communications

IoT communication technologies include the connectivity technologies, standards, and protocols that connect the IoT to the internet. Connectivity is an essential element of any IoT system as it allows users to control "things" and makes data extractable and manageable remotely. Today there is no single communication protocol or standard that allows the exchange and integration of sensor data across devices, users, and domains (i.e., the data has a meaning that is independent of the device or format in which the information is used). This is because some devices require communication chips that consume very little power and only need to function over short distances, while others are less constrained in their power requirements and need to transmit data many miles away.

The situation has led various technology companies, organizations, and consortiums to work on the development of standards that allow interoperability between devices and systems that are very heterogeneous. Some of them like Bluetooth, Wi-Fi, and GSM are standards already widely used. Others, such as WeMo, ZigBee, and Thread, are used by IoT device manufacturers but have failed to gain a foothold in the market. There are also communication standards, particularly in the 5G environment, that are still under discussion and development and have not been released yet. Finally, to achieve the ubiquitous connectivity that IoT applications require, many more features and functionalities will need to be added to the current broadband approach, which will require new developments, such as the heterogeneous networks (HetNet) approach that allows multiple types of wireless access nodes.

13.6 IoT Applications

Although IoT is in its infancy, the number of applications is growing exponentially as more and more devices gain the ability to connect, opening new opportunities for innovation with smart products and services. Basically, IoT applications can be categorized into two types: consumer IoT and industrial IoT, each with different implications for underlying technologies and business models (Shim et al., 2019).

Consumer IoT

Consumer IoT (cIoT) applications involve the interconnection of consumer devices and everyday "things" located in the users' environment, be it the home (e.g., refrigerators, televisions, and appliances), the workplace, or even cities, aimed at improving people's quality of life, saving time and money, and making decision-making easier. cIoT applications interconnect physical "things" through digital means of communication (wired or wireless) and allow them to exchange information with each other to help the user in their daily tasks or make intelligent decisions on their own. For example, in automated homes and buildings, the temperature and brightness of lights can be automatically adjusted according to the context and the presence of people in the room, thus improving people's comfort. A thermostat in a room and the air conditioner are two "things" that can communicate with each other through IoT, in such a way that the thermostat exchanges the sensed temperature with the air conditioner so that it regulates the temperature.

User acceptance of cIoT technologies and services depends on whether they find it useful and easy to use (Lu et al., 2018). Some studies have also found that fun and pleasure are additional factors influencing the acceptance of IoT technologies and that should be considered when designing IoT products and services. Price is another important factor. Many of the business models of IoT companies are based on subscription models in which users pay periodic fees for the value-added services that the products deliver. Therefore, as long as fee schemes do not share the customer's logic and are difficult to understand, prices will continue to be a key conditioning factor that affects the rate of adoption by consumers.

Additionally, emerging innovative technologies like IoT often come with risks. When these are known or perceived by users, they directly influence the intention to adopt the IoT. In addition to issues related to privacy and security, the intangibility and intensive use of different technologies in IoT systems often make users perceive higher risks, thus discouraging its adoption. Last but not least, there are the factors associated with social influence (i.e., when an individual believes that someone influential thinks he/she should be using the technology). Social influence is a more important factor than it seems at first glance and can significantly persuade users to adopt IoT. To address this challenge, firms' business strategies should focus, for example, on offering financial incentives to lead users who recommend IoT products or services to others, as well as encouraging peer-to-peer promotion and diffusion of IoT.

Industrial IoT

Less visible to most users than the cIoT, IoT business and industrial applications (iIoT) have a far greater economic impact than consumer ones (De Saulles, 2016). Indeed, it can be said that iIoT technologies are the real driving force of the so-called Industry 4.0, in which industrial processes and machines become more intelligent and modular, providing greater flexibility to meet customers' demands.

The iIoT focuses on the integration between industrial technologies (i.e., production and delivery technologies) and ICT, so that the resulting IoT systems can monitor and collect data, and exchange and analyze information to intelligently change their behavior or reconfigure business processes. Some examples of iIoT are hotel process monitoring, vehicle fleet tracking, and beverage vending machine monitoring, control, and replenishment. iIoT has important implications for the management of organizations and their decision-making processes, as well as their ability to create new services and revenue streams. Today, iIoT applications are varied and can be found in all types of activities and industries, from public services, to transport and logistics, manufacturing, smart cities, smart buildings, industrial automation, and tourism, where they can improve the productivity and save costs through process optimization.

By collecting and analyzing IoT-based Big Data with data analytics, firms can enhance their understanding of consumer behavior and decision-making, including predicting what is going to happen, all of which has relevant implications for marketing management, innovation, and the management of business strategies. This explains why tourism firms should start equipping themselves with infrastructures and new capacities that allow them to process and analyze the Big Data generated by the IoT (Jesse, 2018; Sestino et al., 2020).

Both cIoT and iIoT applications share common features and have clear differences. For example, communications in cIoT are from machine to user (client–server type) and seek that the connected "things" have low energy consumption and are easy to install, integrate, and maintain. On the other hand, iIoT applications use machine-to-machine communications for automation and/or monitoring of control processes. Additionally, cIoT and iIoT specific communication requirements often differ in terms of reliability, quality of service (i.e., latency, throughput), and privacy protection (Palattella et al., 2016).

Marriott International IoT

Marriott International has partnered with companies such as Samsung and Legrand to develop a smart hotel room. The project aims to improve the guest experience through more efficient room design and construction, thereby contributing to Marriott's sustainability goals. Some of the expected benefits of the smart room include guest access to their own data and information, voice- and mobile-controlled systems, and enhanced personalized service. The project features various IoT systems, devices, and applications for device-to-device communications. Samsung develops end-to-end IoT services from intuitive lighting to voice-activated room controls powered by the ARTIK platform and SmartThings Cloud. Legrand offers a suite of power, light, and data solutions that bring power and connectivity to previously untapped locations. Marriott has well-established private data centers that extend into the public cloud and more than 20,000 devices are equipped with enterprise mobility management software that ensures its mobile workforce can reliably serve customers anytime, anywhere. IoT-enabled building management, room management, energy management, and lighting control software and hardware reduce the environmental footprint of water/carbon/food waste and contribute to the general goal of reducing the environmental footprint.

Source: Own elaboration based on Lee (2019)

13.7 Challenges of IoT

Before the IoT can reach the levels of adoption seen with other technologies, there are a number of significant barriers and challenges that need to be resolved, and they are not just technical. As ICTs become more and more integrated into our daily lives, at work and at home, so do the challenges and potential adverse consequences, such as theft of sensitive personal data from devices, or hacking of IoT systems in industrial settings to take control of power supply systems. The controversial "always on and listening" IoT encompasses a growing range of devices. Google's "Ok Google" search command for smartphones and PCs, Apple's Siri, and Amazon's Alexa are examples of how tech companies are trying to make their digital assistants a natural part of our daily lives. As machine learning and voice recognition improve, they will become even more integrated into our homes and the amount of data that can be coveted by criminals will grow.

Due to the rapid growth of the IoT industry, the demand for professionals with the necessary skills to create IoT-enabled devices and manage the layers that make up the IoT architecture has increased exponentially. Firms need staff with high-level technical skills in today's programming languages who also understand the business environment and infrastructure in which IoT devices operate. Software developers are always in high demand, but the growth of IoT will increase that demand even more.

In many ways, these challenges are nothing new. In fact, we have seen similar challenges arise in the life cycle of other technologies, such as when the internet or personal computing began to grow. However, it is the sheer scale of what the IoT promises that raises concerns among consumers, firms, developers, and regulators. Below are some of the main challenges affecting IoT adoption, including IoT Big Data, privacy and trust, security, regulation, and business models.

IoT Big Data

IoT systems generate a large amount of data that comes from highly dispersed and heterogeneous sources. For firms this means a great challenge, since they must deal with a large variety and noise in the data. Added to this is the fact that conventional data mining algorithms are not ready for Big Data, therefore the problem of how to modify them to operate in a Big Data environment arises. In this challenging context, it is especially important that firms develop a Big Data mining framework that is efficient and effective for exchanging data between different applications and systems, and that guarantees data security and privacy.

Privacy and trust

People are increasingly concerned about how firms use their personal data. Despite the many applications and benefits that IoT can have in everyday life, if privacy concerns become too great, consumers may lose their willingness to interact with IoT platforms and business opportunities will not arise. This situation becomes even worse when we think about the dispersed nature of the IoT and location-based or geo-referencing tools. Personal data has become a new "asset class" that can create value for individuals, firms, and society. As the volume of data grows, firms must take steps to encourage the kind of trust that fosters privacy, increasing the transparency with which personal data is used and balancing business interests with user privacy concerns (Sestino et al., 2020).

Blockchain technology can help overcome some of these obstacles by providing a solution for authenticating the identity of nodes in an IoT network and verifying that only authorized nodes can access data, thus maintaining data privacy and access control (Shim et al., 2019). IoT devices could send data to a blockchain-based database that would only allow authorized parties to access and contribute data without the need for centralized control and management. Eventually, the combination of IoT with blockchain can reduce the privacy and security risks of IoT data and turn it into a trusted source of valuable data. These benefits could also be extended to the management of payments in the form of micropayments that would help monetize data from IoT networks. For example, hotels could send information about the flow and characteristics of visitors to visitor control centers, then this data could be traded (or auctioned) on Big Data marketplaces to facilitate analysis, marketing, and market research as value-added services. In summary, the convergence of IoT and blockchain is at a very early stage and still offers numerous technical challenges given the immature state of these technologies and the speed at which they evolve, however it is worth keeping an eye out for further developments.

Ultimately, firms are expected to develop new trust models especially focused on 5G scenarios where there will be a huge number of different devices connected, each with its own security scheme. This will raise new expanded requirements for authentication and protection of personal data between different actors that will reinforce those currently used for 4G. Furthermore, firms will be required to develop new ethical practices in the handling of personal data that involve users in the process. These practices will be critical for people to gain a high level of confidence in using IoT, and include issues such as data ownership, accessibility, and accuracy. At the same time, it is to be expected that as people get used to new IoT environments and realize its benefits, some privacy concerns will diminish and people will gradually become more willing to share their personal data.

Security

Wherever a technology can be used to exploit a weakness in a system for profit, it is inevitable that someone will take advantage of it. In other words, if a system can be hacked, it will be hacked. Being aware of this maxim, security has become a primary concern for firms focusing on adopting IoT. IoT technologies open the doors of firm systems and data to everyone and increases security threats. Leakage of personal data through illegal hacking has become commonplace, and until reliable and secure systems are in place, many people will be reluctant to share their data with third parties. Interestingly, it often happens that firms that have been hacked never know, or underestimate, the scope of such illegal attacks. In 2012, it was thought that 6.5 million passwords from the LinkedIn social network had been stolen. However, it was only in 2016 that the company revealed that more than 117 million passwords had been taken and that this data was being sold on the internet.

IoT-enabled devices are vulnerable to hacking, and it is feared that they are an opportunity for hackers to steal data and harm people. These risks are further heightened as there is no government regulation of IoT and no organized effort by manufacturers to make these devices more secure. In addition, as firms increase their use of AI, machine learning, and increasingly sophisticated algorithms to give their products and services a competitive edge, the amount of personal and sensitive data circulating on the internet will increase. So, what can firms do to ensure that people's personal information is secure?

Incorporating strong security at all levels of IoT is crucial to preventing fraud and building the necessary trust among users for mass adoption. It is the responsibility of owners and managers to implement methods to protect information in business systems and in interactions with customers. It is also essential that security is built into products and services at an early design stage and not as an afterthought. IoT vendors have addressed some of these security challenges by testing a variety of query protocols, jamming, and other techniques (Attaran, 2017). The Federal Trade Commission of the United States of America also makes these recommendations to firms that want to take their first steps in the IoT (FTC, 2015).

1. Conduct a privacy and security risk assessment.
2. Minimize the data collected and retained from IoT initiatives.
3. Rigorously test IoT products before launching them on the market.

Any firm developing IoT products should seek to maximize the strength of its security strategy. This means thinking about security from the customer's perspective (i.e., in terms of what the solution will do, who it will report to, and how data will be protected), as well as applying a multi-layered approach that incorporates identity, access, encryption, and network security and monitoring. It is also important for firms to set up the controls that will govern the management of the IoT system throughout its life cycle, how issues will be resolved, and when the system will be retired and upgraded.

There is no doubt that security is more of an issue in some scenarios than others. Hacking someone's smart thermostat and turning off their home heating can be inconvenient, but taking control of a self-driving car and crashing it on purpose is something quite different. Another concern with IoT security is the number of new firms entering IoT with no security background. For example, a small business that has amassed experience designing and marketing tours of historic places probably doesn't have the knowledge or skills to create products that can resist hackers. Therefore, new technologies and methodologies must be developed that meet the highest requirements in terms of reliability, security, and privacy, and that can be adapted to the particular needs and capabilities of tourism firms.

Regulation

A prerequisite for IoT to become a practical reality is building trust among users. As with other ICTs, regulators play an important role in establishing legal frameworks that allow firms to operate with confidence and deliver trust to users. In some cases, new regulations are going to be required that take into account the new challenges posed by the IoT; in other cases, the existing regulations may be sufficient or simply require an update.

Today most IoT regulatory frameworks are still very vague or non-existent. There are also clear differences in regulations concerning personal data between the two world leaders in this subject: the US and Europe. In 2016, the European Commission adopted the General Data Protection Regulation (GDPR), which included some modifications related to data portability and the "right to be forgotten" that recognized how much technology had evolved since the 1990s. The European regulation states that when data is kept in a "structured and commonly used format", individuals have the right to request that it is transferred to another service provider. Meanwhile, the United States has a more fragmented approach with a series of specific federal and state laws, each addressing different aspects of personal data. This reflects the importance of state legislation in the United States, but perhaps a more relaxed attitude among many US citizens regarding how organizations use their personal data. The lack of regulatory consistency between Europe and the United States with respect to data protection has also been a source of friction with some EU member states due to the restrictions placed on firms operating in these countries and the extent to which they may move personal data outside the European Union. The data protection directive and the GDPR allow the movement of data to firms in the United States as long as they are subject to the same data protection treatment that they have in Europe. Furthermore, IoT has also become the topic of power politics due to the risks of war and cybercrime.

Although there is no clear consensus on how to legislate on privacy, privacy protection policies should guarantee a series of key principles, including data anonymity; notification to users when information leaks and privacy violations occur; that the data collected is treated with the sole purpose of providing services; transparency in data collection; and accountability of data collectors (Lu et al., 2018). Especially challenging is the massive collection of data in the IoT. Soon it will be almost impossible for a person to avoid being monitored and recorded by sensors in public spaces. What's more, once this information is generated, it may be retained indefinitely, and its disclosure may not be controlled on a personal level. Therefore, when designing IoT applications, information protection becomes critical as it directly affects the user experience and trust in IoT services. The implementation of functions such as "forget me" by service providers, which allow user information to be deleted on request, may be a possible solution to this problem.

In the same vein, there is the issue of security in access to personal data and the problems with organized crime and cyberterrorism. Due to its connected nature, the IoT generates a large amount of data that is collected and transferred between the virtual and physical worlds, and this can be a never-ending source of security problems. Some studies suggest that 70% of IoT devices are vulnerable due to lack of encryption, inadequate software protection, insecure web interface insecurity, and insufficient authorization levels. Furthermore, security threats appear at every layer of the IoT architecture making it even more difficult to analyze security issues, such as phishing and routing attacks, viruses, trojans, and spam messages in data transmission, to name just a few. In the end, solutions to IoT security issues need to involve multiple stakeholders, making it clear who is responsible when a system fails or is attacked by cybercriminals. Improved accountability in IoT (i.e., the obligation of each person responsible for the layers of the IoT system to explain and justify their actions and decisions to users), reinforces security and confirms the need for a stable legal framework for all firms.

Business models

The IoT can impact the business models of tourism firms, both by transforming traditional business models and creating superior value, and by leveraging new business models based on new competitive advantages. IoT-based business models can help improve knowledge sharing by providing firms with a wealth of contextual information about their customers and product/service usage patterns. This abundant knowledge is highly valuable in business terms to increase the innovation capacity of organizations and create new value (i.e., facilitate the decision-making and marketing process, engage the consumer, set prices, etc.). IoT provides opportunities for tourism firms to design new strategies, optimize their business process flows, and offer personalized products/services to customers (Lu et al., 2018). In short, IoT-based business models can improve the firm's competitiveness by better satisfying its customers in the product/service delivery process.

Two new but powerful IoT-specific business models have recently emerged: 1) digitally loaded products, and 2) Sensor as a Service (Fleisch et al., 2015). The idea of digitally charged products is that classic physical products are loaded with a package of new digital services based on sensors, so that they acquire more relevance based on new value propositions. One usual approach to this business model is the physical freemium model, which refers to a physical product that is sold at a very small margin together with a free digital service (i.e., a product that is sold with some digital instructions for installation, operation, and maintenance). Over time, customers can activate (purchase) more digital services based on their needs, thus delivering higher margins to providers. Physical products can become a kind of repository for electronic services as they can embed digital advertising, accrue and redeem loyalty points, and record what is happening around it through a smartphone. For example, the Amazon app already offers features like these for products with an Amazon barcode.

The Sensor as a Service model provides the ability to collect, process, and sell sensor data for a price. This model is based on the idea that measured data from the physical world no longer needs to be vertically integrated into a single application, but can be integrated into a wide range of applications. Unlike the digitally uploaded product business model, the focus here is not the product that generates data or the resulting services, but the data itself. An example of this business model is Streetline (http://www.streetline.com), which installs sensors in municipal and private spaces to detect free parking spaces and then sells the collected data to interested third parties.

A drawback when developing IoT-based business models is that for each IoT-based service, several different parties must cooperate (and thus receive benefits) to support the technology and standards that the service requires (Shim et al., 2019). As IoT devices become cheaper and their complementary software services mature, new applications and business models will emerge based on the data they generate.

Discussion Questions

- What type of IoT products and services have the greatest potential for the tourism consumer-oriented market (B2C)? And for the business-to-business (B2B) market?
- What relationship is there between the development of IoT, Big Data, and AI technologies? Is it realistic to foresee an integration of these technologies in the future?
- What factors related to privacy, security, and regulatory issues are most influencing the development of IoT in the market?
- What factors may be having an impact on the adoption of IoT-based technologies and services by the end consumer? And by firms?
- What professional profiles are most needed by tourism firms to benefit from the IoT? Does the tourism firm have them?
- What management practices should the tourism firm implement to accelerate the adoption of IoT technologies in the organization?
- What role does business leadership play in the adoption of IoT technologies in tourism firms?

14 Cybersecurity

As tourism firms embrace smart technologies they become increasingly vulnerable to security risks (Paraskevas, 2020). Cyber-attacks are highly damaging to customer trust and the reputation of firms, not to mention the serious financial and legal consequences that can result. Kaspersky Lab (2018) estimates that firms pay an average of US$551,000 to recover from a security breach (US$38,000 in SMEs). These are direct losses only, that is, the money firms must spend on IT recovery services, business losses and downtime, as well as legal and PR services. Indirect losses or the costs of hiring and training additional staff, infrastructure improvements, etc., are estimated on average at US$69,000 (US$8000 in SMEs). Sometimes there are even fines to pay. Data breaches in the Wyndham Hotels and Resorts computer network between 2008 and 2010 compromised the records of more than 600,000 guests with US$10.6 million in fraudulent credit card charges. The event led to a lawsuit from the US Federal Trade Commission in 2012 and from a shareholder in 2014. Both lawsuits were settled for undisclosed amounts. Many other high-profile tourism firms have made negative news headlines because they did not pay adequate attention to these risks and came under attack from cybercriminals, some examples include Marriott, Hyatt, MGM Resorts, Carnival Cruise, British Airways, and Cathay Pacific. This happened because, despite the warnings they received from the ICT industry, they adopted a fragmented and clearly insufficient cybersecurity approach.

Deliberate cyber-attacks are becoming more frequent and their impact has drastically changed from paralyzing the operations of organizations to threatening the security of more critical tasks, such as medical care for a patient or the financial viability of a firm (Reagin & Gentry, 2018). The firms' response to this new threat environment has been quick to materialize: firms are spending

DOI: 10.1079/9781800620261.0014

huge amounts of money on cybersecurity. In 2004, the global cybersecurity market was worth US$3.5 billion and it is expected to grow to US$155.83 billion in 2022 and reach US$376.32 billion in 2029 (Forbes Business Insights, 2021). Cybersecurity is no longer an option, it is a strategic asset that every organization must address to ensure business continuity, even more so in the new era of smartization. Cybercriminals are continually improving their techniques and escalating their attacks with higher levels of sophistication, thus causing damage that has more impact for organizations. For this reason, and since preventing cyber-attacks is increasingly difficult, tourism firms need to approach cybersecurity decisively and strategically, quickly adapting information security and developing cyber defense programs that are up to the challenges that lie ahead.

14.1 Cybersecurity in Tourism

Despite the billions spent on cybersecurity, cyber-attacks are considered the biggest threat to the deployment of smart technologies around the world. Furthermore, the travel and tourism sector is specifically one of the sectors most susceptible to cybercrime; the hospitality sector ranks third in terms of the number of security incidents, only behind retail and financial firms (Trustwave, 2020). With the size of the global online travel booking market projected to reach US$2 billion by 2028 (Facts and Factors, 2022), cybersecurity threats can only grow. Customer data stored on countless computers, servers, and networks in hotels, airlines, car rental and reservation firms, and even in bars and restaurants, is a juicy steak for cybercriminals around the world.

Most international hotel groups, from Marriott and Starwood to Wyndham, Intercontinental, Hyatt, Rezidor, Mandarin Oriental, and Omni, as well as smaller independent hotels, have been victims of cyber-attacks multiple times in recent years. But also other firms in the tourism sector such as in airlines (e.g., Air Canada, British Airways, Cathay Pacific, Delta), tour operators (e.g., Thomas Cook), travel websites (e.g., Expedia, Orbitz, Rail Europe), third-party booking platforms (e.g., Sabre Hospitality Solutions), holiday camps (Butlins), trade travel groups (ABTA), chain restaurants (e.g., Arby's, Checker and Rally's, Cheddar's), and even bakery-cafes (Panera Bread), to name a few. They have all been hacked in the last few years and many of them not just once (Paraskevas, 2020). Analysis of data breaches and attacks suffered by tourism firms shows that they are highly vulnerable to cybercrime due to the millions of customers who interact with them in cyberspace, weak defenses at points of contact and sales, and the inherent complexity of travel booking and payment networks, which involve a large number of actors and third-party service providers in a distributed scheme. As a result, even substantial cybersecurity investments often yield very low returns, leading to a "cyber paradox" that involves entering an endless cycle of massive cybersecurity spending with no evidence of security risk reduction.

Many times what really fails in tourism firms is that owners and managers think that cybersecurity is only about implementing a technology, and forget that it is also about training people and (holistically) guiding processes, technologies, the culture, and the physical environment towards the same goal, which is to achieve organizational security. Tourism organizations must create a digital environment that is secure for their employees, customers, and suppliers, and ensure that they will resist in the event of an attack. That is why they need to adopt measures to prevent and respond to cyber threats by adopting an approach based on risk measurement and management.

14.2 Types of Cyber-Attacks

Cyber-attackers employ a wide range of attack techniques to circumvent the cybersecurity measures put in place by firms to protect their systems and data. Most attacks involve high-level technical skills in programming to detect network and organizational security vulnerabilities and inflict damage on organizations, or earn money for data theft or to recover and return to normal (Fig. 14.1). Next, in this section, some of the most frequent types of cyber-attacks that occur in tourism firms are examined (Paraskevas, 2020).

Malware attacks

This is perhaps the most common type of attack and involves compromising an organization's sensitive data and systems by infecting them with malware such as viruses, worms, Trojan horses, keyloggers, and other spyware. There are literally thousands of different malware variants that, once they infiltrate systems and hardware, spread to other systems leaving a trail of disruption and destruction. These attacks typically use a command and control server that allows cyber-attackers to communicate with infected systems, leak sensitive data, and even remotely control the targeted device or server. In the case of tourism firms, one of the most typical types of malware is the Carbanak/Anunak attack.

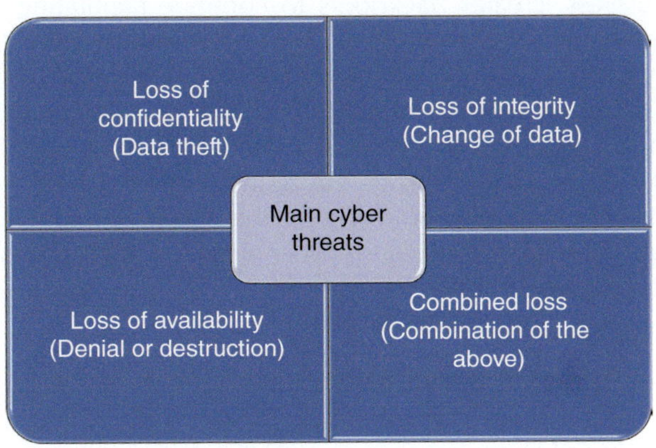

Fig. 14.1. Main types of cyber threats. Source: own elaboration based on Maalem Lahcen *et al.* (2020).

Point-of-Sale attacks

Point-of-Sale (POS) attacks are a very common type of malware among tourism firms, whereby cyber-attackers obtain valuable data, including the identity of credit card holders, card numbers, and personal identification numbers (PIN). For example, when a customer pays at a point of sale with a credit card, personal and transaction data is temporarily stored in the RAM of the POS terminal while it is transmitted to the payment processor. If a cyber-attacker has managed to install some type of malware on the payment terminal, they can copy the card data and transfer it to an external server. Many hotel groups (e.g., InterContinental, Mandarin Oriental, Radisson) and restaurant chains (e.g., Arby's, Checker and Rally's, Cheddar's) have suffered from these types of attacks, most of which have resulted in customer data being leaked through their POS systems.

Although the number of methods used by cyber-attackers is large, they typically look for vulnerabilities in external systems by injecting SQL into a web server or finding a computer on the internal network that is still using the manufacturer's default password. Alternatively, they may identify someone within the organization through a spear-fishing campaign or find a vulnerability among digital service providers (usually the weakest link in the chain) to be their gateway. Legacy systems that have not been updated with the latest security patches or are completely out of date are sometimes easy targets for cyber-attackers to infect with their malware. Another variant in the sector is the attack on mobile travel applications through which cyber-attackers can steal the personal data of mobile application users.

Ransomware attacks

The goal of this attack is not to steal data, but to deny access to the owner's data. While other types of malware have a harder time monetizing (i.e., the stolen data has to be first extracted and bundled and then sold on dark web forums or marketplaces), the goal of a ransomware attack is to make the target pay the attacker directly to regain access to data. The ransomware identifies the most sensitive or valuable data for the organization, corrupts the backups to the point of making them useless, creates backdoors to continue facilitating future infiltrations, encrypts the data and then sends the victim a message with some ransom conditions, usually the payment of an amount in cryptocurrency. It is quite common for firms that are the target of these attacks not to disclose them for fear of retaliation and the impact the attack may have on their brand image, and if they do, they usually deny having paid any ransom. Given the unpredictability of these attacks and the negative repercussions they have for the business, many firms take out insurance that protects them from these cyber-attacks. There are many firms that even have cryptocurrency reserves to cover demands in case of ransomware attacks.

Botnet/Distributed Denial of Service attacks

Botnet/Distributed Denial of Service (DDoS) attacks use thousands or even millions of computers or smart devices (called botnets or "zombie armies") connected in networks to conduct massive spam attacks, malicious login attempts, or bring down networks, devices, or websites. Some computer or devices are pre-infected with some type of malware, enslaving them so that they can be controlled by cyber-attackers and thus have access to the computing power they need to launch these attacks. Once the attack is started, the botnets send large numbers of requests/queries ranging from simple pings to mass emails. The sheer volume of traffic flowing to attacked networks, devices, or websites literally blocks the capacity of the targeted network or server to the point that other legitimate users cannot access or are denied access. The objective aimed by these attacks is to interrupt normal business operations or degrade the level of service of the organizations targeted by the attack, which is why they are also known as Distributed Denial of Service attacks. Many tourism websites are attacked in this way causing further slowdowns and downtime of networks and websites, and even increased GDS transaction costs when attackers ravage ticket sales.

Wi-Fi/website compromise

Cyber-attackers are good at hitting unsecured public networks in hotels, cafés, airports, visitor centers, etc. to infiltrate tourists' devices, infect them with their malware, and steal personal data, or use them as unintentional insiders for their next targets. In addition to exploiting Wi-Fi vulnerabilities, cyber-attackers also use a technique known as the "evil twin" attack, which consists of locating near a genuine Wi-Fi access point (i.e., the public Wi-Fi network of a tourist office) and discovering its Service Set Identifier (SSID) and frequency. Once this information is obtained, they send out a radio signal using the same frequency and SSID that appears to visitors as legitimate access with the same name. By the time visitors connect to the "evil twin", cyber-attackers have already taken control of the device, so all they have to do is collect personal data and start monitoring every activity that takes place on the device. "Evil twins" are also known as "honeypots" or base station clones and are very common threats in the travel and tourism industry.

Tourism firm websites are another attack vector through which to steal valuable customer data, including payment data. The websites of major hotels, museums, and transportation firms often provide detailed information about guest reservations, including reservation code, name, address, telephone number, last four digits of credit card, etc. to third party advertisers, social media, and other partners. Cyber-attackers can access and use this data to log in to a reservation, view personal details, and even modify or cancel the reservation.

Cognitive hacks

The continuous improvement and sophistication of cybersecurity defense systems has led many cybercriminals to review their tactics and focus on easier targets: the weak points of the human mind. Unlike regular cyber-attacks where cyber-attackers exploit vulnerabilities in IT systems, cognitive hacks are designed to exploit people's psychological vulnerabilities, using deception to manipulate their perception and circumvent security. Phishing is the most common type of cognitive attack and involves massive email campaigns that attempt to trick employees of organizations into downloading attachments containing some form of malware or providing their login details. Another more sophisticated tactic is spear-phishing, which targets specific members of staff or specific roles, and tries to make them believe that the malicious email they are receiving has originated from a legitimate source. A large number of hotels and guest houses listed on online booking platforms have been targeted by phishing campaigns in recent times. Customers of these platforms receive emails that appear to come from a legitimate source asking them to provide payment details and personal information such as name, address, phone number, etc. Other times, users are asked to log in to their accounts or sign in to troubleshoot a technical issue.

In the Mind of Cyber-Attackers

To anticipate the movements of cyber-attackers, it is important to understand not only the hacking techniques they use, but also the reasons that move them to act and the prize they can achieve. Not all cyber-attackers think the same or follow the same methods as defenders, so defenders must be open-minded and interdisciplinary in order to foresee various defense techniques. It is important that the people in charge of providing cybersecurity solutions in the tourism firm periodically try to answer the following questions to predict or try to anticipate the next move of cyber-attackers:

- What are the most critical assets of the organization, the most vulnerable, and the most damaged if they are attacked?
- How can the firm protect its access points?
- By what methods can cyber-attackers gain access to the firm's "crown jewels"?
- Where are the "crown jewels" located? Do they have sufficient protection measures to repel a cyber-attack?
- Is there an updated inventory of authorized devices and the level of access granted?
- What kind of malware defenses are being used by the firm? What defenses are planned?
- How effective and robust are the intrusion detection systems used by the firm?
- How is the firm notified of a potential threat or breach? Is there a plan to combat the insider threat?
- Are operating systems of the firm properly configured and up to date? What about the rest of the software installed on the computers owned by the firm?
- Is there a process or system to identify stolen credentials or compromised user accounts?
- Are employees aware of the risks of cyber-attack in the firm? Do they receive any kind of training or awareness program on the subject?

Source: Own elaboration based on Maalem Lahcen et al. (2020)

14.3 Cybersecurity in SMEs

One of the main problems of tourism SMEs with cybersecurity is that they tend to underestimate the threats they face, thinking that cybersecurity is not an issue they should worry about but something more typical of large firms, since their assets are much more lucrative for cybercriminals. This attitude is unprofessional and shows little knowledge of the subject. Quite often the result is a headache for the organization and its stakeholders.

SMEs have become one of the most valuable targets for cybercriminals. In most cases, the techniques used by SMEs to defend themselves are weak and involve little experience, inferior skills, and old or weak security methods. It does not matter that SMEs think that they have few assets with no apparent value for cybercriminals, it is enough that they are suppliers or partners of a larger business group to become a target through which cybercriminals will seek to catch the biggest fish. If we add to this the fact that they have fewer resources to defend themselves and are not willing to take costly legal action against cyber-attackers, SMEs are more attractive to many cybercriminals who see them as easy prey that they can attack without serious consequences.

Currently, poor risk management and the lack of adequate and sufficient measures to protect against cyber-attacks have become major issues for SMEs. This means that SME business strategies often pay less attention to risk management implications, which can have very unfortunate consequences affecting both tangible and intangible assets or, worse yet, lead the firm to bankruptcy (Alahmari & Duncan, 2020). SMEs are more likely to adopt a low-cost cybersecurity approach with a minimal cybersecurity budget and very basic standards far from their strategic priorities, thus becoming more vulnerable to cyber-attack threats. The development of cybersecurity practices as a key strategic factor of the SME could be a solution that would save owners and managers a lot of trouble and mitigate risks. For example, strengthening employee cybersecurity training is an important factor in improving their behavior within the organization and preventing them from being the cause of many of the cybersecurity threats that affect SMEs (i.e., when they ignore policies on information management and access to information). Other times, it will suffice to implement the practices contained in the General Data Protection Regulations (GDPR) and carry out campaigns that make the organization's decision makers aware of the consequences of a bad cybersecurity policy and the need to invest more to reduce the biggest cyber risks.

Today the use of IT technologies that are reliable and secure is a crucial factor in supporting the growth of tourism SMEs in the context of the Smart Revolution. Owners and managers should keep in mind that when cybercriminals identify vulnerabilities in the systems of an SME, attacking them will become their top priority, and they will not hesitate to exploit them sooner rather than later. Therefore, without technologies that are reliable enough, up-to-date, properly integrated into a managed IT framework, and well protected from the threats that surround them in cyberspace, cybercriminals will exploit the disadvantages and inflict serious damage that will cost a lot of time and money to repair. At the end of the day, it is not "if" the firm will one day be the victim of a cyber-attack, but "when"; therefore, leaders must strive to adopt a risk-based approach that allows them to defend their organization from all kinds of cyber-attacks.

14.4 Key Challenges in Cybersecurity

Stories of cyber-attacks have become routine, with cyber-attackers showing increasing levels of sophistication and damage inflicted. The race to discover new ways to exploit weaknesses in the defenses of firms' IT systems does not stop, so it is important that owners and manager are aware of the magnitude of the challenges they face and understand what is at stake. Below are some of the most important challenges that every tourism firm must know about to ensure the security of its operations in cyber space.

What matters most?

A cybersecurity strategy must provide the best possible protection for the key assets and operations of the firm. In this vein, organization must understand which IT systems are mission critical for its operations, as well as being clear about which information assets are the most valuable to cyber-attackers in order to apply enhanced protection to them. In addition, the firm must know where are and who has access to critical assets, and know how these assets are related to each other and what vulnerabilities (organizational and technological) can expose them to cyber-attackers. This process is known as OCTAVE (Operationally Critical Threat, Asset, and Vulnerability Evaluation) and is a methodology for evaluating information security risks that allows establishing a set of risk profiles for critical information assets and systems (Fig. 14.2). These profiles will allow business owners and managers to make informed decisions about the allocation of resources that strengthen the organization's cyber defenses and implement an adequate cybersecurity strategy for the firm (Caralli et al., 2007).

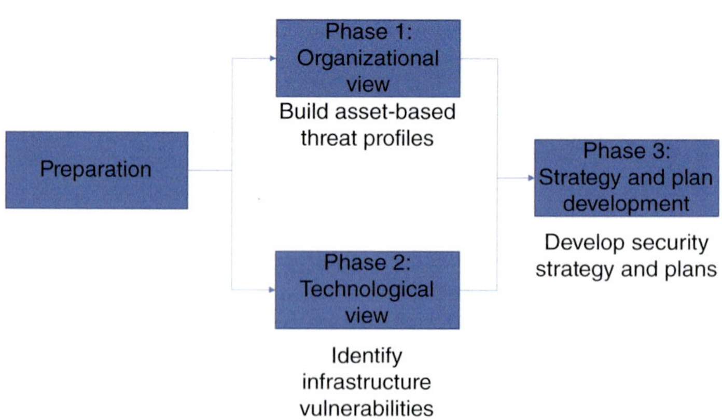

Fig. 14.2. The phases of OCTAVE methodology. Source: own elaboration based on Alberts et al. (2003).

Risk management frameworks

Tourism firms must properly manage risks if they want to succeed in cybersecurity, so choosing a risk management framework from the many available on the market is an essential task. One of the most popular cybersecurity frameworks is the US National Institute of Standards and Technology (NIST) Cybersecurity Framework. This is a high-level, easy-to-read set of guidelines for mitigating organizational cybersecurity risks, which organizations can customize to suit their needs. The framework consists of a core divided into five functions (identify, protect, detect, respond, and recover) further subdivided into 23 categories. For each category, a series of subcategories of cybersecurity results and security controls are defined, with 108 subcategories in total.

ISO/IEC 27005 is another good option for tourism firms. It is a set of standards developed by the International Organization for Standardization (ISO) and the International Electrotechnical Commission (IEC) that provides owners and managers with system-wide guidelines and techniques for implementing and managing information security risks in the organization. With the ISO/IEC 27005 standard, firms can acquire the ability to identify, analyze, evaluate, and treat various risks to information security, as well as prioritize risks and take appropriate actions to mitigate them. Firms that wish to improve their information security management systems do not need to apply the entire methodology of the standard but can focus on a series of factors such as the information security risk assessment, or the information security risk treatment.

Firms may also decide to use the OCTAVE methodology, a framework developed at the CERT(R) Coordination Center of Carnegie Mellon University in the United States widely used by organizations of all kinds to assess cybersecurity. OCTAVE helps organizations identify and classify key IT assets, assess threats to those assets, analyze their vulnerabilities and impacts, and develop security priorities that help reduce IT asset risk. At the core of OCTAVE is the concept of self-direction whereby a small interdisciplinary team of people from within the organization's departments take responsibility for directing the assessment process and setting the organization's security strategy.

The CMMI Institute has developed the Cybermaturity Platform, which shares a risk management approach similar to that of OCTAVE. It is a proactive, evidence-based cybersecurity approach to assess, optimize, and report on cyber capabilities. A distinctive feature of the Cybermaturity Platform is that it allows organizations to prioritize a risk-based roadmap, which is a personalized list of actions based on the most relevant cybersecurity risks that have been identified.

Global security and aerospace company Lockheed Martin has also developed the Cyber Kill Chain framework, which is part of the Intelligence Driven Defense model for identifying and preventing cyber intrusion activity. The model consists of seven steps through which the organization identifies what cyber-attackers would do to achieve their goals, thus enriching an analyst's understanding of an adversary's tactics, techniques, and procedures. The model proposes the measures that the defender must take to break the "chain" at an early stage, as well as at later stages.

Security of data and IoT

Data cybersecurity is an unavoidable problem that must be seriously addressed by all tourism firms transitioning towards smartization. If the problem is not adequately addressed or resolved effectively, cyber-attackers will exploit flaws and weaknesses in the organization's data systems to hijack them, distort data, or simply disrupt the organization's service. The damage inflicted by an attack on the firm's data will far outweigh any gains the firm has previously made.

Cybersecurity attacks are occurring at unprecedented levels on IoT systems and are causing a variety of significant issues for people and having damaging consequences for the reputation, compliance, finances, and operations of tourism firms. Furthermore, according to data provided by a recent survey, IoT-based threats have become a major problem: 8 out of 10 firms have suffered some type of cyberattack on their IoT devices in the last 12 months, and 26% of organizations that had been attacked were not even using security protection technologies (Irdeto, 2019). This increase in the level of threat to IoT systems is due in part to the phenomenal growth of IoT devices and technologies. As the number and variety of connected devices in IoT networks grows, security threats grow exponentially. Advances in 5G and related IoT developments are expected to make cyber-attacks even more common. This highlights the sheer magnitude of the threat and points to the need for tourism firm owners and managers to understand cyber risk management processes and be proactive in investing in IoT cybersecurity.

The purpose of IoT cybersecurity is to reduce the risk of a cyber-attack for users and organizations operating in the ecosystem through the protection of IoT assets and data privacy (Lee, 2020). Due to the dynamic and temporary nature of the connections between IoT devices and the diversity of actors acting in IoT systems, this is undoubtedly a big challenge for firms operating an IoT system. Furthermore, traditional security protocols and mechanisms are no longer adequate, and new methodologies and technologies are needed to meet IoT security, privacy, and reliability requirements.

The very complexity of IoT architectures is another factor that makes it difficult to defend against attacks, as each layer of the IoT architecture has its own challenges and security issues that are unique and interact with other layers. This requires dealing with security measures in a holistic way for the entire IoT architecture and not for any specific layer. Fortunately, the cybersecurity industry is very dynamic and new cybersecurity technologies are continually emerging that provide greater opportunities to withstand attacks unscathed, but also bring with them new challenges for managing IoT cybersecurity risks that organizations must prepare for.

Despite the significant investments that are being made in cybersecurity, owners and managers should keep in mind that there are still no standards that can help them choose the most appropriate cybersecurity technologies. This becomes particularly evident when it comes to deciding on cybersecurity solutions that must protect systems that handle large amounts of data (i.e., IoT), and for which there is an urgent need for secure products and services.

There are currently three open lines of work focused on achieving a security standard in the IoT, to which managers of tourism firms will have to pay attention in the coming years to see the results. These efforts are carried out by the Global Standards Initiative on Internet of Things (IoT-GSI) of ITU-T, the ISO/IEC JTC 1/SWG 5 IoT of the Joint Technical Committee ISO/IEC JTC 1 of the ISO and the IEC, and the TS 103645 standard of the European Telecommunications Standards Institute (ETSI) for the security of internet-connected consumer devices and their associated services.

Hackers' behavior

Since most cyberattacks are human-caused, it's important for cybersecurity managers at firms to understand who these criminals are, why they do what they do, what they're after, and how they operate. Knowing the behavioral aspects surrounding cyber-attackers is very useful to better understand the challenge facing the organization and try to align the behavioral aspects with the technical aspects of cybersecurity. For an organization to understand what its cyber risk landscape looks like, it suffices to consider that cyber-attacks are broadly classified into three main categories: war, activism, and crime. In other words, hackers have geopolitical, ideological, and financial motivations, and can be sponsored by states, be activists, or be criminals (Paraskevas, 2020). Figure 14.3 shows the cyber-risk landscape, that is, the connection between the types of hackers, the threats that occur, and the type of impact they can have on businesses.

There is no IT system that is 100% secure, but it is also not possible to maximize security without considering the human factor. In general, cyber-attackers will attempt to penetrate all levels of the organization's defense system after having gained access to the first level. Consider that it takes most hackers an average of 15 hours to gain access to an organization's system and

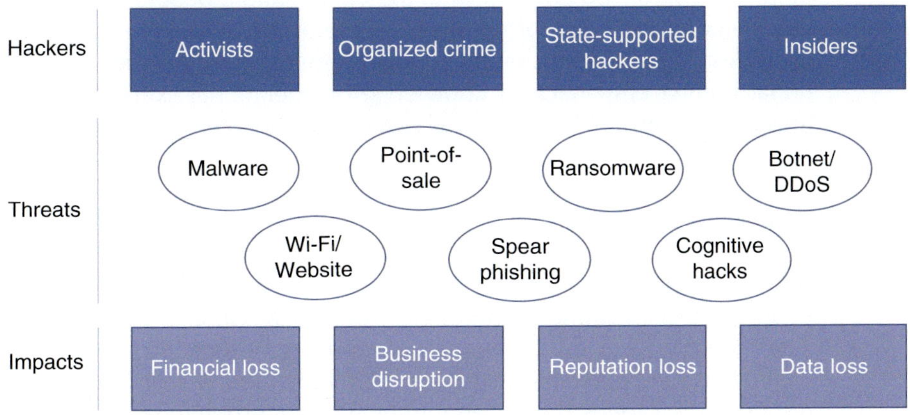

Fig. 14.3. The cyber-risk landscape. Source: own elaboration based on Paraskevas (2020).

map and detect its most valuable data to compromise it (Pogue, 2018). There is no doubt that cybercriminals have an advantage over defenders, which is why defenders are obliged to analyze security at all levels using tools that allow them to discover vulnerabilities before attackers do. Here the famous slogan "Trust, but Verify" (which became internationally famous when it was used by US President Ronald Reagan in the context of negotiations on nuclear disarmament with the Soviet Union) is perfectly valid, which means that even if the firm achieves a certain level of trust in some domains of the organization, it will always be necessary to carry out continuous verification.

Insiders

An insider is a hacker within the organization who has access rights, privileged information, and is behind firewalls. The threat of insiders has long been recognized as a major problem for the management of cybersecurity in organizations (Theoharidou et al., 2005). The Insider Threat Spotlight report indicates that firms are more concerned about data breaches stemming from unintentional insider threats than malicious data breaches (Information Threat Group, 2017). According to this report, 74% of organizations feel vulnerable to insider threats, albeit 42% have adequate controls in place to avoid an insider attack. Privileged users, such as administrators who have access to sensitive information, pose the greatest internal threat to organizations, followed by contractors, external consultants, and regular employees. Most security professionals confirm that insider threats have become more persistent in the last 12 months and that they expect their budget to deal with these threats to increase over the next year. Moreover, 75% of organizations estimate that internal violation remediation costs could be as high as US$500,000.

A troubling issue is that organizations often do not want to admit that they experience internal incidents of this type, preferring instead to fire the insider and protect their reputation. Nonetheless, insiders exist and are more frequent when there are disgruntled employees in the organization, people who suffer stressful events (i.e., sanctions), or people who are prone to this type of behavior (i.e., drug use, aggressiveness, etc.). Sometimes these issues are the result of unintentional human error (lack of knowledge or operational skills), intentional error (misuse of assets), or malicious behavior (with the intent to cause specific harm).

14.5 Success Factors

Any successful cybersecurity program should consider three related factors: people, processes, and technologies (Reagin & Gentry, 2018). Many organizations emphasize the technical components of their program, such as firewalls, intrusion prevention systems, and vulnerability scanning. However, the responses that tourism firms should give to the problem of cybersecurity must address much more than the technical components. We discuss below how each of these factors should be considered.

People

It is vital that organizations form a cybersecurity team and have a professional management framework dedicated to cybersecurity risk assessment that the team can align with at all times. This requires harnessing the leadership of the firm's executive team to develop robust risk management processes and empower all firm personnel through ongoing training. It is important that senior management view cybersecurity as an organization-wide issue and mobilize all employees to be accountable for risk management and oversight of cybersecurity as they would do with any other significant risk to the organization. Establishing an information security oversight committee, chaired by the firm's CEO or COO and involving executives from all business units of the firm, is a proven best practice in many organizations. The committee should meet regularly to review newly identified cybersecurity risks and update security policies and procedures.

The development of a cybersecurity program naturally requires qualified personnel to take on new tasks. It will not be an easy task for a tourism firm, especially when it is an SME, to find the right personnel with which to launch a cybersecurity program given the current high demand for this personnel and the shortage of professionals. Another option available to SMEs is to partner with a Managed Security Service Provider (MSSP) and create a hybrid team (i.e., combining internal and partner resources to bring together the key components of a security program, rather than developing the program completely themselves). This will help the firm gain agility as well as reduce costs.

Once the team is formed, establishing the right links between people, processes, and security technologies is another key success factor for the organization's cybersecurity program. This is the time when the cybersecurity team will need to decide on the cybersecurity framework to use (e.g., NIST, ISO/IEC 27005, OCTAVE). The chosen framework should allow the team to document and assess its ability to prevent, detect, and respond to cyber-attacks, while applying good cybersecurity practices that integrate both a compliance-based approach and promote anticipation through proactive threat behavior. The selection of the key controls to be implemented in the firm is an important step that must be decided based on the probability and impact that the threats may have, even more so considering the large number of providers that currently exist and the large number of technical controls that are available.

Processes

The process framework establishes the model to be followed by the organization's cybersecurity program. It contains the set of policies, rules, and procedures to be followed by the organization that guarantee that the cybersecurity measures and tools will be used consistently and effectively, and that they will be updated periodically. It also forms the basis for a cybersecurity auditor to examine the firm's security framework and assess whether the program is achieving its objectives. To evaluate and compare the effectiveness of the cybersecurity processes developed by the organization, the firm should develop key performance indicators (KPI) that focus, among other things, on the following indicators (KPMG, 2015):

- The probability and impact of the main threats.
- The degree of compliance with internal and external standards.
- The metrics from previous security incidents (i.e., the number, business impact and source of serious incidents, average incident detection/response time, etc.).
- Cybersecurity awareness and culture indicators (i.e., clarity of rules, exemplary behavior, practicability, visibility, organizational openness, etc.).
- The degree of progress of the key security initiatives with respect to the plans.
- The level of current threat.

When considering a cybersecurity framework, business owners and managers should keep in mind that excessive application of security controls can also be a drag on innovation, without bringing additional benefits to organizational performance. For this reason, it is convenient to evaluate both the probability and the impact that each security threat has on the organization and thus decide which ones should be prioritized and will require an additional investment for the implementation of technical controls.

Technologies

When it comes to determining which cybersecurity technologies the firm should invest in, owners and managers should try to optimize the elements that make up the organization's cybersecurity program. Basic controls, such as firewalls and email filters, are essential elements that every tourism firm should have, but there are also other essential components to consider.

- **Anti-spam/anti-virus software:** It remains a vital piece of the cybersecurity puzzle in organizations, especially for blocking known malware on a large scale. However, it is no longer the only component that can be relied upon to keep the business safe from cyber threats. Organizations need to ensure their anti-spam/anti-virus software is always up to date to protect themselves from online threats.
- **Automated patch management:** It is a process dedicated to automatically assessing which updates are critical (and which can be ignored) and applying them to the systems that need them, thus ensuring that risks are not increased by running older operating systems or software.
- **Perimeter security:** It is about using firewalls (network devices that block certain types of network traffic) to protect the organization's network and prevent unsolicited and insecure traffic. A good firewall must also

provide visibility into any intrusion attempts and allow the firm to block access to unwanted websites and applications immediately. The firewall must also support remote access VPN networks to securely allow access to external users.

- **Data backup technologies:** Data backups are the only certain guarantee against data loss. Data loss can be caused by human error, a cyber-attack, or a local disaster such as fire or flood. That is why it is important for the organization to ensure that the heart of its business (its data) is safe and sound no matter what. Once server-level and desktop-level backups are captured, the organization should periodically test them to ensure they are working properly.
- **Multi-Factor Authentication (MFA)** (also known as dual-factor authentication or two-step verification): It creates an additional layer of security that forces cyber-attackers to have not only a username and password to gain access to the system, but also a token to authenticate remotely.

Once the organization has the above controls covered through foundational-level technologies, there are much more advanced technologies that tourism firms will need to invest in to counteract the cyber threats that are becoming more sophisticated and damaging. The following advanced security technologies are designed to provide greater visibility and control so the organization can successfully address today's threat landscape.

- DNS security, to prevent the organization's systems from resolving and connecting to malicious domains.
- Managed Detection and Response (MDR), for advanced threat hunting and incident response through continuous visibility.
- Mobile Device Management (MDM), so that the organization's control extends to mobile devices that have access to firm data.
- Honeypots, or custom lures that simulate the organization's data and intellectual property, including spreadsheets, documents and files, etc.

14.6 The Future of Cybersecurity

Traditional cybersecurity strategies and tools have become increasingly ineffective as the cyber threat landscape has continued to grow. Organizations have no other option but to continue responding to the new security needs that are emerging. The mentality of tourism firms regarding cybersecurity will need to change from being mostly reactive to being proactive. Instead of only detecting malware and other types of threats days after they have been installed on the organization's systems or after they have caused damage, firms will need to actively search for malware and all types of attacks lurking in cyberspace before being activated and causing damage within the organization. This approach that seeks to anticipate threats is already possible thanks to advanced Endpoint Detection and Response (EDR) solutions and AI technologies.

With so many cases of security breaches made public in renowned firm and that have a high media impact, privacy and data protection will become competitive factors to be considered by tourism firms. Those firms that treat cybersecurity as a simple compliance exercise will eventually lose the trust of the market (Paraskevas, 2020). Tourism firms will need to seek solutions that offer flexibility in identity and access management to respond to the increasingly fluid needs of customers. The adoption of as-a-Service models (i.e., IDaaS, PAMaaS) that enable passwordless authentication, single sign-on, and user-friendly multi-factor biometric authentication are options that will greatly facilitate the management of cybersecurity in tourism firms.

Finally, as more tourism firms move their IT infrastructures and data centers to the cloud, it will be essential to adopt a distributed, multi-cloud approach that includes reservation, revenue, call center, channel/distribution, and content management, as it will bring more flexibility, resilience, and cost advantages for the tourism firm. In this vein, two of the main challenges for tourism firms will be: 1) to guarantee compliance with card payment data security standards (Payment Card Industry Data Security Standard, PCI DSS) when personal identification data crosses multiple layers of the cloud; and 2) design an integrated prevention–detection–response strategy automated to threats.

Discussion Questions

- What are the most critical and vulnerable assets of a tourism firm that would be most affected if they were the target of a cyber-attack?
- What are considered the "crown jewels" of a tourism firm from a cybersecurity point of view? How are tourism firms protecting attackers' access to the "crown jewels"?
- What non-technological measures can the tourism firm take to better protect itself from cyber-attacks?
- Identify a publicly known case of a cyber-attack on a tourism firm and describe what type of damage was caused. Was the viability of the business compromised?
- What impact does the adoption of cybersecurity measures have for the firm's employees and customers? Is there any impact on productivity and/or customer experience?
- How should a tourism firm react when it is aware that it has suffered a cyber-attack? What internal and external communication actions should be carried out?
- What capacities and professional skills do tourism SMEs currently have to manage cybersecurity processes and decision-making? Are they enough?

15 Enabling Technologies

The technologies examined in the previous chapters form the technological hard core that is shaping the Smart Revolution and its potential to bring benefits to tourism firms. These technologies are responsible for acquiring the data, processing it, analyzing it, and converting it into useful knowledge for the firm. However, on many occasions these technologies do not come alone but are part of or integrated with other technologies that connect them with the specific needs of organizations and those of consumers and users. Sometimes even the power of core technologies is substantially enhanced or increased thanks to their integration with so-called enabling technologies, whose growth is exponential. Ultimately, enabling technologies are instrumental in achieving more and improved smart applications that better fit the needs of firms and consumers. Together core and enabling technologies are powerful tools to generate the competitive advantages that tourism firms seek. In this chapter the main enabling technologies are characterized, more specifically, mobile applications, the combination of robots, AI, and service automation technologies (RAISA), augmented and virtual reality, and blockchain.

15.1 Mobile Applications

The use of smartphones and mobile applications has become a daily habit for most consumers around the world and, by extension, their use for tourism activities is increasing exponentially. In 2021, the number of mobile users worldwide stood at 7.1 billion and is forecast to increase to 7.49 billion by 2025 according to Statista. In line with these figures, it is a proven fact that consumers rely more and more on their mobile applications to travel and make decisions before, during, and after their trip. This means that mobile applications have a very significant influence on the development of smart tourism.

DOI: 10.1079/9781800620261.0015

Today there are thousands of online applications available to help tourists plan and organize their tourist activities. There are apps to help tourist find hotels, choose restaurants, check the weather, research a destination, book tickets, check exchange rates, check in for flights, and perform hundreds of other tasks. Mobile applications overcome the barriers associated with navigating websites that are not optimized for mobile use, their use can be personalized, and they allow a higher level of convenience, since consumers can use them to compare prices, obtain discounts, get information about products and services, locate services and places of interest, etc. and everything on the fly. It is estimated that, on average, travelers access about 16 app categories for each trip (TravelPort, 2017).

From a business standpoint, mobile applications can positively impact customer loyalty, increase the effectiveness of trade promotions, and bring mobile users and customers closer anytime, anywhere. Furthermore, the constant development of new mobile information services, such as electronic tourist guides and application-based mobile guides, or payments through near field communication (NFC) technologies, is the consequence of the widespread adoption of mobile technologies by consumers for travel-related purposes. Tourism firms may also take advantage of GPS embedded in mobile apps for location-based marketing (Law et al., 2018). In short, mobile devices and applications have drastically changed consumers' behaviors and business processes in tourism firms.

Mobile apps and the consumer

The use of mobile phones by consumers for tourism purposes is becoming more and more popular. Data provided by the annual survey of travel retail platform Travelport (2017) seems to prove it: 50% of travelers rely on their smartphones for searching and booking during a trip, 60% say they would be lost without their smartphone and 45% of travelers say they can't live without a map app when traveling. What's more, nearly two-thirds of US travelers rely on mobile apps while traveling, 61% have booked and paid for trips through their smartphone in the past year, and 64% use their smartphone en-route to their destination. These figures make clear the enormous potential that the mobile application market represents for both consumers and tourism firms.

The reasons why tourists use mobile applications are many and varied. Consumers are primarily looking to save time, be more efficient and effective in their travels, and deepen their travel experience, but there are also these other benefits.

- Search for and obtain information on tourism products and services and on places.
- Make reservations (e.g., lodging, events, transportation, attractions, restaurants).
- Buy products and services.
- Communicate with friends and family and engage in social activities.
- Find activities and entertainment.
- Locate themselves on the map.

- Feel more secure and be able to solve problems when traveling.
- Keep track of loyalty programs.
- Access digital boarding passes.
- Improve the travel experience at the destination.

In general, the trend is for consumers to adopt mobile technologies and applications when they find them useful, easy to use, and compatible with the other technologies they already use. There is also a direct relationship between consumers' greater ability to use the internet and their more favorable attitude towards mobile applications and their intention to use them. In addition, the higher the frequency with which consumers use their smartphone at home and the greater the availability of Wi-Fi at the destination, the greater the use they make of mobile applications (Mang et al., 2016). Another key factor is the perceived usefulness of the recommendations obtained through mobile applications, so that the greater this perception, the greater the value and satisfaction that the consumer obtains from the use of mobile applications.

What is needed for consumers to increase the use of mobile applications? It is important that these are easy to access and navigate, that allow consumers to control the navigation and the transactions they carry out and that the purchase environment is reliable and safe. Other very important aspects not to be overlooked are the role of psychological factors such as enjoyment and inspiration when adopting mobile applications, as well as the differences that exist between young and old users, since the latter are less influenced by social media, less dependent on the opinions and advice of others, and less likely to spend time looking for inspiration than the former.

Mobile applications and tourism firms

Mobile applications allow tourism business owners and managers to stay connected with their customers and prospects anytime, anywhere, speeding up the entire service delivery process. Mobile applications are, in short, an important source of competitive advantage (Law et al., 2018). If we pay attention to what some authors say, the introduction of mobile applications in tourism firms can increase the profitability of shareholders up to 1.32% in the case of hotel and airline companies (Adukaite et al., 2013). So, it seems reasonable to think that any tourism firm that wishes to continue improving its competitiveness should take the process of implementing mobile technologies and applications in its business very seriously and do so strategically.

However, there is still not enough insight as to guide tourism firms in the successful implementation of mobile technologies in the organization, nor is the impact that mobile applications have on the organizational and financial performance of the firm well known. What we do know is that the decision of tourism firms to develop a mobile application usually depends on factors such as the degree of compatibility with the systems already in place in the organization, the technological resources available, the marketing objectives pursued, and the expectations of use by users and customers.

A critical aspect to be considered by tourism firms when implementing and adopting mobile applications is that of the functional characteristics that the mobile application must have. Mobile applications for tourism must provide customers and users with reliable and abundant access to information, but also personalized experiences and smartphone-enabled services that are only available on mobile platforms (i.e., route recording in real time, the "shake" function to select restaurants at random). These services must be aimed at providing new experiences to customers that allow them, in turn, to make innovative consumption decisions. In this spirit, context-aware functionalities such as information sharing or geo-tagging of products and services, along with connection to data repositories and data mining capabilities, can substantially increase the value that firms provide consumers and users through mobile applications. Furthermore, by capturing data through mobile applications and analyzing and exploiting it, tourism firms can start thinking about attracting their target consumers in a more informed way and engaging them with increasingly personalized services. However, finding a way to motivate customers to download mobile apps remains a major challenge for tourism firms. This does not only depend on having good promotional information, but also and above all on the information content and services offered by the mobile application. Recent studies reveal that the main reasons why users download mobile applications in the hotel industry are to obtain information about the firm and carry out transactions (i.e., reservations, check-in, payment of services).

Trends related to mobile applications

Tourism firms, together with the mobile technology industry, are continuously innovating and developing new solutions that add value to consumers, either through the development of new technologies or by improving existing ones. This is the case of technologies as popular as quick response (QR) codes, which have been used in the hotel industry to digitize traditional loyalty cards and in augmented reality games to experience tourist places or attractions. Emerging issues that have recently emerged in relation to the use and development of mobile applications and that will affect tourism firms include the following:

- The rapid growth of NFC technology in tourism. NFC technology is a set of protocols that enables communication between two electronic devices at a distance of 4 cm, which facilitates simple and secure communication between them. By integrating NFC technology in mobile phones, users can perform tasks such as transfer photos, videos, or music files, complete purchases and payments, manage their access control identities, or download information (Liebana-Cabanillas et al., 2020).
- The development of mobile applications for disaster situations and crisis management (e.g., earthquakes, COVID-19) is proving very useful to provide relevant and real-time information on evacuation, and mobility, and share security alerts that affect people and tourists.

- Travel recommendation applications that take advantage of the GPS capabilities of smartphones to help tourists discover and select points of interest in a place according to their spatial and temporal preferences. These apps also save them the effort of making decisions and prevent them from being overwhelmed by a barrage of information.
- The impact of mobile technologies and applications on the segmentation of the firm's customers. Since each segment of mobile app users has different usage patterns and is motivated by different benefits offered to them, this should allow the tourism firm to segment its customers in a precise way.
- The potential to adopt mobile technologies and applications to conduct employee training activities, thereby overcoming the limitations of traditional training methods, and increasing the effectiveness of employee training.
- The perception that employees have about the use of mobile technologies and applications at work. Many employees believe that the use of mobile technologies can increase their self-efficacy and job performance, which in turn can be a factor that contributes to higher levels of job satisfaction and commitment to the organization.

The risks of mobile applications

Perhaps the number one risk faced by users of mobile applications and tourism firms is that related to privacy and the protection of personal data. Mobile applications constantly collect a lot of personal information about the consumer in order to deliver the accurate and personalized information that the consumer wants. At the same time, consumer confidence in the treatment and protection of their personal data through mobile applications is a critical factor that affects their perception of risk in the mobile environment and their intentions to use the applications. These concerns negatively affect consumer travel behavior and lead them to continually weigh the risks and benefits of using mobile apps. In general, it can be said that the risks associated with the lack of privacy and the possibility of theft of personal data is an issue of paramount importance for the tourism firm and of great concern for the development of smart tourism (Dorcic et al., 2019).

Finally, although we live in a world in which being connected with family and friends at all hours and accessing social media is a fairly common habit, if not a necessity for many, in the back of the tourist mind there is still the need to disconnect once in a while and relax by the pool and forget about real life for a moment. Yet, it's somewhat surprising to see that while 80% of travelers find it important to relax on a trip, an almost identical number (75%) agree that keeping in touch is very important (TravelPort, 2017). With these figures in hand, one wonders about the effects that habits associated with being always connected have on people's mental health (i.e., feeling like you're lost without your smartphone, getting anxious when you can't get online, worrying when technology fails or batteries die) and whether travelers really get to live authentic and real experiences when they are only thinking about the next opportunity they will have to reach for their mobile phone. Without a doubt, this is an issue on which both consumers and tourism firms should continue to reflect and delve into the pros and cons.

15.2 RAISA Technologies

The progressive massive adoption of robots, AI, and service automation RAISA technologies is driving ever-widening layers of society and firms into a new economy, which some call "robotomy". Robotomy poses a qualitatively different economic scenario than the present one that may lead to greater automation of the production of goods and services, to the point where most goods and services are delivered by RAISA technologies and not by human employees. In this "robonomic society", some authors believe, RAISA technologies will do most of the work and only a small portion of humans (probably less than 10%) will do any work. Once the need to work is eliminated and a universal basic income is established, people will free up a lot of time that can be used for leisure and recreation. Consequently, the tourism market will expand substantially. It's also not hard to predict the profound implications this will have for the nature of work, the consumer experience, and customer-business interactions, to name just a few.

Whatever happens in the future, these changes will not be exempt from significant social, economic, and political tensions that will affect developed societies and also those aspiring to become one (Webster & Ivanov, 2020). Developed countries with zero or negative population growth and aging populations may end up accepting RAIA technologies as a potential solution to labor shortages that threaten their ability to grow and well-being. This will most likely alter migration flows, as most industries in developed economies will have been automated. In short, it should not be ruled out that society and business firms are heading towards quite radical changes that will be a consequence of the growing incorporation of smart technologies and that the lower labor needs across all economic activities will accelerate and will have a high impact on tourism firms.

RAISA in tourism

Tourism is no exception to the adoption of RAISA technologies. In many tourism activities, human workers have begun to be replaced by robotic concierges, chatbots, and information assistants, to the point that these new inventions are giving rise to a "robotic tourism" based on a wide variety of devices powered by Big Data, AI algorithms, mobile internet, and IoT technologies, and new models to design and serve human consumers.

Chatbots are already part of the communication between tourism firms and customers. Tourists use chatbots not only to search for travel information and find out about offers, but also to receive service attention and make reservations. They also use virtual reality to see in advance the attractions of the destination and the hotel in which they are going to stay. At the airport, travel is made easier by self-service check-in machines, self-service bag drop, and automated passport control with facial recognition. Tourists can arrive at their hotel in an autonomous vehicle, be greeted on arrival by a robotic doorman, check in at a self-service kiosk, and enter their room using a mobile app on their smartphones. Once inside the room, tourists can control the temperature, the amount of light, and the audiovisual services they prefer through technologies available on their mobile device. They can place their room service order online and have it delivered moments later by a robot knocking on their door. The same robots clean the floors and mow the lawn. And so we could continue describing a good number of radical changes that occur when the tourist decides to go to a restaurant for lunch or go out to visit the attractions of the destination.

Overall, from the consumer's perspective, all these new technologies contribute to improving their experience due to reduced waiting times, convenience, and ease of use, while improving their overall satisfaction. However, owners and managers must not forget that the use of these technologies by consumers is not guaranteed and that it varies according to the culture, the category of services, and the target market segments. From the perspective of tourism firms, RAISA technologies not only help them reduce costs and improve productivity, but also allow them to streamline operations and design new service experiences, which in turn can have profound implications for their business models (Ivanov & Webster, 2019). By reducing service errors and eliminating unnecessary tasks that entail costs for the firm, RAISA technologies also allow firms to be more efficient and improve their financial results. Notwithstanding all these benefits, before implementing any process of robotization and automation of services, firms should carry out a cost–benefit analysis and evaluate the financial and non-financial consequences of implementing RAISA technologies, as they would with any other technology.

RAISA technologies

RAISA technologies include a wide variety of service automation innovations that enable customers to obtain service without the need to directly involve service employees. Among these technologies there are a long and varied list of technologies, such as check-in and information kiosks (and applications) in hotels, smart tourist offices and transport stations, ticket machines in stations, food and beverage vending machines, bag drop counters, biometric scanners at airports, etc.

A common feature of service automation technologies is that they shift responsibility for service delivery from employees to customers, so that these become prosumers of the service process. Service encounters conform to processes that are predetermined (i.e., the steps that a customer must follow to check into a hotel or check in their luggage at an airport) and are not very flexible to deviations. However, despite the rigidity that characterizes robotic/automated services, they are more efficient and less expensive than those provided by human employees, which explains why they are used to provide fast and cheap services to mass tourism.

While service firms have been using self-service and automation solutions for decades, robots are only now making their first steps in service firms. Robots are programmable mechanisms that can move and manipulate their environment in two or more axes and that have various degrees of autonomy, interactivity, and intelligence to perform some intended tasks. To carry out their tasks, robots have sensors allow them to obtain data from the environment (i.e., for the identification of objects, sounds, distance/location, pressure, temperature, etc.) and actuators (e.g., motors, lights, speakers, arms, etc.) that allow them to act on their environment. In tourism there is a great variety of robots that are dedicated to very varied tasks, from the most basic in which they practically do not interact with humans (i.e., mowing the lawn, cleaning surfaces), to the most sophisticated in which the robot (with humanoid appearance) can actively communicate with humans (i.e., for room service, robotic waiters).

When deciding what type of robot to use and what tasks it should perform, it is important that tourism firms consider the attitudes that customers will have towards RAISA technologies. Very often these attitudes are shaped by their previous interactions with these technologies and the extent to which robots are involved in the service delivery process. Recent surveys show that the services that customers perceive as most appropriate for the use of robots are those related to the provision of information, the delivery of food and various items, as well as cleaning services; while services that require tourists to entrust their bodies to a robot (i.e., massages, childcare, hairdressing) encounter greater resistance from customers (Ivanov & Webster, 2019).

Although the behavior of robots is perceived by users as a bit clumsy at the moment and many times they create even more work for human employees or cause problems for customers (the Henn na hotel in Japan had to "fire" half of its 243 robots after being declared the first robotic hotel in the world), it is to be expected that advances in robotics and AI will gradually make robots increasingly capable of serving humans and perform, tasks beyond 3D tasks (dirty, dull, and dangerous), so its use by tourism firms will increase.

Adoption of RAISA

The adoption of RAISA technologies in tourism firms is influenced by several macro factors, among which the demographic factor is one of the most important. For the richest and most developed countries in the world, such as European countries, Japan, and South Korea, which have a large aging population and very low birth rates, replacing jobs has become a challenging endeavor with negative impact in the labor market. In other words, in the not too distant future it is very possible that in developed countries there will not be a sufficient number of workers to cover the needs of the productive sectors. Since the solution of "importing people" does not seem feasible, both because of the time it would take to bring in people from abroad, train them, and put them to work in qualified jobs, and because immigration is a source of social tensions in many countries, it seems that the only viable option in the short term is to replace (or reduce) the labor factor. Hence, instead of trying to find more human employees, RAISA technologies can help firms substitute the human factor and thus fill employee shortages through technology. However, as promising as this solution may seem, for RAISA technologies to become increasingly accepted by firms, several conditions must be met (Ivanov & Webster, 2019).

1. First of all, there must be RAISA technologies that are practical, work well, and are affordable for firms, in other words, that add value to the firm. In the coming years, although the active population will continue to be concerned about how technologies can replace human labor, the need for automation and the greater technical capacity to achieve it will be powerful incentives for tourism firms to automate more tasks and processes.
2. Second, given the inability of governments to offer long-term solutions to the economic, social, and political puzzle that developed societies face due to the stagnation and aging of their populations, it will be necessary to adopt political measures that consider RAISA technologies as a solution that can help solve the problem and replace many jobs that will hardly be filled in the future.
3. Third, there is the problem of the speed with which consumers and tourism firms are going to adopt RAISA technologies in the coming years. Consumers will adopt the use of RAISA technologies as long as their problems are solved and they obtain some financial advantage for their use. Initially, it is possible that firms could pass on part of the savings achieved by replacing workers to consumers, which could encourage them to adopt RAISA technologies more quickly. In this initial phase of the process, it will also be important for firms to do a lot of pedagogy, showing consumers the real benefits of using these technologies. Once RAISA technologies are introduced and the benefits of their use become more apparent, consumers are more likely to accept them naturally.

Although firms are still in the early stages of incorporating RAISA technologies, forecasts indicate that there will be a growing use of these technologies in tourism activities. The time is surely not far off when consumers forget what hotels were like when they didn't have robotic services. However, despite the benefits that these technologies can provide to use human labor more efficiently, there are still many challenges that tourism firms will need to address, such as how they will adapt to the transformation of the labor market, the new profiles of workers that they will demand, and the development of a new work culture.

Impact on business processes

The adoption of RAISA technologies is beginning to have a significant impact on the main business processes of tourism firms and in all functional areas, including operations, human resource management, marketing, and financial management. These impacts are reflected in changes that affect the competitive capacity of firms and their financial performance. Next, some of the most significant impacts that RAISA technologies are having on the key business processes of the firm are examined (Ivanov & Webster, 2019).

Operations

Since the use of RAISA technologies implies that a service is provided by a non-human agent (e.g., a robot, a chatbot, a kiosk, a vending machine, etc.) that can serve a large number of customers simultaneously at any time of the day, this allows organizations to have a greater service capacity, which in turn leads to productivity gains. The use of RAISA technologies allows for easier and more efficient programming and planning of operations, given the high availability of these technologies compared to human employees. On the other hand, the lower use of resources by RAISA technologies can contribute to improving the environmental sustainability of operations, since it reduces the production of waste and eliminates unnecessary activities. However, the use of RAISA also has some drawbacks at the operational level. Among them is the decrease in the flexibility of the service delivery system, and the need to create facilities in firms that are friendly to robots and positively reinforce the customer experience. No less important is the reengineering of the service delivery processes that the firm will almost certainly have to carry out when it introduces RAISA technologies and that should be accompanied by new processes for controlling and monitoring activities.

Management of human resources

The implementation of RAISA technologies can help save human employees the time they spend performing dirty, dull, and dangerous tasks (3D tasks), i.e., tasks in which the human factor does not provide a differential advantage over machines, such as repetitive tasks in which precision and speed prevail, tasks that pose a risk to the physical and mental integrity of workers, those in which the emotional component is not valued, or in which it is not required to make decisions in a flexible and adaptive manner. Thanks to RAISA technologies, employees can dedicate their time to more creative activities where human intelligence and the emotional component are key, and those that improve the customer experience and create value for the business. RAISA technologies can help make the work that was previously done by humans more efficient, productive, and sustainable, and solve the problem of temporary employment and high churn rate in low-skilled jobs, of great concern to tourism firms.

The introduction of RAISA technologies has opened a highly speculative debate in the business world, and in tourism firms in particular, about what should be the role and capabilities of the human being in business, and if these technologies should focus on improving the work of human employees or simply replace them. Although the debate has defenders on both sides, it is no coincidence that the evolution of the number of human employees in tourism firms follows a downward trend, some experts even assure that we will see hotels with zero employees. Employees are well aware of this situation. Many of them perceive RAISA technologies as a threat to their jobs and turn against the adoption of these technologies. Notwithstanding the foregoing, owners and managers should be aware that RAISA technologies will require reorganizing human resources within the organization, sometimes creating new departments and sometimes new jobs, and that employees will need different skills than those required today.

Marketing

When a firm's competitors use RAISA technologies and this gives them a sustainable competitive advantage over time, the firm has only two choices: either the firm decides to start using RAISA technologies to create similar advantages itself (and evolves towards a high-tech firm) or decides to reposition itself as a "high-touch" firm focused on the personalized relationship with the customer. In the real world, business alternatives are not black and white, but there are various shades of gray in between. Regardless of the path that the firm chooses, what is clear is that it will have important consequences not only on its business model, but also on the way the customer will perceive the quality of the service and on how the firm will communicate and interact with customers.

RAISA technologies can improve the quality of service perceived by the customer through new attractive and interactive ways of providing services, communication, and interaction. For example, robots, chatbots, and service kiosks offer the ability to communicate with customers in different languages 24/7. In many cases, these technologies can create value for customers by making the service delivery process fun and entertaining. These technologies can also affect the pricing of tourism products and services, since they allow discriminating between the prices of mass "high-tech" products (with lower prices) and the prices of personalized/individual "high touch" products (with higher prices). In this "robonomic" future that awaits us, an increasingly smaller segment of tourists will continue to demand "high-touch" service experiences, which will be satisfied through products and services specifically designed for high-level niches. Meanwhile, the rest of the tourists will enjoy a highly automated tourist experience at lower prices. In the latter case, pricing will take into account not only the costs of the product or service, but also the quality perceived by customers of RAISA services and their willingness to pay for tourist services not provided by humans.

Besides discriminating between customer segments, RAISA technologies could allow owners and managers to set prices automatically without human intervention (i.e., based on moment in time, purchase history, degree of loyalty), as well as assign automatically the available capacity by distribution channel, taking advantage of the analytical, predictive, and AI capabilities available through virtual assistants and smart channel managers. The image and positioning of the firm that adopts RAISA technologies would also be impacted. In some cases, the firm would benefit from reinforcing its image as a highly innovative high-tech firm and positive word of mouth, but in other cases, if the firm automates the communication with customers (i.e., through chatbots, voice assistants, or robots) it could suffer negative publicity as it would be perceived as a firm that puts profits before humans.

Financial management

RAISA technologies can be assimilated to a workforce that works 24 hours a day, 7 days a week and can serve a large number of customers simultaneously. In addition, the firm is not required to review RAISA's salary every year, nor to pay attention to its working conditions. RAISA allows tourism firms to increase their presence in the market thanks to its 24/7 availability with the same labor costs (or even lower) than before, offering the opportunity to increase sales volume at a very low cost. However, it is not gold everything that shines. RAISA technologies entail significant financial costs associated with the acquisition, installation, maintenance, and updating of hardware and software, as well as costs for the creation of friendly environments and facilities for the use of robots, upskilling, and a long list of other costs associated with robot damage and insurance costs.

15.3 Augmented and Virtual Reality

Augmented Reality (AR) is a technology that allows the superimposition of data and synthetic images generated in 3D by a computer on the real world, interactively, and in real time (Azuma, 1997). AR represents the first step towards the so-called virtuality continuum (VC), at the extremes of which is the virtual reality (VR) on one side and the real environment on the other (Roxo & Brito, 2018). Therefore, AR is a form of Mixed Reality (MR), i.e., a combination of elements from the real world and VR. AR has the potential to expand human perception and quickly adapt to different contexts, giving users greater awareness of their surroundings and making tasks more user-friendly and effective (Wei, 2019). This technology also makes it possible to create more immersive experiences for customers and users, aware of the context and transparent for people and firms. As such, AR is the basis behind new platforms to deliver content and services to a global audience.

On the other hand, VR consists of a computer-generated environment that involves total immersion in the digital world (Guttentag, 2010). VR technology provides unique capabilities to simulate real situations and allow users to navigate in a virtual environment through their senses. Today, there are a wide variety of VR devices and programs, such as wearable head-mounted VR displays and online 3D virtual tours. Although owners and managers should not confuse AR with VR, they can treat them in combination by considering AR a type of VR. Compared to an all-digital and artificial 3D immersive experience such as that provided by VR, AR technology presents virtual information overlaid on the real-world view that users can interact with (i.e., a mobile application that provides overlay of text, audio, 3D animations, or avatars on a real-world environment). This key distinction implies that AR is generally better suited to enhance (and co-create) rather than replace the tourist experience (Neuhofer et al., 2014).

Key elements of Virtual Reality

One of VR's greatest advantages is its ability to visualize spatial environments. This is especially relevant for tourism products and services which, as we know, cannot be tested by consumers in advance. Putting on a VR headset and being able to compare different destinations or tourism products and services can be of great help to consumers when making informed decisions. The three main elements that characterize VR are the following (Yung & Khoo-Lattimore, 2019):

1. **Visualization:** or the user's ability to look around, usually through a head-mounted display, and perceive the "virtual" world around them.
2. **Immersion:** which consists of imaginatively transferring the user to an artificial representation of the real world or an imaginary world that they can manipulate or interact with. This is made possible by a combination of sights, sounds, and technologies, including gyroscopes and motion sensors, which track the position of the head, body, and hands; small high-definition screens for stereoscopic displays; and computer processors that are small, light, and fast.
3. **Interactivity:** by allowing the user to have a certain degree of control over the immersive experience provided, generally through the use of sensors and an input device such as a joystick or keyboard.

Two terms that are related to VR and worth explaining are Virtual Environment (VE) and Virtual Worlds (VWs). A VE is a computer-generated environment where the user can move and interact with and modify the digital objects that compose it according to their will and interests. VR itself involves the user's immersion in a VE. There are three main types of VE used in VR today, each with its own level of immersion and features: non-immersive, semi-immersive, and fully immersive VEs.

Non-immersive VEs are computer-generated environments that allow the user to be aware of and maintain control of their physical environment. Non-immersive technologies include a computer or video game console, a display, and signal input devices such as keyboards, mice, and other controllers.

Semi-immersive VEs are a partially virtual environment where users perceive that they are in a different reality but still connected to their physical environment. Semi-immersive technologies provide realism through 3D graphics and depth of field. This type of VR is often used for educational or training purposes and relies on high-resolution displays, powerful computers, projectors, or simulators that replicate the functionality of real-world working mechanisms, such as a flight simulator.

Fully immersive VEs provide users with the most realistic and comprehensive simulation experience. Users can experience and interact with VR using a high-resolution VR headset with a wide field of view, or a head-mounted display (HMD). The screen is typically split between the user's eyes, thus creating a stereoscopic 3D effect combined with input tracking to establish an immersive and believable experience. This type of VR is commonly used for gaming and entertainment purposes, although use cases in other sectors such as medicine, education, or tourism are increasing and seem limitless (e.g., Mechdyne CAVE VR system).

On the other hand, VWs are virtual environments open 24 hours a day, 7 days a week, which simulate an artificial world or environment (inspired or not in reality) and that allow people represented by avatars to create, play, and interact with each other and with virtual objects or goods in real time. It is therefore a metaverse, or an environment where humans interact socially and economically as avatars without the limitations of the real world. Today one of the most active VWs in the world is Second Life, a platform founded in 2003 where avatars socialize, interact, and create their own virtual spaces. In 2021, Second Life reported around 64.7 million active users on its platform and since its inception has created an in-game economy valued at approximately US$500 million. The rise in popularity of VWs has not gone unnoticed by the tourism industry and countries like Sweden, the Maldives, Estonia, Kazakhstan, Serbia, and Italy have already created their own virtual embassies in Second Life, as have international hotel brands, such as Starwood, Hyatt, STA, and Crowne Plaza. Popular world tourism attractions such as the Eiffel Tower and the Arc de Triomphe in Paris, and the Maasai Mara villages in Kenya, also have their virtual site in Second Life. In VWs, the social and promotional aspect predominates through virtual marketing, with the avatars being able to travel to these attractions in groups and interact with other avatars present on the site.

Key elements of Augmented Reality

Following the success of the Pokemon Go AR game in 2016, there was a surge in public interest in the use of AR applications, leading the world's leading destinations and tourism firms to explore the potential of this innovative technology. Advances in mobile computing have accelerated the development of AR applications in tourism, taking advantage of the geolocation capabilities provided by mobile devices to deliver users with additional relevant information about their environment.

For the user, the main difference between AR and VR is the level of immersion. In AR, most of what the user sees belongs to the real world, while with VR the user is completely immersed in an artificial environment.

In addition, AR's geo-location capabilities, personalization, and multilingual capabilities allow for more accurate and personalized marketing messages to be delivered to consumers, fostering a more positive attitude, more trust, and, ultimately, higher purchase intent (Javornik, 2016). AR and VR technologies offer users a realistic pre-experience of the potential destination and of the products and services, which opens the door to new ways of promoting the services of tourism firms. As with mobile applications, to be successful in the development and adoption of AR applications in tourism, it is important to consider not only the utility and ease of use, but firms must also educate internal stakeholders about what implementing AR entails and what the expected results are.

AR/VR applications

AR and VR-based applications are being increasingly adopted and implemented in various areas of tourism and hospitality, such as hotels, tours, museums, attractions, cruises, heritage, and destination marketing. Today there is a wide variety of VR and AR applications that provide tourism firms with innovative ways to revitalize their value offerings, attract more visitors, and improve the service experience through more immersive and interactive environments. AV/VR technologies offer tourism firms the ability to reach a broader audience, influence their customers' attitudes, experiences, and emotions, as well as consumer purchase intentions and word of mouth. Through AR/VR, tourists can access valuable information and improve their knowledge about tourist products and services and attractions, while enhancing their tourist experience and enjoying higher levels of entertainment in the process (Kounavis et al., 2012). All of the above can be done in a personalized way, delivering services and multimedia content tailored to the needs of users.

Among the factors that influence the user experience and are useful in predicting the degree of satisfaction with AR/VR applications are the advantages/benefits perceived by the user, the functional and practical attributes that the applications have, the perceived enjoyment, and the degree of emotional involvement of users. Another important element that influences the experience is the aesthetics and the visual appeal of AR/VR applications (Roxo & Brito, 2018). The Boston Consulting Group (BCG) has estimated that more than 80 million US citizens use AR, meaning that about one-third of smartphone users use AR technology at least once a month, and many firms are convinced that AR applications will begin to affect sales, purchase intent, and customer engagement within two years. All the above are arguments that can make business owners and managers consider the implementation of AR/VR technologies in their operational and business plans. Some of the AR/VR applications that are expected to have the greatest potential for tourism firms are presented below.

Marketing

Technological advances in AR and VR are transforming the way tourism firms and destination marketing organizations (DMOs) communicate, deliver content, and interact with consumers and suppliers. AR/VR applications are used to influence tourists' positive attitude towards products and services and places, reinforce the brand, and as an innovative means of developing interactive advertising to increase the number of customers. In essence, AR/VR applications allow consumers to learn about and experience products and places in new ways (Wei, 2019). Regardless of the type of AR/VR application in question, one of its greatest strengths is the ability to visualize spatial environments and provide rich and enhanced information to users during the planning stage of their trips. AR/VR offers an indirect experience to users by allowing them to experience sensory and even spatial aspects of the destination without actually being there. This makes advertising through VR/AR more effective when compared to other media (i.e., brochures), given the greater richness and interactivity of the information delivered.

Destination organizations such as Tourism Australia (http://www.australia.com/, accessed September 2022) and Destination BC in Canada (https://www.hellobc.com/, accessed September 2022) have launched interactive VR experiences through their websites. In other cases, cities have been recreated, such as the case of Valladolid in Spain, which allow the user to tour the city on the Second Life platform and learn about it in a virtual context. These initiatives can be very useful in the initial phases of the consumer decision process, since tourism products are intangible and consumers cannot try them in advance. At the same time, several studies have found that those users who interact with AR and VR manage to feel a greater degree of commitment and participation, and this contributes to generate greater positive feelings towards tourism firms and places (Yung & Khoo-Lattimore, 2019). Although much remains to be learned and researched, it should not be ruled out that AR and VR technologies could become tools that allow marketers to revolutionize interaction with customers through information.

Meetings, Incentives, Conferences, and Exhibitions

The ongoing transition of the Meetings, Incentives, Conferences, and Exhibitions (MICE) industry to virtual mode, recently accelerated by the COVID-19 pandemic, is having important implications for the roles of event organizers and travel managers, as well as for the global business travel market. The substitution of many business trips for virtual events is nothing new as this need began to arise to save time and money and feel more secure. However, this trend is now transforming the business strategies of the main players in the MICE industry and readjusting the contractual conditions with travel agencies and the relationships with customers. Faced with this reality, AR/VR technologies emerge as tools with the potential to provide innovative solutions that strengthen the competitive capacity and growth of firms in this industry. Although the capabilities provided by AR/VR technologies may still seem insufficient to justify their widespread adoption in the industry, their use has begun to attract the interest of many of the industry leaders. AR/VR technologies offer new and innovative ways to show customers previews of an event hall set up for trade shows, or social events. Moreover, creating VR experiences in virtual events is a way to provide attendees with engaging and immersive experiences to make them feel more involved and engaged. Therefore, it would not be surprising if its adoption became widespread in the coming years in specific niches of the MICE industry where the replacement of face-to-face events with virtual ones makes more sense (i.e., in professional events, product demonstrations, brand presentations, networking meetings, etc.).

Gamification in museums and attractions

Gamification is an effective strategy that uses game mechanics, such as points, leaderboard rankings, achievements/badges, levels, stories/themes, objectives, feedback, and rewards, etc., to influence user behavior and affect user engagement, loyalty, and satisfaction (Pasca et al., 2021). The implementation of gamification in digital platforms can help to attract and retain certain types of visitors, making available a large amount of data to service providers on the preferences, perceptions, opinions, and itineraries of these tourists. With the exponential growth of digital platforms, social media, and mobile technologies, gamification has become a pervasive innovative tool in service design. It is increasingly used by tourism firms to promote consumer interaction and participation in the co-creation of experiences and services. Gamification has also spread to other contexts not strictly related to games, such as in firms that are looking for new tools to create more meaningful connections with users and influence and improve their behavior and that of their employees.

One of the applications of AR technology that is gaining more and more ground is as a tool for disseminating information and as an interactive guide in museums, attractions, art galleries, and heritage sites. Its use is justified by the more mobile nature of AR compared to VR (the latter requires the user to be static in one place and more processing power). The use of AR in contexts where it is necessary to disseminate large amounts of information is an opportunity to improve the tourist experience before, during, and after the visit. Since AR is also a technology that is easy to use and very convenient for the average user of a mobile phone, it makes it possible to incorporate game elements in mobile applications that serve as an incentive for the public to have fun and improve their knowledge and perception of places and history (Jingen Liang & Elliot, 2021; Pasca et al., 2021). For many specialists, this combination of AR technologies with gamification has important practical implications and represents one of the opportunities with the greatest potential to attract heritage tourists.

Additionally, the idea of socializing the experience when visiting heritage sites (i.e., sharing opinions, scoring points, badges, etc.) can also contribute to the joint creation of business-to-consumer (B2C) and consumer-to-consumer (C2C) value, since it motivates visitors to return attracted by new challenges. As a next step, visitors could create their own "treasure hunts" in venues, museums, and attractions to be solved by other visitors (Jung & tom Dieck, 2017), thus becoming co-creators of content and activities in the AR environment. However, given that AR environments and experiences are computer-generated reconstructions, it is not uncommon to find managers of heritage sites who feel that the use of these technologies undermines the objective authenticity of the sites and trivializes them, so they are not willing to accept the use of AR or gamification. These attitudes can be expected to change as the benefits (and limitations) of using AR technologies become more apparent to users and organizations over time.

Mobile AR tour guides

AR mobile tour guides can display on-demand content as tourists travel through the city, exploring the cityscape or visiting points of interest. These applications allow adding new layers of information and knowledge to the real world, delivering new, more interactive, and dynamic experiences. In addition, since most users access these applications through smart mobile devices with GPS, tourists can take advantage of the benefits of geographical positioning.

Information within AR mobile apps is delivered in various media formats, from images, videos, and sounds, to 3D models and hyperlinks that can direct the user outside of the app. This combination of AR technology with multimedia content and the careful design of the functionalities of the mobile application allows tourists, for example, to create lists of favorite points of interest linked to a map. Since AR mobile apps can work like a tour guide that provide layered and on-demand information, the effect of information overload can be minimized. Each user can be given information according to the level of knowledge and interest they have and personalize their tour or visit, thus resulting in a much more memorable experience (Kounavis et al., 2012). Furthermore, users of mobile applications can take advantage of the capabilities provided by mobile devices and mobility, and use AR/VR technologies as social media and interact with the rest of the users who access them. Tourists can instantly exchange information and tips, or share their opinions with others inside and outside the app. Eventually, tourism firms can use this information to profile the wishes and expectations of tourists and personalize their visits.

Challenges of AR/VR

Tourism firms can reach out to new markets and create a competitive edge by implementing AR/VR technologies. However, there are a good number of challenges that affect both users and firms and on which the promised benefits depend. From the users' perspective, the main challenges lie, as with other smart technologies, in the effort required to learn how to use them, in the compatibility of the applications with the devices that are already being used, and in the cost of use. AR/VR applications often require an internet connection, but not every city or location is fully covered with Wi-Fi networks that offer free internet and 4G connection. Furthermore, data roaming charges remain a considerable expense for many tourists, especially younger ones. Ease of use, simplicity, accessibility, and app functionality continue to be major challenges faced by AR/VR app developers and firms. It is key to ensure that users have a positive experience and that tourists do not feel overwhelmed when using unknown technologies (Yin et al., 2021). Users also expect AR/VR apps to help them with post-trip retrospection by providing functionalities related to remembering, taking photos, and recording. It is therefore important that there is an interaction between application developers and tourists and that the former take into account the needs of the latter when developing and implementing AR/VR and guarantee the creation of value for the tourist.

For many tourism firms, the development of AR/VR applications is perceived as very risky, especially when a reliable proof of concept is not previously available. AR/VR content development, device provisioning, and, in some cases, 3D printer purchases require significant investment and, as the pace of AR/VR adoption accelerates, smaller firms are afraid of making large investments in these technologies. Additionally, there is still the issue of a lack of interoperability between mobile platforms, which means that most AR/VR apps cannot be used on all operating systems, increasing development costs for firms. Despite the pitfalls, the widespread use of smartphones allows tourism firms to offer basic applications for visitors' own devices, with very little investment, and thus providing an interactive experience to their customers and users.

15.4 Blockchain

Blockchain is one of the 21st-century technologies with the potential to change the way tourism business is done, payments are made, and management processes are executed. Based on the Distributed Ledger Technology (DLT), blockchain applications provide innovative solutions based on the security they provide to the actors of the network, the efficiency in the maintenance of records and transactions, and the automation in the execution of contracts when certain conditions are met (smart contracts). Although most of the literature on blockchain focuses on the financial industry and cryptocurrencies, interest is growing in how this technology can be applied in other industries and management processes, including tourism. Blockchain can radically disrupt current business models in tourism and tourism marketing. Not surprisingly, many leading tourism firms have already announced their intention to test and adopt blockchain technology in their booking and payment systems.

How it works

The blockchain is an open source software that allows the creation of a distributed database of records (ledger) organized in a block structure, which is decentralized and securely records all the transactions that take place between the participating members of the blockchain. The goal of the blockchain is to ensure that digital information is securely recorded and distributed, but not edited. Hence, the blockchain is the perfect basis for recording transactions that should not be modified or deleted.

Unlike a typical database, transactions on the blockchain are collected in groups called blocks. When these blocks are filled, they are closed and linked to the previously filled block, thus forming a chain of data known as a blockchain. Blocks are chained together with cryptographic hashes, so that each block (except the first) contains the hash of the previous block. The blockchain ensures integrity by chaining blocks of transactions in such a way that altering any one block breaks the link to the next block. Furthermore, all the blocks together form a ledger which is an auditable record of all transaction history. In the ledger, each transaction has a specific user code or pseudonym attached. The blockchain allows all actors to maintain their own copy of the ledger, which can be decentralized and replicated without any central institution maintaining a single authoritative record.

A ledger is stored in a distributed system made up of nodes on a computer network. Nodes are the infrastructure of the blockchain and can be any type of device, from a laptop, a desktop computer, to even larger and more powerful servers. All nodes in a blockchain are connected to each other and constantly exchange the latest blockchain data so that everyone is up to date. Blockchain nodes can be regular (i.e., only interact with the blockchain), and complete (i.e., contain a complete copy of the blockchain transaction history and validate transactions). A third type of node present in many blockchains is the blockchain miner, whose mission is to create new transaction blocks from full nodes.

The blockchain can be public or private. The former is also known as open or permissionless blockchain, and anyone can freely join and establish a node. As it is open, this blockchain must be protected with cryptography and a consensus system, such as proof of work (PoW). Private or permissioned blockchain requires that each node must be approved before joining, so the security layers do not need to be as strong as in the public blockchain.

Blockchain technology has six main characteristics that make it suitable for a number of applications in the field of tourism, these are: decentralized nature and operation, transparency of data records, open source access, autonomy and trust, immutability, and anonymity (Antoniadis et al., 2020). These features provide greater security and auditability to the transaction recording system, while the absence of intermediaries builds trust in the system, reduces the costs of transactions and the maintenance of a central database, and mitigates the risks of network collapse from malicious attacks.

Applications

Blockchain applications can change the way tourism firms interact with customers in a secure environment and validate the content they generate about their products and services (i.e., fake reviews on social media). Token-based loyalty programs can change the experience offered to customers of a hotel or a destination and make it much easier to manage these types of programs. Digital payments with customers (B2C) and with suppliers (B2B) can be greatly facilitated through cryptocurrencies, and when combined with smart contracts, operations can be made more agile and secure, which generates more trust between the participating parties (Antoniadis et al., 2020). Figure 15.1 shows some of the main applications that blockchain can have in tourism.

Cryptocurrencies

The growth of blockchain technologies and the emergence of new applications are largely due to the success of the Bitcoin cryptocurrency. Bitcoin was one of the first applications of DLT technology to emerge as a peer-to-peer cryptographic payment system in response to a lack of trust in global finance created after the 2008 financial crisis. In the Bitcoin ecosystem, any digital interaction can be tracked through a transparent and secure structure that is also resistant to both external and internal hacking attacks. Although there are many different types of cryptocurrencies today, the most important in terms of transaction volume and recognition are the following (Çapar, 2020):

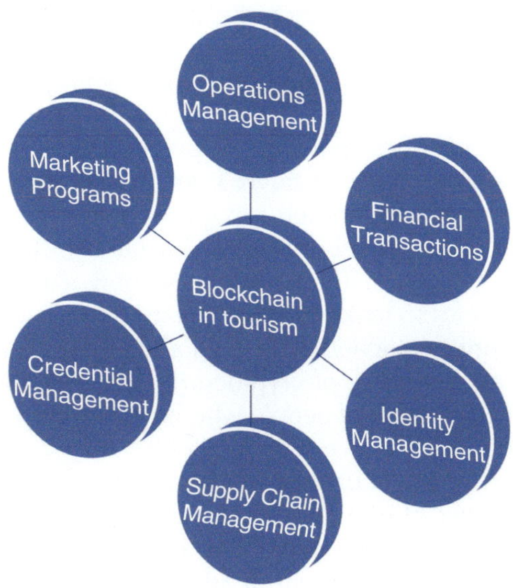

Fig. 15.1. Main applications of blockchain in tourism. Source: own elaboration based on Antoniadis *et al.* (2020).

- **Bitcoin:** It is the most popular of all cryptocurrencies. Released in 2008 by an unknown person using the pseudonym Satoshi Nakamoto, it can be traded like a physical currency. In January 2021, 400,000 daily Bitcoin transactions were reached.
- **Altcoins:** These are alternative cryptocurrencies to Bitcoin, of which there are currently more than 10,000 on the market (e.g., Ethereum, Ripple, Litecoin, Cardano, Dogecoin, etc.), 200 of which support active trading between customers. Although most are different versions of Bitcoin, some are actually quite different, since they use very different algorithms from Bitcoin (e.g., Factom). According to the "dominance ratio", which measures how strong Bitcoin is compared to the rest of the "altcoins", the Bitcoin market was slightly above 50% at the beginning of 2021.
- **Tokens:** They are quite different from the other two types of cryptocurrencies, as they do not have their own blockchains. Tokens are used with DApp applications, usually to buy a product or get a discount.

Like other new technologies, the adoption of cryptocurrencies depends on the acceptance by consumers, after all tourism firms will not see compelling reasons to invest in Bitcoin if they do not find a critical mass of consumers who use them (Ozdemir et al., 2020). Among the factors that may explain consumers' reluctance to accept cryptocurrencies are lack of knowledge about cryptocurrencies, the low number of retailers that accept them, as well as lack of trust and the high risk of losing value of the cryptocurrency. On the contrary, consumers value the advantages of paying with cryptocurrency, such as flexibility, convenience, and anonymity, compared to their traditional counterparts. As credit card fraud continues to increase at a high rate and this negatively affects the development of electronic commerce in the world, the use of cryptocurrencies and bitcoin in particular, appears as an emerging technology that is gaining more acceptance every day.

Some recent statistics show that in post-COVID-19 times there is a strong appetite for spending more on travel from US travelers with 22% planning to use crypto to pay for part of their trip (prnewswire.com, 2021). Of those people planning to use cryptocurrencies, 70% choose Bitcoin, 45% USDT, 37% Ethereum, and 35% AVA. These estimates show a record in the intention to use cryptocurrencies to make travel reservations and payments. It is also worth noting that the prices reached by Bitcoin and Ethereum are at an all-time high, with the total market value of cryptocurrencies exceeding US$2 trillion, and that the estimated number of people who have any cryptocurrency in the world has reached 106 million.

A large part of the tourist products and services marketed in the world involves the transfer of money between people and firms located in different countries. This requires the intervention of trusted intermediaries to carry out the transactions and the payment of additional high commissions – just think about the 10–25% commission that Booking.com charges hoteliers – and that logically affects the price paid by the final consumer. However, by using cryptocurrencies it is possible to ensure money transactions without intermediaries.

One of the fastest-growing tourism activities in the adoption of cryptocurrencies is medical tourism. Many medical tourism operators use high-quality medical care as a competitive weapon, and also that safe, transparent, and quick in response. On the other hand, patients (medical tourists) prefer to minimize risks in the medical care process and carry out as few transactions as possible. By using a cryptocurrency, patients do not transfer any medical information to third parties, thus maintaining confidentiality in the provision of medical services (Barkan & Tapliashvili, 2018). This has led many clinics and hospitals to accept cryptocurrencies as a means of payment for their services, especially from their international customers.

There are many experts who believe that cryptocurrencies will change the way in which tourists and firms exchange money, thus avoiding the intermediation of third parties in transactions and reducing costs. The high level of anonymity provided by cryptocurrencies such as Bitcoin is another reason for optimism, as "user wallets" are not directly connected to their owners through shared personal data, meaning that the chain does not require personal data to register an address.

Smart contracts

Another key feature of blockchain technology that has become very popular since Nick Szabo introduced in 1998 the idea of a decentralized mechanism to create a digital currency called "bit gold" and later the open source platform Ethereum was created in 2015 by Vitálik Buterin, are smart contracts. Smart contracts are programs written in a programming language that are registered in a blockchain and automatically executed when there is a consensus between the parties involved that predetermined conditions are met (Caddeo & Pinna, 2021). When the automated procedure is activated in the block chain, the intervention of intermediaries is not necessary, thus reducing the costs of the transaction and reinforcing the trust between the parties involved.

The programmability of smart contracts and their automated execution without human intervention opens up a field of great opportunities for tourism firms. For example, transactions could be paid through instant payment systems, which would improve collaboration between accommodations and travel agencies; in hotels, rooms assignment to guests upon arrival could be done through digital keys in the block chain; and in airlines, in the case of delay or cancellation, smart contracts could automatically settle the compensation provided for as established in the insurance. By eliminating the need for intermediaries to establish contracts between the parties, new opportunities are also created for tourism SMEs. Larger firms, such as big tour operators and hotel groups (and even consumers), may be interested in diversifying their supplier networks and focusing more on smaller firms as transaction costs fall and confidence is strengthened (Antoniadis et al., 2020).

IoT

One of the key challenges for smart tourism firms is the growing number of devices connected to the internet through IoT systems and the need to have an infrastructure that provides security against the multiple threats affecting the layers of the IoT architecture. As the number of connected devices increases, so does the number of devices with outdated or inconsistent firmware versions, the amount of corrupted or compromised data, and the number of fake integrations. These risks are compounded by the increasing volume and variety of data, and the different types of infrastructure and users that are connected to the internet. The key question that arises here is whether it is possible to develop a data management platform in the context of smart firms, based on a generic data model and using reliable technology.

The decentralized nature of blockchain technology could help minimize the risks of fraud and manipulation by participants, by effectively managing identities through immutable block-based access systems. Therefore, the IoT could use blockchain to secure many of its applications and devices by publishing and storing data in the form of a public ledger that cannot be changed, as each user or node in the network maintains the same ledger as everyone else (Lee, 2019). Hence, some of the key challenges related to IoT security could be addressed. However, while blockchain technology provides decentralized privacy and security capabilities, it also comes with high power consumption, delays, and heavy workloads for computational resources that are still not suitable for most resource-constrained IoT devices. To overcome these barriers, several alternative implementation studies are underway, such as dividing the IoT network into several levels and adopting blockchain technology in each of them to provide security; or creating a distributed cloud infrastructure that reduces response time, while still achieving high-performance and economical computing to respond to attacks in real time on the IoT network (Brandão et al., 2018).

DApps

Decentralized applications (DApps) are applications or programs created to run on a network of blockchain computers. DApps are outside the reach and control of any single authority and are often created through the Ethereum platform for a variety of purposes, including gaming, finance, and social media. For example, a developer can create a DApp similar to Instagram or Twitter and put it on a blockchain where any user can post messages. Once there, no one, not even the creators of the app, can delete the messages. DApps allow users to interact with blockchain technologies in a regular and "friendly" way through smartphones and web browsers. As DApp developers seek to make blockchain technology easy to use, new consumer-oriented business models are emerging from them. For tourism firms, this means they can develop their own DApps to better connect and engage with their customers. At the moment, there are several smart tourism DApps in the works focused on online review systems, trip planning, direct communication with owners, personalized marketing, etc. Some examples are Nocturus (a blockchain-based hotel reservation firm that offers direct-to-customer hotel reservations as well as sales channel consolidation for hotels), and Further Network (a blockchain-based P2P network to help airlines and the travel industry to complete the billing, settlement, and payment (BSP) process in real time).

Airlines

In airlines, the blockchain can be used to carry out authentication and identity management, allowing the identity of a person to be unequivocally determined. For example, biometric information, such as fingerprints and facial recognition can be added to the traveler's registered information, thus simplifying the work of hotels, which would only have to register the arrival and departure dates of the guests in the blockchain. Tests are being carried out with the virtual (digital) passport, which uses a unique token that contains biometric data and other personal information stored on mobile devices to reduce the control of documents during trips. Blockchain technology is also being used to enhance customer loyalty programs, such as to automatically connect the digital identity of travelers with loyalty programs and hotel reservations (i.e., for flight crews). These use cases often require the integration of multiple actors into a single system, and since there is no middleman, the system needs to establish trust between the multitude of nodes that are part of the blockchain network, in addition to complying with privacy and security. Other use cases in airlines include: the implementation of tokens for payments (replacement of credit cards); the tracking and monitoring of baggage at key points (ensuring the security of the data of all the actors involved in the service, including passengers, airlines, airports, and insurance companies); and automated compensation through smart contracts.

Hospitality

The blockchain can be used to keep track of guest check-ins and run loyalty programs that issue tokens as rewards to travelers and even connect with credit cards. For a blockchain implementation to be successful, customer privacy must be always guaranteed. For example, the start-up firm Loyyal (https://loyyal.com/, accessed September 2022) uses proprietary blockchain technology and smart contracts for businesses to manage their loyalty programs by offering various types of reward systems. Blockchain technology can also be used in inventory management systems (i.e., available rooms in hotels, or the number of seats available on airplanes) to provide information on availability and coverage rate, thus replacing Property Management Systems (PMSs) and getting rid of intermediation costs.

Disintermediation not only concerns monetary transactions, but also directly affects the delivery of tourism products/services. Examples of these applications are the WindingTree platform (https://windingtree.com/, accessed September 2022), which allows hoteliers and airlines to list their availability and tourists to book it; and the BedSwap project, from the tour operator TUI, which allows firms to assess demand and move inventories between different points of sale in real time, making sales margins more flexible based on demand.

Blockchain can have an impact on the structure of online travel platforms, which may shift from relying on centralized intermediaries to having decentralized governance (Calvaresi et al., 2019). This would mean creating new business-to-consumer (B2C) or consumer-to-consumer (C2C) markets for tourism products and services through the adoption of blockchain and cryptocurrencies. An example is the TripEcoSys project (https://www.tripecosys.com, accessed September 2022), which aims to be a decentralized travel ecosystem based on the Ethereum blockchain, through which to find all travel service providers, from flights, to accommodation and excursions, and where the traveler can earn cryptocurrencies and rewards by sharing his or her own experiences.

Blockchain technology can also revolutionize the quality of online customer reviews through reliable systems for evaluating tourism products and services (e.g., hotels, restaurants, flights, events, etc.). Its true potential extends to the possibility of certifying that user opinions are original and that they cannot be manipulated by tourism firms or by consumers. Blockchain-based evaluation systems would ensure people have identities that are traceable by unique private keys, thus avoiding manipulation in reviews and making them more reliable. A decentralized, impartial, and transparent system could be created that would guarantee that the opinions are authentic and reliable and that once entered into the blockchain it would no longer be possible to modify or eliminate them. In such a system, the privacy of users would be guaranteed. Furthermore, users would surely be more willing to do reviews in exchange for financial rewards in the form of tokens or cryptocurrencies (Caddeo & Pinna, 2021).

Restaurants

In restaurants and food and beverage firms, food traceability is of the utmost importance. It is also for consumers, as 33% of US consumers place trust among the three main drivers for the purchase of food/meals (Accenture, 2017). Through blockchain, consumers can know everything about the ingredients that are part of a menu and what its status is, from the origin until the dish is served on the table. This brings transparency to the process and increases consumer confidence in the food supply chain, which in turn helps reduce unnecessary waste and prevents costly reputational damage to restaurants. For example, the Foodchain platform (https://food-chain.it/, accessed September 2022) uses blockchain and the IoT to track food products from origin to the final consumer, making all information accessible through a dedicated QR code on the product packaging (Grecuccio et al., 2020).

Benefits

Tourism is one of the industries leading the investment in blockchain and its growth is expected to continue to be very high in the coming years (PWC, 2017).

However, while blockchain has garnered global attention and its promises have captivated many, tourism firms are still struggling to understand what blockchain can do for them. Tourism products and services are intangible services that rely heavily on trust and reputation management, which is traditionally centralized and delegated to trusted third parties. Since the strengthening of trust is one of the potential benefits of the application of blockchain in tourism, blockchain can mean a leap forward for tourism firms, especially in a period of great uncertainty such as the one we are currently experiencing (Caddeo & Pinna, 2021). For example, renting a room on an online platform requires multiple levels of trust. On the host side there is the issue of trust in the guests. On the guest side, it is key to trust the host and the accommodation, as well as the services they offer. Both parties must trust the ability and integrity of the platform to handle the booking and payment processes. In a scenario like this, the online platform is, in addition to the provider of the technical infrastructure, the user interfaces, and the reservation process, responsible for establishing and maintaining trust between users (Calvaresi et al., 2019). Unfortunately, current technologies cannot prevent malicious behavior or misleading information, therefore neither a centralized organization nor an intermediary can address these issues. This is exactly where blockchain technology can unfold its full potential.

Blockchain has the potential to transform the tourism value network into a well-coordinated, self-contained, distributed network, thereby providing an enhanced experience for tourists. Likewise, blockchain technology has the potential to cause substantial changes in the business models of tourism firms in three main ways.

1. Blockchain helps build trust between contracting and transacting parties by providing a trusted platform through which traded goods/services are authenticated and transactions validated. Blockchain technology enables the adoption of cryptocurrencies and tokens by tourists and firms, and help improve the tourist experience by simplifying transaction processes through smart contracts.
2. Blockchain can increase direct access to markets for both tourism firms and consumers, limiting the need for intermediaries. This could lead to changes in the structure of many markets in the tourism industry.
3. Finally, blockchain can improve operational efficiency and help reduce costs for tourism firms, thanks to the greater speed, elimination of errors, and organizational agility achieved.

All the above, however, will only be possible if tourists understand how blockchain works and consciously perceive the benefits that its use brings. It will also be necessary that the local tourism actors accept it and avoid limiting its use to a restricted group of experienced users. Furthermore, full blockchain implementation will require not only to create interoperable networks within the different tourism activities, but to create and integrate those networks with other tourism-related industries. Initiatives such as this require a substantial level of collaboration between tourism stakeholders (e.g., tourists, firms, government, destination marketing organizations) to innovate and reinvent the tourism experience. The good news is that they can also provide a competitive edge to the actors who decide to take the step forward.

Challenges

Despite the growing interest in the use of blockchain technology and the great benefits it promises to tourism firms, it can be said that owners and managers are mostly unaware of how it works and in which cases blockchain solutions can lead to real improvements in their businesses. In addition, the technology is still at a very early stage of development, which means that there are still important challenges ahead before achieving its massive implementation in tourism firms. The mechanisms for creating new blocks require a large computational capacity, which entails high energy consumption and creates difficulties in scaling to a large number of transactions per second. Blockchain mining requires complex algorithms to be solved that consume a lot of computing power, so more and more electricity is going to be needed. Today, blockchain miners are using 0.5% of the total electricity consumed in the world (*New York Times*, 2021), but if the blockchain continues to grow at the current rate, this could mean that miners would need more energy than the world can produce. Also, as the number of users on the network increases, transactions take longer to process, and the number of users on the network is restricted. Processing an entire transaction could take days, significantly reducing the benefits that blockchain adoption brings to firms and consumers. For example, a long period of time is required to confirm financial transactions with Bitcoin, in addition to a great deal of computing power. Some alternatives based on a peer-to-peer network of virtual machines to create and run smart contracts and decentralized applications, such as Ethereum, are complex and very difficult to implement (Brandão et al., 2018).

In the context of tourism firms, the problem of scalability and the lower speed of transactions due to the high consumption of processing resources are the most important challenges. If we compare the number of transactions carried out with a centralized system with conventional standards and those of an equivalent blockchain solution, the number of transactions of the latter is usually exceeded by the former. Furthermore, the benefits and effectiveness of implementing a blockchain solution still depend on the adoption of the main players in the tourism industry and a standard is not yet available (Calvaresi et al., 2019).

The evolution of the blockchain from version 1.0 (mainly used to send and receive value transactions and with the capacity to process 5 transactions per second) to version 2.0 (which incorporates smart contracts and the capacity develop DApps and with a capacity of 25 transactions per second), and version 3.0 (aimed at the scalability of applications and with a capacity to process between 2000 and 4000 operations per second) have specifically sought to overcome some of the original limitations of the blockchain. The number of transactions per second has been increased along with the possibility of combining various types of records and native exchange with different types of currency. However, there are still key legal aspects that require attention, especially those related to privacy and data security, and the authentication of people's digital identity. Moreover, there are challenges related to the investments that need to be made to implement a technology that is still very complex. Many firms, especially SMEs, may not have the necessary financial resources and knowledge to properly exploit blockchain technologies. This may lead many owners and managers to think that blockchain technology is for now a tool that can only add value to those firms that are in a position to exploit it for their own benefit (Caddeo & Pinna, 2021), and that they are still far from being able to do so.

Discussion Questions

- What impact can the technologies described in this chapter have on the skills and abilities of future employees of tourism firms?
- Which of the technologies described will most influence the customer experience? How will these technologies affect the design of tourism service delivery processes?
- What is the relationship between the technologies focused on data management and exploitation and the technologies focused on increasing interaction and commitment with the customer? Can they work separately?
- To what extent is the customer experience affected by the use of RAISA technologies? Are they having an effect on the reach of tourism products and services?
- What variables are decisive for generating trust between firms, and between firms and consumers in the tourism ecosystem?
- To what extent can the role of large online platforms be threatened by the adoption of blockchain technologies?
- How can the tourism firm meet the financing needs arising from the implementation of new smart technologies and infrastructure?

Part 3 Smart Management Practices

One of the biggest concerns of owners and managers of tourism firms, and their stakeholders is how to successfully tackle the transformation process towards a smart organization. The answer is clear: the conventional management practices used until this point are no longer useful for a world dominated by data and the high speed at which things happen, so it is necessary for tourism firms to implement new practices according to the environment in which they operate today. Tourism firms need smart leaders with a vision that everyone can buy into and which becomes a roadmap that guides the organization's efforts and culture toward change. A new culture is needed that motivates the organization to be agile, innovative, and to act swiftly whenever customer preferences change or the environment transforms. This chapter provides owners and managers with a synthesis of the key management practices and skills that should help them successfully drive organizational smart transformation. By learning about these practices, business leaders will be able to assess how close their organization is to realizing the benefits pledged by the Smart Revolution, as well as reflect on the strategic nature of smart transformation and the need to set an agenda for change.

16 Introduction to Smart Management

In light of the strong impact that the smart technologies are having on tourism firms, there is no choice but to find new ways to transform organizations to remain competitive. Smart technologies offer endless opportunities that can be game-changers for many tourism activities, but they also present daunting new challenges that can threaten the very existence of many businesses. Therefore, the need to transform is not only an endogenous phenomenon that arises in response to the opportunities offered by smart technologies, but it is also a broader exogenous phenomenon that, by altering the competitive environment of the firm, calls for bold responses from tourism firms to stay alive in the market. Technologies such as Big Data, data analytics, artificial intelligence (AI), the Internet of Things (IoT), cloud computing, and social media are making new smart organizations possible by offering new capabilities focused on consumer needs and innovation, and streamlining response processes to environmental changes. The role of these data-centric technologies has reached such relevance that how they are integrated and managed by the organization has become a vital issue for tourism firms of all kinds, regardless of their activity, size, ownership, or nationality.

Tourism firms see that smart technologies offer unprecedented opportunities to: 1) create value from products and services that are delivered to consumers right when, where, and how they need them; 2) develop new business models that exploit new and innovative sources of revenue; and 3) access markets that were previously too difficult or expensive to enter. However, in the new smart era, business owners and managers must be aware that firms will no longer be competitive by the simple fact that they implement this or that smart technology, thinking that this will make them extraordinary. Smart transformation goes far beyond mere technological change. Technology is just one part of the intricate puzzle that organizations must learn to solve to stay competitive in the new smart era.

DOI: 10.1079/9781800620261.0016

Organizational transformation becomes a key driver when it comes to increasing the competitive capabilities of the firm and improving its profitability. Be aware that transformation is aimed at generating deeper knowledge of the changing needs of the customer that can be leveraged into faster and more accurate business actions. The secret sauce lies in the organization's ability to continually learn from data, extract meaning from it, and inform decisions. In short, the smart organization is one capable of adopting a data-centric culture and using information technologies to accelerate the production of knowledge and engage customers in new and innovative ways, learning in the process. Therefore, it will not be enough for tourism firms to use new technologies and think of new strategies to do things digitally, but they will need to emphasize changes in their organizational structure, in their core processes, in their business models, and in their own culture to find new ways to create value (Mahraz et al., 2019; Vial, 2019). This is called deploying a smart transformation strategy, which encompasses the set of far-reaching responses that the organization will need to give in the functional, operational, and technological fields simultaneously to address the challenges posed by the Smart Revolution. The end result is the creation of value to both the consumer and the organization itself.

16.1 Transformation and Value Creation

One of the most significant consequences of the growing incorporation of smart technologies in tourism firms is the change that occurs in the way organizations create and capture value. The creation of value has become a crucial factor for the sustainability of tourism firms, on which the generation of profits, customer retention, and the achievement of goals depends. However, value creation is a difficult and quite complex process, especially when it involves internal and external factors to organizations, and one that most tourism firms find difficult to tackle when smart technologies come into play (Mirarab et al., 2019). The smart tourism firm can create value through two main data-driven modalities (Zeng & Glaister, 2018):

- **From internal data:** This mode allows the firm to focus mainly on the analysis of transaction data to generate differential advantages that only the organization can enjoy.
- **From open data:** This allows the firm to focus mainly on the analysis of the data obtained through the collaborative relationships it maintains with the outside world to generate differential advantages that the firm and its partners can enjoy together.

Firms can be very different when it comes to harnessing data, putting a data strategy in place, and deploying smart technologies to create value. But, as far as a firm creates value from data (i.e., is able to collect, process, analyze, and use insights from data to create effective business value) it will outperform the rest. Furthermore, the value generated by sharing data with external partners reinforces the value generated by internal data, further driving the opportunities that result from smart transformation. The value created from smart transformation does not depend solely on the firm's IT and analytics capabilities. Strategic and organizational practices are also critical to maximizing business value, hence the need for the tourism firm to develop a smart transformation-specific value creation framework that takes into account key factors such as the dynamic capabilities and ambidexterity of the organization. Figure 16.1 encapsulates the main management vectors that come into play when driving the smart transformation of the tourism firm.

Dynamic capabilities

In the new smart era, the consumer is king when firms must decide on technologies and products/services. Tangible resources (including products) lose relevance with respect to intangible assets and services, and the firm's ability to open up to the outside world and interact with other players in the tourism ecosystem becomes more important. These new rules of the game are having a strong impact on the very configuration of the value networks, which are increasingly wider and more complex. Consequently, tourism firms must operate in a much more turbulent, uncertain, and hypercompetitive environment. To manage all the complexity that surrounds them, firms must devise mechanisms to sense what is happening and quickly adapt to the changes that occur inside and outside their competitive environment, optimizing the organization's key processes (i.e., with robotization, automating tasks), creating new sources of revenue (i.e., personalized services), or developing new business models (i.e., Uber, Airbnb). In fact, these circumstances are not much different from previous times when disruption was also the general trend, for example, when the internet revolution began. What is different this time is the different way firms must act to achieve their goals, and the much larger size of the opportunities that lie ahead for organizations that know how to seize them.

A long-term challenge facing tourism firms is how to stay relevant in front of increasingly digital and savvy customers, and how to build deeper and more lasting engagement with them that will earn firms more pocket share (Bala, 2018). A consistent way to achieve this is through dynamic capabilities, i.e., by

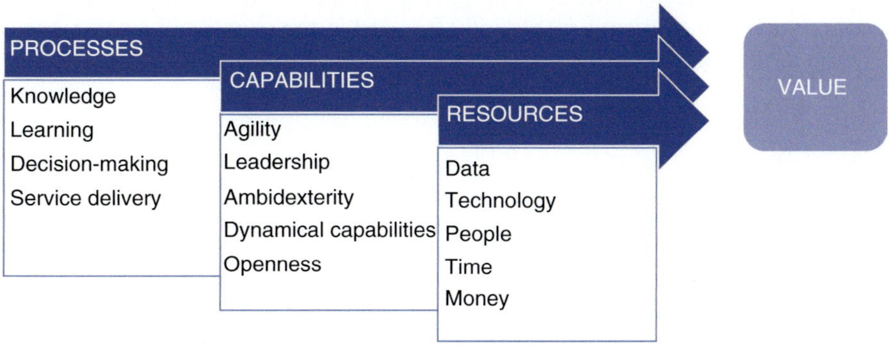

Fig. 16.1. Management drivers of smart transformation. Source: own elaboration.

strengthening the ability of firms to deliberately modify their resource base, and by acquiring ever-higher levels of digital maturity through successive waves of digital and organizational innovation. Dynamic capabilities support the way in which firms create, implement, and protect the intangible assets that drive superior long-term business performance, helping to develop mechanisms that allow them to increase their adaptation to the environment and ensure their competitiveness. These dynamic capabilities are difficult for other firms to replicate and make all the difference when it comes to adapting to customer changes and harnessing the potential of new smart technologies. Dynamic capabilities development can be accomplished through three main mechanisms (Teece, 2007):

- **Sensing and detecting opportunities and threats:** Smart firms must constantly explore their environment in search of new technologies and business opportunities, both local and distant. This requires informational resources, analytical and organizational skills, and a culture aimed at acquiring, analyzing, and using data. The speed in acquiring information assets and converting them into useful knowledge becomes the key to maximizing value for the firm.
- **Seizing the opportunities:** Once the firm detects a new (technological or market) opportunity it should be addressed through new products and services and agile processes. Maintaining and enhancing technological competencies and complementary assets is also key to seizing opportunities. Later on, when the opportunity is more mature, the firm may decide to invest heavily in the particular technologies and value offerings that are most likely to gain market acceptance.
- **Sustaining competitiveness:** Enhancing, recombining, and protecting the organization's tangible and intangible assets (including organizational structure) as the firm grows and markets and technologies evolve.

However, in most cases, firms do not have the capacity to develop dynamic capabilities and innovate on their own, and must rely on multiple parties to design new mechanisms to create and capture value (Vial, 2019). In such cases, firms need to open up and start collaborating across the platforms and ecosystems they operate in, whether with customers, partners, or even competitors. The integration of smart technologies, provided or co-developed with partners, can move the tourism firm steadily forward in the process of smart transformation while meeting its business objectives.

Ambidexterity

Addressing smart transformation involves developing capabilities to quickly adapt to predictable and unpredictable environmental changes and create and capture new value. Events such as the appearance of new technologies and disruptive/innovative business models, the shortening of product life cycles, and a more dynamic competitive landscape require new strategic responses from business firms. Those organizations that take a step forward and are committed to smart transformation do so by adopting different combinations of information technologies, processes, and resources that end up impacting the entire organization and leading to profound changes in its structure, culture, and behavior.

These transformation initiatives deliberately introduce organizational changes that are global and pursue ambidexterity, namely, the simultaneous combination of two different modes of operation within the firm (Jöhnk et al., 2020; Olszak & Zurada, 2019). On the one hand, the exploitation mode, focused on the most efficient and flexible use of the existing organizational knowledge base. On the other hand, the exploration mode, which seeks to go beyond the current base of knowledge and skills of the organization and focuses on the search for new sources of knowledge, resources, and the acquisition of new skills that encourage speed, experimentation, and innovation. Both modes differ in terms of the strategy, structure, processes, and culture, so the smart organization will need to find an adequate balance between the two, avoiding overweighting one mode over the other.

There are many approaches to achieving ambidexterity in the firm that usually encompass different initiatives at various levels, these basically can be grouped into the following:

- Cultural change initiatives.
- New organizational structures, including new roles and responsibilities.
- New processes aimed at continuous innovation and collaboration with partners.

In general, firms use different approaches in their exploration and exploitation approaches. For exploitation, firms typically implement traditional strategies that establish the strategic direction to be followed by the firm, allocate resources, and set up priorities. This highly programmed way of introducing change into the organization can slow down (if not prevent) the exploration of new transformation scenarios. For this reason, firms usually use open and agile strategies for exploration that guide the discovery of new possibilities.

Organizations sometimes create dual structures to make ambidexterity structural, meaning that they simultaneously maintain a traditional organizational structure for the exploitation mode and another open and agile structure for the exploration mode. On other occasions, firms establish hybrid configurations when justified by the need to combine both approaches. Very often this means that multiple transformation initiatives come together in the same organization, adding even more complexity to the transformation process. In these cases, it is crucial that each transformation initiative is aligned with the overall strategic vision of the firm. To do this, the organization can create a specific unit with people dedicated to coordinating all initiatives and ensuring synchronization and collaboration in accordance with established business objectives.

Most of the time managing multiple concurrent transformation initiatives and the consequent structural and cultural changes is a challenge for organizations, even more so the smaller the organization. This explains why, in today's highly complex business context, it is crucial that the firm has a governance framework in place for transformation initiatives that regulates resources and responsibilities and manages to seamlessly integrate them with the business, IT, and people functions. This is maybe the best way to reduce resistance to change and overcome the inertias that often impede organizational change, replacing them with new innovative and more collaborative capabilities instead.

Ultimately, owners and managers should not forget that the transformation of a business organization is made by and designed for people, so employees must always be at the center of any smart transformation strategy. Interpersonal and cultural aspects play a crucial role in a successful organizational transformation strategy, which must be based on communication and leadership as tools that facilitate the collaboration of employees in the business and IT areas. In this vein, maintaining continuous personal interaction at all levels, understanding the objectives pursued, and sharing the necessary knowledge not only helps to reduce the economic and social costs of organizational transformation, but also contributes to accelerating the whole transformation process.

16.2 Value Creation

The creation of new value is the leitmotif that drives all the efforts that the tourism firm dedicates to achieving smart transformation. However, this is not a quick or easy path; rather there are numerous challenges that run through the entire value creation process, including creating, delivering, and capturing value, and to which owners and managers need to pay careful attention. Let's see below the main challenges that can be found in each of the value processes.

Challenges for value creation

In the context of smart transformation of tourism firms, value creation is about providing tourists with a "smart experience" that is fast, accurate, personalized, and can be tracked in real time. Making an experience smart depends on the firm's ability to collect information, process it, analyze it, and make decisions quickly and effectively. The Big Data life cycle management thus becomes one of the main engines for value creation in the smart firm. This requires a set of skills associated with data literacy and interpretation of results that should enable employees to make valuable contributions to the business, and that includes data management, analytical tools, and business and decision-making processes (Brinch et al., 2021). In addition, value creation must be driven by the firm's business strategy, which makes smart technologies coordinate with the rest of the firm's ICT to support continuous improvements.

To harness these sources of potential value for the business, the tourism firm must be able to manage mixed offerings made up of products and services based on smart technologies and "conventional" products and services (Vaska et al., 2021). This may confront firms with important decisions about whether to cannibalize "conventional" products and invest in the most advanced smart products and services, or maintain a balance between the two value propositions. These decisions may also affect the speed at which the firm must transition from one value proposition to another, and the impact on the existing business model.

Another key challenge is related to the impact that smart technologies have for the co-creation of value with customers and stakeholders, and the new opportunities that arise as data analytics is integrated into consumer devices and tourism products and services. While these implications remain to be seen, tourism firms need to remain vigilant about new media and emerging opportunities to co-create value to quickly deliver the best value propositions to customers.

Finally, incorporating greater organizational agility is another important challenge that the firm must address to achieve a faster and more accurate operational framework to respond to changes in the environment. By implementing agile practices, tourism firms can increase their ability to offer solutions to customers quickly and in a personalized way, creating new value for the customer and the organization. However, gaining organizational agility in the context of highly complex ecosystems such as tourism is a crucial challenge that tourism firms must learn how to address.

Challenges for value delivery

Value delivery describes how the firm's processes and activities are used to deliver the promised value to the customer. As smart transformation challenges conventional core business competencies, activities, processes, and technologies, and brings the threat of new players to market, the way organizations deliver value to customers and stakeholders is deeply impacted. In the smart firm, the activities and processes focused on delivering value change and become governed by the data and knowledge that the firm generates from customers and the environment. This new way of operating internally requires the firm to work in an agile manner and take advantage of all its internal knowledge, extending it to all levels of the organization to ensure that it can act quickly and accurately.

However, delivering value in new tourism ecosystems dominated by increasingly complex value networks is a significant challenge. Firms seeking to transform must relearn how to deliver value using the new relationships, roles, and activities emerging in smart ecosystems, orchestrating their resources accordingly, and making smart decisions about what skills and capabilities to develop and what investments to make first. Additionally, tourism firms must be open to continuously incorporating new smart technologies and adapt to the new business models that emerge. This makes it necessary to delve deeper into the knowledge of smart technologies and better understand the opportunities they offer to deliver value to the customer in a close, direct, and more efficient way.

Challenges for capturing value

Smart transformation also brings significant impacts on the way firms capture value. For example, there are impacts on revenue models, cost structures, and the financial viability of tourism firms, as well as on the capacity of the firm to reduce the uncertainty associated with decision-making. Through the implementation of smart technologies, tourism firms can optimize their work processes and increase the quality and efficiency in the delivery of products and services to the customer. Tourism firms capture value when they can offer experiences to customers that respond to their preferences, in the place, time, and through the channel of their choice, or when they personalize their products and services quickly and seamlessly. Technologies such as Big Data and data analytics help firms identify new sources of value in the markets and design superior value propositions, while reducing the costs of delivering customer value over time. Smart technologies also enable tourism firms to capture value through platforms, leveraging the abundant knowledge they generate about customers and the close and direct relationship they allow with the customer.

Notwithstanding, tourism firms must make investment decisions in smart technologies with caution and intelligence. They must carefully weighup the pros and cons of investing in certain smart technologies and undertaking deep organizational changes, as well as assess the impact that all these changes may have on the value to be captured by the firm and its business model. Business owners and managers must be aware that many times the value captured from investments in new technologies does not reach the income generated and may not compensate for the risks assumed by the investments made. This is perhaps more evident in ecosystems dominated by platforms where each actor will surely end up capturing part of the value of the investments made by the firm. That is why it is key for tourism firms to build a well-thought-out revenue model, in which the value captured is adjusted to forecasts throughout the life cycle of technologies and products and services.

16.3 Creating a Transformation Strategy

Creating a smart transformation strategy based on the organization's dynamic capabilities and ambidexterity through the implementation of multiple concurrent projects is an effective way to reduce risks and costs. In a scenario like this, organizations are constantly changing, adapting their structures, resources, and objectives as they evolve and overcome the inertia that impedes change. This is different from those organizations that perceive the need to transform as a result of episodic external events (or disruptions) that alter a previous equilibrium situation and that affect the internal structures of the organization, which are then replaced by others.

The continuous change approach entails a micro-level perspective (as opposed to the macro-level of the episodic change approach), according to which transformation does not only respond to external changes that abruptly shake the organization at specific moments (i.e., when a new technology appears), but is substantially driven by the continuous evolution of the contextual conditions in which the firm operates (i.e., by the continuous interaction with customers, the evolution of customers' consumption expectations, etc.). It is precisely the constant evolution of the contextual conditions, and the complexity arising from the growing interconnection between the actors of the tourism ecosystems, that determines the degree of transformation that the firm will require, from continuous incremental micro-adaptations to more profound selective changes that introduce new organizational configurations. Ultimately, the organization should always remain aligned with the contextual conditions in which it operates, seeking an adequate balance between continuous and episodic change.

Unlike what may happen in other business environments, equilibrium situations are not expected to be reached any time soon after external changes or disruptions occur in the tourism ecosystem. Rather, in tourism changes often occur in cascade and their consequences are unpredictable. For this reason, owners and managers should better approach the smart transformation strategy from the paradigm of continuous change, which means that the organization and its structure become malleable and are always prepared to face endless micro changes that are going to occur continuously over time (Hanelt et al., 2021). However, despite the predominance of the continuous change approach, this does not rule out the need for the tourism firm to respond to episodic or disruptive changes that may occur in its environment, even more so when these episodic changes can be the precursors of continuous changes.

All activities aimed at implementing a smart transformation strategy and creating value for the tourism firm must be converted into concrete actions. Hence, owners and managers should always remember that deciding to implement technologies is not going to add direct value to the organization on its own. Only when owners and managers are aware of the specific context in which these technologies operate and pursue specific and ambitious transformation goals can the firm discover new ways to create value. Therefore, a successful smart transformation strategy will depend on the emphasis that leaders place on preparing the organization's value creation processes for change, acting more specifically in the five areas that are discussed below (Vial, 2019):

1. Value offerings.
2. Value networks.
3. Digital channels.
4. Business model.
5. Organizational agility.

Acting in these areas can be a source of inspiration to create innovative products and services, optimize key business processes, improve decision-making, and evolve to an agile culture that leads to a smart data-driven organization.

Value offerings

Smart technologies provide the ability for tourism firms to create new value offerings from reviewing or expanding their existing portfolio of products and services. These new value offerings are determined by the firm's ability to collect and analyze data on customers from their interactions with the products and services, as well as by incorporating increasingly innovative elements into them. A paradigmatic example of the creation of new value offerings through smart technologies is Airbnb, whose business model is based on the ability to collect information from customers and partners and use data insights to better understand how customers travel and what accommodations they prefer when visiting a travel destination. Putting this knowledge into action means that Airbnb provides customers with real-time solutions in the form of accommodation recommendations that best suit their preferences. When tourism firms equip themselves with the right smart technologies and capabilities, they are opening up new possibilities to better understand customer needs and preferences and deliver the experiences that customers want, how, where, and when they demand them. Furthermore, firms are much closer to creating personalized value offerings and implementing value co-creation initiatives aimed at deepening customer engagement.

Value networks

Smart technologies make it possible to redefine value networks across the tourism ecosystem. On the one hand, smart technologies give customers and users the ability to become co-creators of value, facilitating continuous communication and interaction between them and firms through online platforms, communities, social media, or online channels enabled by the firms themselves. On the other hand, smart technologies facilitate interaction and direct exchanges between value network participants (i.e., customers, users, firms) bypassing intermediaries and favoring new collaborative relationships between the different actors focused on value creation.

As the collaborative ties and coordination between the actors are strengthened, increasingly complex and omnidirectional relationships emerge between the stakeholders of the ecosystem that can end up creating more value for the customer. In view of these positive effects, it is key that business owners and managers encourage stakeholders to engage with smart technologies that promote the joint creation of value.

Distribution channels

Smart technologies have a significant impact on distribution and sales channels, through which many tourism firms are redefining the way they deliver value to the customer. Tourism firms use smart technologies to create new direct and bidirectional touch points and communication channels with their customers and users (i.e., social media, chatbots, IoT, communities) that foster continuous dialogue with them and close the gap between the physical and digital world. Likewise, the use of sensors and technologies based on IoT, although still at an early stage, could contribute to making the operation of supply chains and the provision of tourism services more efficient through the automation of tasks and data-driven decision-making.

Business models

Business model transformation can create new value for the tourism firm, especially when the new models are based on emerging technological developments and the provision of innovative services to customers. In some cases, the new business models completely transform the way firms behave and compete until they practically become technology firms, such as those that offer to connect firms with users, or users with other users. In other cases, business models provide value offerings aimed at satisfying certain customer needs, such as those with lower incomes or those seeking a very particular or specialized tourism experience.

There are also business models that aim to create new value using disruptive technologies to find solutions to problems related to sustainability or the shared economy. This is the case, for example, of the new business models arising from sustainable mobility (i.e., car and bike sharing), which lead to new tourism products and services that improve the customer experience. In recent years, business models based on digital platforms have also emerged that facilitate communication and connection between the different actors in the tourism ecosystem. An example are reservation and exchange platforms in the accommodation sector (i.e., Booking.com, Airbnb), which create new value propositions for customers based on cost reduction and the integration of value-added services.

Organizational agility

Smart technologies can help tourism firms cope with changes in their environment and adapt more quickly to them. This is possible when firms speed organizational agility, which means developing the ability to acquire knowledge about the environment, detect opportunities, and be able to coordinate responses quickly and accurately. Systems combining Big Data, data analytics, AI, and IoT provide a huge amount of information and processing capacity that can be used by the tourism firm to detect untapped opportunities, increase customer knowledge, and, ultimately, create new customer-centric value. For example, a tourism firm can anticipate the preferences of its customers and offer personalized experiences based on the accumulated knowledge of the customer and the conditions of the context. In this way, smart technologies can drive ambidexterity in the tourism firm, as they encourage exploration through smart innovation and exploitation of existing resources.

16.4 Smart Decision-Making

Smart transformation is becoming a key competitive driver for tourism firms, as the implementation of smart technologies and the transformation of organizations provide real-time insights to make faster and smarter decisions. Decision-making is about choosing among several alternative courses of action to achieve specific goals and is at the core of all business management functions. Therefore, developing a process and systems for smart decision-making is critical to the success of smart tourism firms that need sound decisions to create and maintain competitive advantage (Forman & Selly, 2001).

Effective use of smart technologies, as well as speed of action, are key factors that distinguish smart tourism firms and have a high impact on the decision-making approach adopted by organizations. Yet, despite the wealth of data now available to tourism firms that could support better decisions, many owners and managers continue to rely too heavily on their experience, opinion, and intuition (or a combination of all of these) to make decisions that affect the functioning of organizations. Consequently, many decisions, instead of being driven by creating value and seizing the opportunities for the organization, are too influenced by personal or political interests.

With data growing at an unprecedented rate, new data science techniques are appearing at an accelerated pace together with a renewed interest in AI and machine learning, thus increasing the capacity and power available for decision-making in tourism firms. The days of decisions being subjective and often lacking in evidence seem to be numbered. In this context, although beliefs and intuition may continue to play a role in everyday decision-making, the strategic and operational decisions of smart firms cannot be based solely on them. In this new smart context, conventional tourism will be compelled to adapt existing decision-making structures and processes to seize the big opportunities offered by smart transformation and enable owners and managers to make decisions based on "what they know" rather than on "what they think".

However, owners and managers must bear in mind that it would be too simplistic to say that the quality of smart decisions depends solely on the amount of data. Good decisions are closely tied to the organization's strategies and capabilities to extract value from data. This means that tourism firms need to develop new capabilities to exploit the unprecedented opportunities offered by smart technologies and address the new complexities that arise. Ultimately, firms need new resources, tangible and intangible, including human resources, culture, technology, and managerial and technical skills (Shamim et al., 2019). The objective is for decision makers to improve their knowledge and make increasingly higher quality decisions that lead to the creation of value for the customer, the firm, and the stakeholders.

Decision-making process

The process to implement a smart decision-making framework is not easy, nor does it happen overnight. Not only does it require business owners and managers to decide which techniques and tools must support data processing, but they must also focus on the quality and quantity of data that is needed. In fact, the basis of smart decision-making lies, first, in the ability to identify and understand the decisions to be made and, second, in collecting the data and knowledge (from inside and outside the organization) that are necessary to support these decisions and turn them into actions that create value, for example, decide what product the firm should offer to a specific customer, or what improvements should be made to the next product or service (Berntsson Svensson et al., 2019; Chiheb et al., 2019).

However, many organizations systematically collect more information than they use, and the excess of information, together with the poor quality and irrelevance of the data managed by the organization, negatively affect smart decision-making. At this point, it is important that owners and managers avoid underestimating the costs of collecting information in relation to the benefits, and put an end to the illusion that by collecting more data they will better manage uncertainty (Kaivo-Oja et al., 2015). In real life, information is never neutral, but always reveals some aspect of the issue at stake at the cost of hiding other aspects of the same issue. Therefore, it is highly important that analysts and data teams work on improving data quality and processing techniques. They should also focus on the quality of the visualizations used to present the results to the leaders of the organization and to facilitate the interpretation of the analysis and make its consequences understandable. Figure 16.2 outlines a smart decision-making process based on four stages: intelligence, design, choice, and implementation.

Once the firm is clear about the decisions to be made and has the necessary data and insights to support these decisions, it must continue to develop and analyze possible alternative courses of action to address the problem or the opportunity that requires a decision. After the alternatives have been formulated, it is necessary to evaluate them according to pre-established

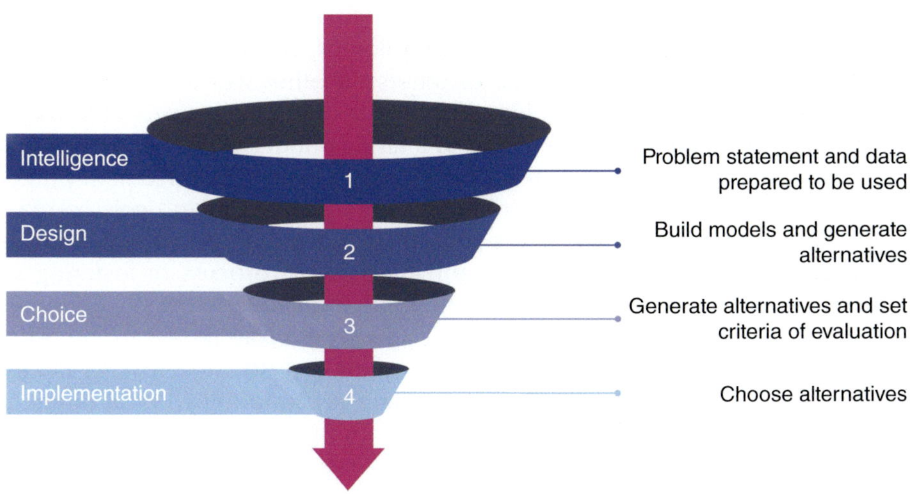

Fig. 16.2. Smart decision-making process. Source: own elaboration based on Chiheb *et al.* (2019).

evaluation criteria to determine the impact that each of the alternatives can have on the business. Based on the results obtained from the evaluation, those people making decisions must choose one of the alternatives (or set of alternatives) that best solves the problem. Finally, the organization must focus on implementing the selected alternative, after it has been communicated and made known by the decision makers to all stakeholders. After this, the decision process will resume once new information has been collected on the results achieved and the real effectiveness of the decisions made has been evaluated.

As can be seen, the smart decision-making process allows the integration of Big Data analytics throughout the different stages of the decision-making process and promotes communication and collaboration between decision makers and the analytics team, with the former establishing the decisions to be made and the latter identifying knowledge that may be useful in supporting the decisions. Both should have a common understanding of the decisions to be made and its requirements, and document the decision-making process to ensure reuse of knowledge from one decision to another.

Decision-making capabilities

Implementing a smart decision-making framework involves significant management challenges, especially when it comes to new capabilities. Smart firms need to reconfigure and upgrade existing capabilities, which depends on effective resource management practices. In this context, leadership, talent management, technology, and organizational culture are among the most relevant capabilities for making smart decisions (Shamim et al., 2019).

Especially relevant to driving smart decision-making is organizational culture, followed by leadership and talent management. The role of technology is comparatively less important, which does not mean that it is not important to have the adequate technologies, but rather that its relevance does not reach the importance of others. In other words, in the new smart era, tourism firms are not going to be successful in the process of transformation just because they have access to a large amount of data or tools, but they need leadership with a clear vision, adequate talent management practices, and, above all, an organizational culture that fosters smart decision-making. Business owners and managers should be those who provide a clear vision and the goals that guide the organization in the process of fostering smart decision-making. They must know how the Big Data life cycle works and the opportunities offered by data analytics, and be a role model in the use they make of data in decision-making. The quality of decision makers is key for the firm to improve its smart decision-making capabilities. As for talent management activities, these must provide the organization with the skills that enable smart decision-making by hiring and retaining experts in Big Data and analytics techniques.

16.5 Knowledge Management

Information and knowledge play a critical role in smart decision-making. Surely this will have been the case since the first tourism firms emerged. However, the advent of Big Data and smart technologies have completely changed things: today they are the basis for scaling decision-making to a higher and smarter level. Through the wealth of information and knowledge they produce, tourism firms can make more informed decisions, which in turn helps improve internal trust and preserve external legitimacy (Kaivo-Oja et al., 2015).

Value creation in smart firms is increasingly the result of the management of information and knowledge assets that improve decision-making and create competitive advantages. In other words, the competitive advantage of the smart firm is based on the ability to effectively apply the organization's new and existing knowledge to create differential advantages, for example, through new products, services, and processes (Thrassou et al., 2012). Knowledge has become an essential asset in terms of organizational competitiveness, and the ability to quickly recognize the meaning of new information, assimilate it, apply it, and integrate it with the rest of the organization's knowledge assets is part of highly valuable management practices. However, there is always a "but", because all these capabilities are only possible when the firms takes knowledge management seriously.

Knowledge management has three key dimensions: processes, enablers, and knowledge management systems (KMS). Knowledge management processes include those that organize and coordinate knowledge for its collective use in the organization (i.e., the creation, exchange, storage, and application of knowledge). Thanks to these processes, the firm develops competitive advantages, speeds up innovation, and increases its ability to respond to changes in the environment. Enablers are those mechanisms that allow knowledge management activities to be carried out, such as the coding and sharing of knowledge between people and teams. These mechanisms provide the necessary infrastructure to improve knowledge management processes, promoting the creation, exchange, and protection of knowledge. KMS are to IT infrastructure (i.e., hardware, extranets, intranets) and collaborative technologies (i.e., shared databases, document repositories, discussion forums) that help manage and apply knowledge effectively in the organization (Santoro et al., 2018). Therefore, a KMS is a facilitator of effective knowledge management as it aims to capture people's knowledge so that everyone in the organization benefits from it.

KMS allow greater freedom and precision in decision-making by combining both the traditional internal view of knowledge management, based on the integration of internal databases that facilitate access to knowledge, and the external view supported by the open innovation paradigm, which suggests that firms must promote new forms of interaction and collaboration with third parties. Implementing a KMS can be a good starting point for collaboration and the exchange of knowledge between the departments of an organization.

They can also create open and dynamic collaboration spaces with external partners where the exchange of information and knowledge drives innovation capabilities.

Business owners and managers must be aware that knowledge management technologies and infrastructure alone are not enough to increase the organization's capacity for innovation. The firm must also strengthen its connections while exploring the environment in search of opportunities and selecting the most suitable partners with whom to establish relationships of different intensity and scope. Furthermore, the organization's leaders need to understand the ICT framework that can support knowledge management and how they will collect the knowledge they need to create value. Ultimately, the multidimensional relationships that the tourism firm establishes between Big Data, data analytics, knowledge management, KMS, and open innovation will depend on building an "open knowledge system" in which information and knowledge flow freely throughout the organization to improve innovation capabilities.

16.6 Recommendations

Some of the main recommendations that can be made to owners and managers that intend to tackle the smart transformation process are summarized in the following points:

- **Dive well below the surface:** Successful smart transformation is not about creating a new web store or developing a new mobile application, while seeking to maintain a balance between traditional business and new digital business. Very often business owners and managers are confused about smart transformation and fail to understand that what has really changed are the basics of doing tourism business. Digital platforms that bring together buyers and sellers have radically changed the very idea of the tourism ecosystem, turning it into a much more liquid and diffuse concept in which control of data and, above all, what to do with data have become the main competitive drivers for tourism firms. Smart transformation goes far beyond scratching the surface, it means reinventing or readapting the entire business model to create real value from data and customer insights.
- **Never stop innovating:** In a world dominated by platforms, learning to orchestrate relationships is much more relevant than owning or controlling tangible resources, i.e., Uber connects drivers and passengers without a car, and Airbnb connects hosts and guests without rooms (Reddy & Reinartz, 2017). Creating value no longer depends on having a large capital base to undertake big equipment, network, or infrastructure projects, but rather on the ability to integrate capabilities around a new idea or innovative project in which the participants collaborate to achieve their individual goals. The ability to innovate has become more important than ever. But if you don't have the resources, people, or ideas to be a champion of innovation, open your mind and let fresh air into your organization from the outside. Avoid the old cliché "I know what the market needs but I don't have the resources to do it" and be open to new ideas and insights from the

outside world. Find in collaboration and open innovation an efficient, faster, and less risky way to innovate and bring your value offering to the market.

- **Don't stop doing things and speed up:** Everything in the new smart era happens at an incredible speed that has never been seen before and which tourism firms are not used to. Speed marks the difference between market participants and is a key factor that determines how organizations approach projects, execute processes, and even hire people. Having the best technologies and data is worth nothing if the firm is not able to process it quickly and deliver new products and services just when and where the customer wants them. A single minute sooner or later can make a difference, so business owners and managers must reflect on the new value of speed and what they stand to gain (or lose) by speeding up the pace of getting things done. Organizations need to break with the old rules of doing things, tear down silos within the organization, and make everything work in a more agile and flexible way. Smart technologies open up a universe of new possibilities for organizations to become more agile and customer-centric, and learn to extract value from data faster and more efficiently. But this does not happen overnight. Firms need organizational agility and agility applied to their data initiatives if they want to keep pace with customers and deliver products and services that their customers will continue to be willing to pay for.

Discussion Questions

- What differences exist in the management of smart transformation between big and small tourism firms?
- If value creation is the main objective of the smart firm, how is the notion of "value" defined? How can it be measured?
- How can the tourism firm make sure that multiple initiatives geared towards smart transformation are aligned into a single agenda?
- What factors can make the tourism firm opt for a transformation approach based on continuous or episodic change? What role do smart technologies play in this decision?
- Is ambidexterity a transitional solution while the organization reaches a higher level of digital maturity, or is it a permanent solution to address smart transformation?
- What is the relationship between the three management approaches described in this chapter: dynamic capabilities development, ambidexterity, and continuous change vs. episodic change?
- How can an owner or manager lead a tourism SME that considers smart transformation as something out of reach or that will not generate enough value for the firm?

17 Managing Transformative Change

Smart transformation initiatives have the potential to structurally change tourism organizations. This requires the implementation of new management practices based on the incorporation of dynamic capabilities, ambidexterity, and greater organizational agility, which means that cross-functional collaboration within the organization plays a crucial role (Vial, 2019). While it is an old recurring theme that has been at the forefront of organizational transformation management discussions for years, the truth is that even today it is difficult to see these forms of collaboration emerging within business firms to facilitate seamless integration of change strategies with the rest of the firm's key strategies, such as IT strategy or human resources strategy.

To overcome the collaborative gap that persists in most organizations, many firms create separate organizational units giving them greater independence. These units often function with greater autonomy and flexibility than the rest of the firm's organizational units, making it easier for them to focus on innovation while maintaining access to the rest of the organization's resources. Alternatively, other organizations create cross-functional teams that remain under the umbrella of the core organization itself. The fact that a number of multifunctional teams coexist in the same organization is a factor that favors the creation of internal networks based on knowledge and skills that transcend the traditional organizational structure, thus strengthening collaboration and the exchange of knowledge within these networks for the benefit of all. This is just one example of the changes that smart transformation brings with it, so let's examine a few others that tourism business owners and managers will need to consider.

17.1 Changes in the Organizational Structure

For smart transformation to be successful, an organization will need to turn its culture upside down. For example, issues such as the traditional separation

DOI: 10.1079/9781800620261.0017

between the business and systems/IT areas, so ingrained in organizations that have been operating for years, will no longer make sense in the context of the smart firm. Instead, the need for the smart firm to work with agility and adapt its value offering quickly and accurately to the changing customer needs and preferences makes it necessary to cultivate a very different culture based on the value of collaboration and shared knowledge. This new "smart" culture must be based on inspiring leadership that attracts the interest of all members of the organization towards an innovation-focused vision, regardless of the department or area to which each individual belongs. Smart firms need smart leaders, who work to build an agile organizational culture focused on innovation and encourage people to cultivate an appetite for experimenting and taking risks, being aware of all the benefits that people can bring to the organization as a whole.

At times, smart firms will have to create new leadership roles that assume the responsibility of scaling the smart transformation processes and technologies to the rest of the organization, while keeping them aligned with the rest of the firm's strategies and the principles of agile work. New roles such as the Chief Digital Officer (or Chief Transformation Officer) aimed at ensuring that digital technologies are used appropriately and in line with the organization's objectives, reveal the strategic value of digital transformation for many organizations. In other cases, Chief Data Officers take on the role of extracting value from data and provide refined insights to improve decision-making by business leaders. Most of the time these roles are created by smart organizations with the intention of transcending their own departmental boundaries and implementing actions that transcend the organizational logic to achieve superior collaboration between the organization's business and technology functions. In the context of the smart transformation, all changes in the organization's structure and culture inevitably lead employees to take on new roles that were traditionally outside their duties. Many of the attributions and responsibilities of employees based on their position within the hierarchy of the organization will become blurred, and organizations will focus more on the individual skills of each employee, their leadership capacity, and the value that their initiatives can have for the organization. For example, employees who were previously not part of the IT function will start leading technology-intensive projects in smart organizations, just as IT function specialists will take on more business-related roles.

As organizational structural changes advance in the tourism firm, it is important for owners and managers to be aware of the need to improve the skills and abilities of employees. For example, people will need to increase their data analytics and technology skills to solve increasingly complex problems that involve a large number of interconnected variables. This also includes the redefinition of the professional profiles that the tourism firm will need to boost transformation in the coming years, part of which will come from new hires outside the firm and others from internal promotions within the organization. Leading the employees in the transition to the smart organization will therefore entail important challenges that will surely involve significant changes in the human resources function of the firm.

17.2 Changes in Strategy

Tourism firms need to establish new management practices to govern the set of changes that are taking place and adapt to the new reality emerging from the Smart Revolution. The way in which firms establish and deploy their transformation strategies depends to a large extent on the competitive position of the organization compared to other firms in the ecosystem (Bala, 2018; Reis et al., 2018). Firms should formulate a transformation strategy plan that serves as a core element around which the firm coordinates, prioritizes, and implements all the actions that are aimed at the firm's smart transition.

However, establishing a strategy in an environment as changeable and unpredictable as the one surrounding the tourism firm can be a daunting task. If conditions are constantly changing, even the best plan is almost certain to fail (Andersen et al., 2018). In addition, smart technologies such as Big Data, analytics, and artificial intelligence (AI) are blurring the boundaries that used to exist between technology and business management, and new concepts and tools are emerging that change the way firms address management challenges, innovate, build relationships, and behave (Vaska et al., 2021). Therefore, tourism firms must approach their business transformation strategy by making it as adaptable and comprehensive as possible.

Organizations that are used to working the "agile way" and that know how to extract value from data will be in a better position to establish a smart transformation strategy. They will find it easier to adopt experimental approaches under conditions of uncertainty and will be able to overcome the traditional barriers between strategy formulation and execution, as they are accustomed to working in short iterative cycles rather than sequentially. These organizations usually establish results objectives, assign a development team the responsibility of proposing solutions, and, after these are tested with the customer or users, they decide to implement the ones that work (Andersen et al., 2018). Once the cycle is complete, the firm starts a new cycle of formulation and execution of the strategy.

A smart transformation strategy must necessarily have a wide scope and include far-reaching initiatives. It should seek to integrate and give coherence to the set of apparently independent or unconnected initiatives carried out by the organization, such as those related to products, customers, markets, or distribution. Due to its transversal nature, the smart strategy will cross the boundaries of all other strategies of the firm, including the business and technological strategy. For example, the firm's IT strategy focuses on managing IT within the organization and setting the roadmap for future business applications, infrastructure, and technologies. As such, it is essential for the development of operational activities in the organization and it should remain aligned with the firm's data strategy, which is responsible for the processes aimed at capturing value from data (Fig. 17.1).

A smart transformation strategy goes beyond the conventional approach of change strategies focused on processes, since it includes a broad repertoire of changes, some of which are disruptive and can affect products and services, resource allocation, and the business model. Unlike a digitalization strategy,

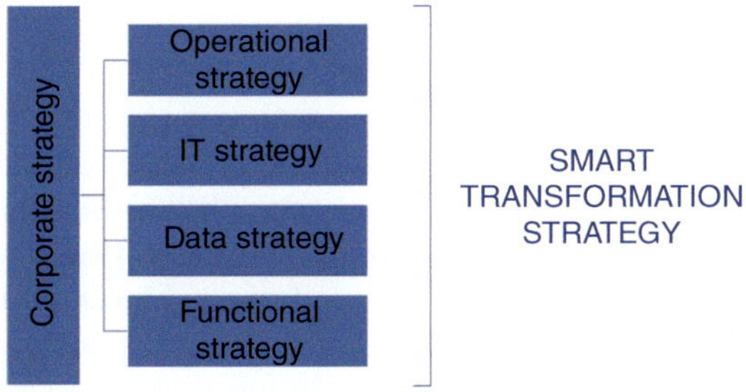

Fig. 17.1. Relationship between smart transformation strategy and other corporate strategies. Source: own elaboration.

a smart transformation strategy has a much broader scope and pursues more ambitious goals. Furthermore, its effects extend to products, processes, and organizational aspects, which is why it requires a governance framework capable of integrating all the initiatives that result from the integrated use of smart technologies with the power of data and shared knowledge.

That said, there are no guidelines that show firms how they should proceed with their smart transformation initiatives in a specific industry. Furthermore, the tourism business environment is continually evolving and there is great uncertainty regarding the assumptions underlying smart transformation strategies. Therefore, it is important for tourism firms to continually monitor and re-evaluate their transformation strategies, testing initial assumptions and the progress made by transformation (Matt et al., 2015). In the event that expectations are not met or deviate from the goals set, the organization must implement processes that connect evaluation with the formulation of strategies, in a way that the credibility of management is guaranteed, and biases are avoided.

17.3 Changes in People

Due to the wide and diverse scope of smart transformation strategies and the impacts they have on organizational, technological, and operational aspects of the business, it is important to ensure that organizations have the necessary human resources, leadership, and responsibilities to address a strategy like this. If the firm approaches its smart transformation strategy without determination, leadership, or commitment, it will almost certainly lose momentum and start to have operational problems everywhere.

Tourism firms have some peculiarities that distinguish them from firms in other industries, including some related to the human dimension. The strong seasonality of the tourism business and the low-skilled and part-time workforce that are so common in tourism firms affect the way organizations design and implement their transformation strategies. Tourism firms often have difficulties in retaining their staff, which can end up affecting the delivery of quality products and services, and make knowledge management and innovation hard, especially if people continually enter and leave the firm and many resources must be devoted to recruiting and training new staff instead of engaging in higher value-added tasks (Pesonen, 2020).

On the other hand, the tourism firm must ensure that the person (or people) in charge of the smart transformation have sufficient experience in change projects and that their roles and responsibilities are aligned with the objectives and processes of the transformation strategy. Besides hiring the right people, owners and managers must decide how to blend the different leadership roles within the strategy, depending on the depth and duration of the transformation, as well as ensuring that the different roles (e.g., CIO and CDO) remain aligned at all times. The ability to delegate leadership to other people is not a given in tourism firms, where most of the time (especially in SMEs) leadership is exercised by the CEO or the owner of the firm, who is the one who makes most of the decisions that affect the future of the organization.

Tourism firms must redouble their efforts in training their employees as they join the new culture of "smartness". Employees need to develop new skills and abilities related to innovation and new technologies (including hardware and software); be excellent at managing customer relationships and handling digital communication channels; be able to search, organize, and process information from scattered sources; and be autonomous in making decisions about how to organize their work to achieve the established goals. Organizations not only acquire these new skills and capabilities by promoting internal staff, but also by outsourcing services, creating alliances with specialized firms, and even through mergers and acquisitions of other firms highly competent in these fields. According to some studies, the firms that attach more importance to the role of people in transformation processes are the ones that invest the most in training their employees and are better prepared to face the challenges of technological change and innovation (Madera et al., 2017). These firms confirm that the knowledge of the organization and the skills and abilities of the employees are closely related to the performance of the firm, its efficiency, and its competitive capacity.

Owners and managers must therefore keep in mind that human resources' function is critical in the new smart organization. It is responsible for improving the quality of the organization's human resources and ensuring that the skills of the workforce are aligned with the new needs posed by the Smart Revolution. The human resources department can shape the skills, attitudes, and behaviors of employees and make them unlock their full potential and improve their contribution to the business. It is also responsible for establishing the incentives available to employees to contribute to the results of the organization. Notwithstanding the foregoing, a large number of tourism firms, especially the smaller ones, do not have a human resources department, or if they have one, their human resource management practices are flawed when it comes to recruiting, evaluating, and rewarding the employees. This affects the way of managing knowledge within the organization and reduces the firm's chances of hiring the right people to tackle the organizational transformation processes required by the Smart Revolution.

17.4 Changes in Business Models

Smart transformation brings with it a new conceptualization of business models and new requirements for capabilities to create, deliver, and capture value. With the Smart Revolution, new opportunities are created so that firms can co-create value with customers, expand their catalog of products and services (making them more flexible and better adapted to customer tastes and preferences), and accelerate the development of new ecosystems dominated by value networks. In this context, tourism firms must rethink new ways of creating and delivering value, as well as developing abilities and skills to capture value generated from a more intimate and lasting relationship with the customer.

One consequence of the above is that the business models of smart firms, that is, the mechanisms put in place to create, deliver, and capture value, are undergoing a radical change that affects each of its individual components. This is due to the different way in which firms approach the creation of value propositions, deliver value, segment customers, and discover new sources of revenue that did not exist until now. In the end, business models are so relevant that there are authors who affirm that the future of digital firms and, by extension, of smart firms, will be based on value architectures able to cope with constant changes in customer value, rather than on business models as we knew them (Brown & Brown, 2019).

Firms' business models have evolved from rigid structures that are difficult to adapt to changes in the environment, to open and permeable frameworks capable of responding quickly to changes of a business, technological, and market nature that occur in the environment. Some of the components that have traditionally been key in the architecture of business models (i.e., customer segments, channels, key partners) have lost relevance in favor of technology, collaborative relationships with stakeholders, the ability to manage organizational knowledge, and the personalization of value propositions. This reflects how radically different smart firms are approaching the processes of creating, delivering, and capturing value. Unlike the conventional business models, those of smart firms consider data flows to be a core asset of the organization and, as such, they play a central role in the processes of creating and delivering value. A direct connection therefore exists between smart technologies, knowledge management, organizational culture, and business models in the smart firm.

17.5 Barriers to Transformation

Although the Smart Revolution and its effects both in tourism firms and in tourism ecosystems are here to stay, the inertia of organizations and resistance to change are factors that can seriously hinder the progress made by smart transformation. Inertia becomes relevant when the organization's own resources and capabilities become a barrier to the transformation of the firm, limiting its own ability to innovate and hindering the adoption of new technologies that can make processes and practices evolve. These issues are even more apparent when firms have been established for years and have had time to develop habits that are deeply ingrained in the core processes of the organization, such as in production processes, customer and supplier relationships, and decision-making. Over time, some of these practices turn rigid and become part of the employee culture. When this happens, the problem is not that the leaders of the firm are disconnected from reality and do not understand the potential benefits of the Smart Revolution for the organization, but rather that the very culture of the organization and the daily practices that make it work inhibit the capacity for change and the innovative power of smart transformation in the organization.

On the other hand, the resistance of employees to change and the adoption of smart technologies is another powerful barrier that owners and managers must consider. Sometimes resistance is a byproduct of organizational inertia that is embedded in daily work processes and employee mindsets due to their distrust (or even fear) of the unknown consequences of change. Other times, resistance is just a reaction to "innovation fatigue" that arises from the sheer number and rate of changes taking place within the organization. In the end, resistance to change cannot be easily overcome by simply asking employees to behave differently or develop different habits. It requires the processes of the firm to be reconfigured to accommodate flexibility and agility.

Both inertia and resistance to change keep the debate open about what should be the right path and the pace at which smart transformation should be implemented in tourism firms – a question for which there is still no optimal answer that can work for all firms. The appointment in many organizations of a Chief Digital/Transformation Officer proves to be a lever of change that ensures that new digital and smart technologies are used in a way that is consistent with the culture and objectives of the organization, and that it is possible to achieve a balance between exploiting its potential and acceptance by the firm's employees.

17.6 The Role of Government

Governments have an important role when it comes to promoting the smart transformation of tourism firms, as they have the capacity to create the right conditions that promote the development of new business models and thriving ecosystems (OECD, 2020). The complexity that the regulatory and normative environment has reached around the tourism firm, together with its unique characteristics, can negatively affect the risk perception of the owners and managers and make it difficult for firms to adopt smart technologies. Tourism firms need a public policy framework that is consistent with the real problems of businesses, consumers, and markets, and that addresses the profound changes that are taking place in the ecosystem, providing them with legal certainty to invest in and reap the benefits of the Smart Revolution.

A well-structured public policy framework implies, according to the recommendations formulated by the OECD, combining short- and long-term initiatives that encourage the adoption of digital and smart technologies by tourism firms, reinforce the commitment of SMEs to transform, and encourage new management approaches oriented towards the development of a culture focused on digitalization and the use of data. Some of the policies that the public administration can implement are as follows:

- Lead the smart transformation policies in tourism. Governments can take the lead in establishing framework conditions that facilitate the development of new business models, value networks, and improved tourism ecosystems through smart technologies. These public policies might include:
  - Create a support framework for tourism SMEs through the modernization of the regulatory framework that fosters innovation and promotes competition.
  - Promote the creation of new entrepreneurial initiatives through technology project incubators and start-up accelerators, the facilitation of mentoring services for entrepreneurs, and the transfer of knowledge through technology demonstrators by universities and public R&D centres.
  - Promote a free market of technological services and knowledge by supporting intermediate agents that serve as a transmission belt for the objectives pursued by public entities.
- Encourage investment in smart technologies by tourism firms. Public policies could help boost investments by tourism SMEs in smart technologies that improve their productivity, efficiency, and capacity for innovation. The following could be put in place:
  - Improve the access of tourism SMEs to quality human resources and market intelligence sources that demonstrate the benefits and opportunities that the adoption of smart technologies entails, thus strengthening the individual capacities of SMEs to participate in new smart ecosystems.
  - Enable the access of tourism firms to digital and smart technologies, tools, and applications in a safe and affordable way, as well as help them with the initial investments and costs of updating and maintaining these technologies.
  - Gather the necessary knowledge that evidences the needs and the impacts that the smart transformation causes in tourism firms. This could help to incorporate into the design of public policies the incentives, aid, or subsidies that best adapt to the real needs of tourism firms.
- Promote innovation in business models and ecosystems through convergence, interoperability, and the adoption of smart technologies. These policies are an important complement to articulating a framework of public policies that effectively supports the smart transition of tourism firms. These actions may include:
  - Extend the access of tourism firms and visitors to high-speed broadband (including 5G) and to data processing infrastructures, in order to promote the creation and delivery of smart tourism experiences and the exchange of knowledge between the actors of the tourism ecosystem.

- o Promote the creation of public–private partnerships between traditional tourism firms, technology providers, and educational and R&D institutions that contribute to improving the exchange of knowledge and open innovation.
- o Encourage the meeting of actors dedicated to complementary activities within the tourism ecosystem that encompass technology, tourism activities, and other activities to create collaborative spaces in which both emerging and traditional tourism firms can prosper.

Discussion Questions

- Make a list of the management practices that can contribute the most to the smart transformation of the tourism firm. Then present the list to your colleagues and discuss to come up with a consensus list.
- How should owners and managers plan for the smart transformation process to be successful (what activities, milestones, responsibilities)? Who should be involved in designing the strategy and what roles should there be?
- What kind of difficulties do tourism firms usually find when it comes to accessing quality human resources? What differences exist between big and small firms?
- What new business models are an opportunity for the tourism firm in the smart era? Describe some examples and indicate the elements that differentiate them from conventional business models.
- What type of agents do tourism firms usually turn to in search of support to spur smart transformation? What collaboration frameworks are established most frequently?
- What intermediate agents do you know that play a relevant role in promoting the smart transformation of tourism firms? Name some examples. What kind of services do they provide that are valued by firms?
- What public policies or programs that support digital and/or smart transformation are you most familiar with? To what extent are these policies helping to drive smart transformation?

18 Organizational Agility

In the uncharted waters of smartization and the uncertainty of the tourism business environment, owners and managers struggle to determine the most appropriate strategies and how to implement them to sustain the firm's competitive edge in the long term. In light of this challenging prospect, what seems clear is that every tourism firm is going to need a new work structure to transform into a smart organization, i.e., become an organization capable of delivering products and services faster according to customer preferences and needs using (internal and external) intelligence. Such a restructuring will undoubtedly have an impact on the strategy, structures, and operational processes of the firm. Ultimately, a new organizational logic is going to be required, which reorganizes the resources and adapts the way of thinking and working of the firm to the world of digitalization and smartness.

In this chapter the concept of organizational agility is introduced as a main transformation management framework that can provide tourism firms the opportunity to embrace a more flexible operational style and respond more rapidly and proactively to the competitive risks and challenges posed by today's turbulent business and economic environment.

18.1 What Is Organizational Agility?

Agile organizations can intentionally increase their flexibility and speed for efficient delivery of customer-centric products in an ever-changing environment. This means that agile organizations are able to swiftly react to predictable and unpredictable changes in their environment and take advantage of them through: 1) a strategic orientation towards sensing and response; 2) a functional

DOI: 10.1079/9781800620261.0018

alignment of knowledge and resources; and 3) agile operational teams that enable members to work effectively and efficiently (Akkaya & Tabak, 2020; Cepeda & Arias-Pérez, 2019; Greineder et al., 2020; Unhelkar, 2017). In other words, agility is a key driver for organizations to gain competitive edge in turbulent and highly uncertain market conditions by pursuing the following four core capabilities:

1. A strategic orientation that revolves around the ability to sense and respond.
2. The ability to deliver value to the customer or the end user by incorporating their tastes and desires in the development of products and services (customer agility).
3. The ability to deploy agile processes and methods for the rapid development and delivery of customer-centered products and services, as well as its continuous improvement through innovation and learning.
4. The ability to bridge the gap between strategy and operations by integrating sensing and responding capabilities with the delivery of customer-centric products and services.

In summary, agility is about the tourism firm being able to comprehend the dynamics of customers' preferences and expectations and improve its performance by developing tailored products that customers value, thus gaining competitive advantage (Attar & Abdul-Kareem, 2020; Burchardt & Maisch, 2019). Agile has been used very often as an umbrella term that ambiguously brings together different methodologies for developing products that meet customer requirements, each with its own process, terminology, and scope. To avoid ambiguity, in this chapter we will use the term Agile as a framework of values and principles contained in the Agile Manifesto, the Agile set of practices, and the Agile methodologies. The Agile framework (Fig. 18.1) thus systematizes the practical application of agility in an organizational context and provides guidelines to successfully tackle smart transformation.

The Agile framework has gained popularity in recent times as a management approach that can help create competitive advantages for firms, as well

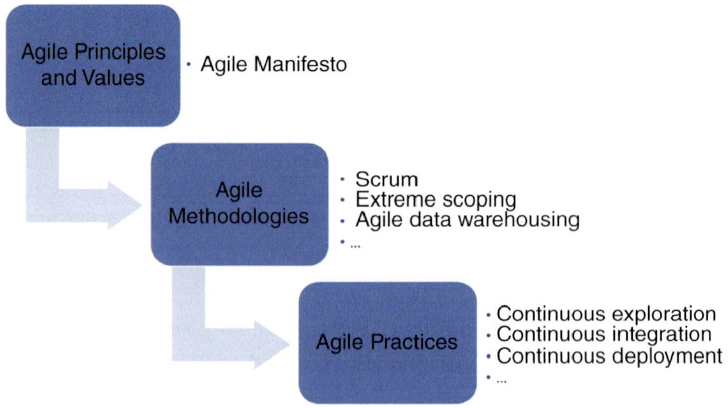

Fig. 18.1. Agile transformation framework. Source: own elaboration.

as drive improvements in customer satisfaction, and operational efficiency and effectiveness. Business firms are beginning to realize that organizational agility can be vital in delivering quality products and services to customers and to differentiating themselves in unique ways in an increasingly complex and uncertain hypercompetitive environment. In this vein, the impact of organizational agility on main business performance indicators, such as customer and employee satisfaction, market share, business success, or competitiveness has been documented by some surveys (Attar & Abdul-Kareem, 2020; Burchardt & Maisch, 2019). A study conducted by Capgemini Consulting (2017) on the agile organization shows that 83% of survey participants had plans for further transformation towards agile practices, with only 17% not seeing the need to move towards an agile environment in their organization. Moreover, 94% of the C-level executives interviewed by Accenture said that their current operating model puts the growth and performance of their organization at risk (Accenture, 2020), hence the need for organizations to move towards agility.

18.2 Strategic, Functional, and Operational Agility

As an owner or manager maybe you are aware that organizational agility started with software development companies looking for ways to create software development processes that would allow them to deliver solutions quickly and tailored more accurately to their customers' needs. In its origins, agility consisted of a manifesto of 12 statements or principles (Beck et al., 2001) that enunciated the characteristics of Agile projects and guided development teams through efficient software development. Over time, these principles have been adapted to a growing number of firms outside the software industry and refined to become an operational framework for organizational transformation around three fundamental dimensions: strategic, functional, and operational (Greineder et al., 2020).

Strategic agility

Strategic agility is the organization's ability to drive faster, more accurate, and less costly business processes, as well as enable the exploration and exploitation of innovation and opportunities. Hence, strategic agility encompasses the critical organizational capabilities needed to sense and respond to changes in the organization's environment that can affect business strategies.

Organizations sense insofar as they can monitor what is happening around their internal and external environment. Sensing involves knowing where to look, what to look at, and what to do with what you are looking at, thereby extracting what is relevant to support action-oriented decision-making. This sensing capacity continually generates knowledge flows that the organization needs to manage in some way. Developing these management skills implies that organizations have a knowledge acquisition and management system that allows them to interact with the actors in their business environment (i.e., customers, users, partners, suppliers, competitors) to acquire knowledge, and to exchange and exploit it. With the knowledge thus acquired, agile organizations are in a position to respond, which means that they can adopt a decision making process in which the knowledge acquired is first evaluated and then opportunities or threats to the organization are identified. Ultimately, agile organizations have the ability to bridge the gap between sensing and acting through people and teams who are empowered and have the autonomy to do so.

Functional agility

Functional agility seeks to align strategic agility with operations through two interrelated capabilities: 1) the alignment and dissemination of knowledge generated; and 2) the flexible and fluid allocation of resources to allow the joint optimization of the organization's strategic and operational tasks. Knowledge alignment is the result of sharing knowledge and understanding between people and business units. Organizations that align knowledge can leverage the tacit and explicit knowledge of employees to create new business opportunities as the environment changes. The flexible and fluid allocation of resources is the ability of the firm to adjust its processes and deploy resources on time, where and how they are needed in response to environmental changes. The greater the knowledge alignment, the more positive impact there will be on the organization's sense and response capabilities. Likewise, the better the flow of resources, the greater the impact on the speed of delivery of customer-centric products and services.

It is recommended that all stakeholders develop such capabilities to ensure that the transformation strategy is as consistent as possible. However, for these capabilities to develop properly across the organization, IT needs to be embedded into the organization's core business processes. IT is a critical enabler for business processes to gather timely and accurate information for proper decision-making and monitor the behavior of customers and users (Cepeda & Arias-Pérez, 2019; Rialti et al., 2018). It is simply not possible to separate an agile organization from strong integrated information systems that streamline and automate business processes. By emphasizing the development of information technology capabilities, the agile firm can strengthen its internal and external networks, reducing costs. These capabilities allow information and knowledge flows between stakeholders to be more dynamic and spread among a greater number of people, enabling the firm to offer innovative solutions more quickly as the context changes. In the end, IT is a great ally for the agile firm to learn, stay connected, and innovate.

Operational agility

Operational agility addresses the optimal use of operational and product delivery systems in accordance with changing environmental conditions. This usually involves reconfiguring the mix of resources of the firm, adapting business processes, and/or redesigning the organizational structure so that the customer can swiftly receive added value. Typically, the agile organization will need to set up a new work environment and implement new work systems that affect factors such as the structural characteristics of the organization, IT, customer interaction, knowledge, employee autonomy, etc.

One of the core transformational practices that agile organizations introduce is the decentralization of decision-making, which means that team members must be able to create their own insights and make their own decisions based on the best information available. Decentralization also means that organizations must provide the staff making decisions with the knowledge and skills necessary to understand the objectives pursued by the processes they lead (Unhelkar, 2017). This requires a clear determination on the part of firm leaders to work on continually improving workforce capabilities, especially when it comes to embedding data and intelligence into business processes. For example, the complexity inherent in managing Big Data requires firms to empower their people with new skills and analytical expertise, as well as tools and techniques to drive the decision-making process. Again, IT plays a central role in leveraging decentralized decision-making processes. Customer agility is another key agile capability, especially at the product development and delivery level. Customer agility relies on agile development methods that bypass plan-driven development processes and rapidly integrate product components as desired by the customer, allowing the organization to control, change, and improve production and delivery processes in real time. Figure 18.2 shows the dimensions and capabilities involved in organizational agility, as well as the effects they may have on the firm's adaptability, speed, and flexibility goals.

Tourism firms willing to move forward as agile organizations will need to strike a balance between each of the components of the organizational agility framework. Unfortunately, not much information is available on how firms can successfully transform into agile organizations and how agile practices can be mastered. Nor do scholars and professionals agree on the conditions under which it is good for firms to implement Agile practices. Therefore, it is strongly recommended that owners and managers first fully understand the smart transformation process globally before they determine what types of job tasks and processes can best be replaced by Agile practices to improve business performance.

18.3 Practices and Methodologies

The principles of the Agile Manifesto lay the foundation for Agile values; nonetheless they are only statements that do not provide owners and managers with any concrete plan or process to guide the smart transformation process. This issue becomes more apparent when organizations embrace agility without properly understanding what the concept is or how they can measure or regulate it. Therefore, it is key that organizations learn as much as they can about how and why they should become agile (Attar & Abdul-Kareem, 2020).

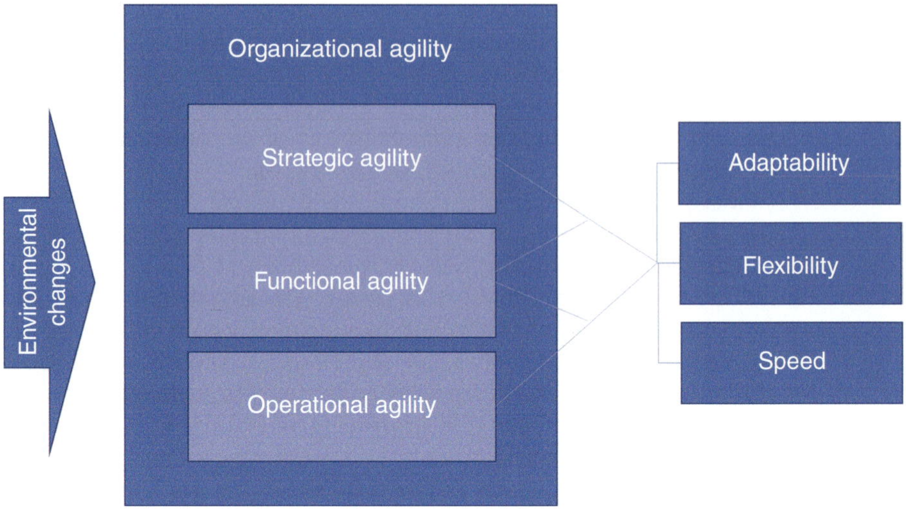

Fig. 18.2. Dimensions of organizational agility. Source: own elaboration based on Greineder *et al*. (2020).

Organizational agility entails proactive management practices that pursue continuous improvement; elimination of downtime and inefficiencies caused by organizational silos, overlaps, redundancies, and unnecessary routine tasks; efficient use of resources; flat organizational structures that avoid bureaucratic bottlenecks; and effective supply chain management. By consistently working on pursuing these objectives, the agile organization shall be better prepared to develop products and services adapted to demand and move swiftly to cope with changes in the environment. But what are the practices that the firm must implement to become agile?

In the absence of a commonly accepted framework that sets the ground for an Agile transition and moving beyond the 12 principles stated in the Agile Manifesto, Rialti et al. (2018) highlight a list of the 7 most relevant principles by which an organization can provide an agile response to a changing environment while exploring and exploiting opportunities.

1. Customer satisfaction.
2. Response to market changes.
3. Organizational flexibility.
4. Technical flexibility.
5. Dynamic business process management.
6. Effective strategic collaboration with partners.
7. Stakeholder accountability.

Organizational agility involves, according to Akkaya and Tabak (2020), the development of four fundamental skills by the organization:

- **Responsiveness:** It is the ability of organizations to grasp the changes that are taking place in the market (i.e., changes in customer tastes and preferences, new digital technologies, etc.) and provide effective and efficient responses quickly and targeted where they are needed. As far as tourism firms can respond effectively to changes in their environment they may be able to gain competitive edge.

- **Flexibility:** It is the ability to respond to changes by offering an optimal combination of internal and external resources available to the firm. To meet changing customer needs, agile organizations are constantly reorganizing their resources (e.g., employees, equipment, facilities, technologies), accommodating new resources and skills, and adapting their product and service delivery to a variable scope.
- **Speed:** It is about organizations making timely decisions when they are needed, and activating the organization's processes smoothly and accurately to deliver some expected results. In short, speed is about delivering products and services where, when, and how customers demand them. Speed and responsiveness are closely related, as it will be useless for firms to respond to changes in the environment when it is not necessary or after the window of opportunity has passed.
- **Competence:** It is the ability to decide when and how to use the other three organizational agility skills (responsiveness, flexibility, and speed), either individually or in combination. Competence is also a process that requires the organization to be aware of all the internal and external competences within its reach, updating them over time, as well as identifying those that are missing and that it would be good to have.

Furthermore, according to Joiner (2019), operational agility is achieved when the organization establishes these three structural and cultural capabilities:

1. Agile management methods in each of the functions of the organization so that work is carried out in short and fast iterative cycles to which feedback from stakeholders is added.
2. Fast and flexible IT systems that improve the organization's speed and responsiveness to changes in the environment.
3. Tear down silos, foster cross-unit/departmental collaboration, and strengthen relationships with external stakeholders.

Agile methodologies

The values and principles of the Agile Manifesto and some successful Agile practices have been systematically packaged in the so-called Agile methodologies. Specific methodologies used for organizational agility include Scrum (for incremental team product development processes); Kanban (for managing workflows that improve organizational bottlenecks and balance demands with available capacity); Extreme Programming (for teams to improve delivery quality and responsiveness to new customer requirements); Test-Driven Development (to convert product requirements into test cases and embed continuous testing in work cycles); Feature-Driven Development (to break down customer requirements into tasks for on-time delivery of products); and

Lean Manufacturing (to reduce production costs by eliminating wasteful processes, getting the right things done in the right place, at the right time, in the right quantity) (Leybourn, 2013). Some of these methods, such as Scrum and Extreme Programming, have become very popular across industries and are considered standard Agile methodologies. They typically include practices such as working in short time cycles (typically a few weeks), continuous product delivery through incremental improvements, and small teams of experts physically working together.

Among the Agile methodologies best known to data-intensive firms are Scrum, Extreme Scoping, and Agile Data Warehousing (Larson & Chang, 2016). Extreme Scoping specifically focuses on data integration within the organization's data and analytics projects. Data integration encompasses data-centric management activities whose goal is the acquisition and subsequent incorporation of data into data warehouses. Some related practices include acquiring and understanding data sources, cleaning data, organizing and modeling data, and preparing data for loading into a data warehouse.

Agile data warehousing is a collection of methodologies and philosophies that systematically encompass end-to-end activities for agile data warehouse delivery. These methodologies stem from the need to avoid the pitfalls that often lead to failed data warehousing projects that cost millions of dollars and take years to implement. These methodologies share a vision of the architecture, the quality of the data, the type of modeling and prototyping of the data throughout a project, the organization of the work by requirements, the documentation, and the participation of the stakeholders.

Scrum is the most popular Agile methodology in software development and business intelligence projects. Some key Scrum concepts that are used in data-driven projects include user stories (informal explanations of the features that a product must have from the perspective of a customer or end user); sprints (small projects that are executed in very short cycles to achieve product improvements that deliver value to the customer or user in a very short time); backlogs (lists of product requirements and/or sprint specifications to be developed from user stories); and Daily Scrum (daily meeting of about 15 minutes to synchronize team tasks and review progress). With Scrum the requirements of a data-driven project are broken down into smaller stories that are then developed, tested, and delivered to the customer or end user with input from stakeholders.

In addition to these market methodologies, some authors have suggested mixed frameworks based on Agile principles, such as in Unhelkar (2017) and the Composite Agile Method and Strategy (CAMS). This approach is aimed at firms that do not operate in the software industry but are interested in improving their capabilities to make fast and effective decisions using Big Data and analytics as support tools. CAMS seeks to balance agile flexibility with planning formality, i.e., an IT governance framework that is complemented by the flexible and collaborative principles of Agile. The same philosophy can be applied to all functions of the organization such as sales, human resources, project management, innovation, etc.

18.4 Knowledge, Teams, and Leadership

Organizational agility is closely associated with the organizational structure of the firm. The organizational structure includes the system of functional and hierarchical relationships contained in the organization chart of the firm, the resources, and the work and leadership styles necessary for human capital and talent to work in the desired direction. The organizational structure of an agile firm is characterized by its flexibility, operational decentralization, and a flatter hierarchy that fosters autonomy in decision-making, interdepartmental communication, and cooperation between teams and individuals (Attar & Abdul-Kareem, 2020). Agile organizations take a more open approach and build stronger connections with key internal and external stakeholders (i.e., they create work teams and alliances). In this way, the agile organization can continuously detect changes in the environment and provide responses. Not surprisingly, interpersonal and communication skills, flexible leadership, entrepreneurial mindset, and self-motivation and learning prevail in the agile organization.

Agile organizations create work teams that are able to swiftly provide solutions to the changes sensed in their environment. Agile teams are self-organizing groups of people who come together because of their shared interest in a project or because they think that a specific team is where they can contribute the most. In organizations with established agile practices, employees freely decide which teams they want to be a part of and how they can contribute most effectively, which in turn can result in higher employee satisfaction as they are more involved in decision-making and build mutual trust and respect (Rigby et al., 2016). Team creation/termination is flexible and depends on the environmental conditions around the firm. The same employee can be a leader in one project, a facilitator in another, and an expert technician in another (Setili, 2015). Agile team members are diverse in terms of age, gender, or training to better respond to the complexity of the environment in which they operate. Furthermore, recent research has found that the smaller and more informal the work environment, the higher the performance.

Agile teams execute projects with a limited scope in short iterations and test them before embarking on large transformation projects that involve higher costs and risks. Therefore, they are often engaged in short-term, fast decision-making processes (Berntsson Svensson et al., 2019). This is especially important when firms struggle to cope with the unpredictable changes in the business environment and when developing innovation. Since firms are very unlikely to anticipate what customers will demand or whether customers will like a new product or service that they have never seen or experienced before, the best thing to do is to experiment again and again. Testing new things quickly to find out with customers and users how they are going to use the innovations provided and whether the products and services are going to be valuable to them is key to the agile organization.

The way agile organizations manage work teams does not mean that they do not have well-defined leadership. What happens is that instead of associating leadership with certain people by their hierarchical position in the organization, leadership is a role that anyone can assume depending on the tasks involved. This emphasizes the autonomous and self-organized nature of the agile organizational structure where collaboration and cooperation are part of the firm's internal work processes (Burchardt & Maisch, 2019). Indeed, leadership style is a critical factor when navigating the transition to the smart organization. Transformational leaders create a vision that the organization must move toward, share that vision with people, encourage them to learn and innovate collaboratively, and make decisions that transform the organization in an ever-changing environment (Akkaya & Tabak, 2020). Therefore, a major challenge for any agile organization is to adapt the leadership style of owners and managers to the changing business environment, so that they can steer the organization in the most effective strategic direction at any time. In the end, agile leadership is about being able to deal with the changing and uncertain nature of today's business environment and foster flexible strategies that take broad perspectives into account (Attar & Abdul-Kareem, 2020).

A final key variable in the transition to organizational agility is knowledge management. Firms must equip themselves with the capacity to manage their knowledge and create the learning mechanisms necessary to exploit it. Without these capabilities, further implementation of knowledge management practices will not make sense. Knowledge management also has a great influence on leadership style, as business leaders increasingly need more and better knowledge to assess and understand their environment. Knowledge is also key for leaders to assess the competencies that firms lack and that are required to implement product delivery strategies that meet customer needs.

However, agile organizations not only make a living from efficient knowledge management and learning capabilities, but also develop decision-making processes that are flexible and seamlessly connected to the firm's knowledge base. In the end, becoming an agile organization is not easy, and neither is staying agile. Instead, it is a never-ending journey for which people must be prepared to proactively deal with the ambiguity and uncertainty of the environment around them. Engaging stakeholders in a joint effort and having a well-designed plan that can be implemented are key to successfully accomplish this process.

18.5 Customer Agility

An essential component of organizational agility is the ability to sense and respond to changing customer needs to improve business performance. This is known as customer agility and requires that sensing provides enough accurate and timely information to the organization so that it can respond effectively to changing customer demands. Ultimately, customer agility is about achieving a higher level of customer satisfaction, thus laying the foundations for customers to buy a tourism product or service again (Fig. 18.3). Given that the factors that determine customer satisfaction and loyalty in tourism are highly variable and depend on the context, tourism firms must be equipped with the ability to constantly detect customer preferences and respond to them quickly and

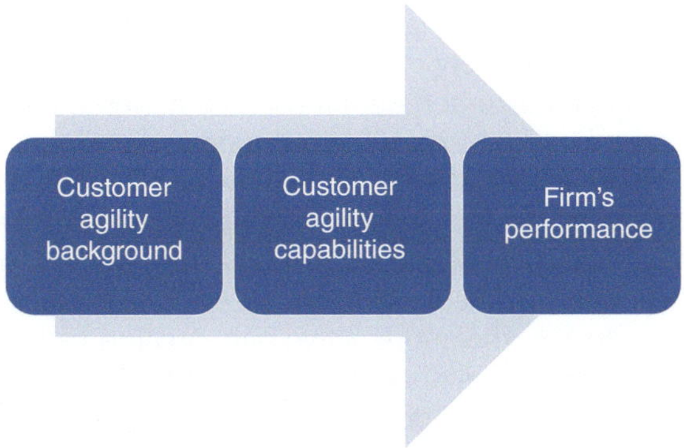

Fig. 18.3. Customer agility framework. Source: own elaboration based on Ngo & Vu (2020).

flexibly as they change. That said, there are a number of steps tourism firms can take to develop customer agility, the first of which is to align business strategy with customer strategy (Ngo & Vu, 2020). Firms must encourage employees to share the common goal of detecting changes in customer demands and responding effectively to them. This will typically require the commitment of the firm's top management, as well as the involvement of the human resources department. With the active engagement of both, the firm will be able to develop a customer-centric culture, build cross-functional customer service teams, and equip employees with the skills to spot customer changes and know what to do to respond to them.

Another key step for customer agility, very often overlooked by business owners and managers, is knowledge management. Generating knowledge and making the firm learn from it (i.e., developing an organizational culture and processes focused on continuous learning of all members of the organization), is a proper way to ensure that the smart firm knows how to proactively reconfigure its value offering. Knowledge is also a lever to unfold the firm's innovation capabilities, which are essential for building an organization that learns and gains competitive edge. More specifically, deep and accurate knowledge about the customer and the market is key for the firm to identify their needs and be able to respond to changes in their preferences. However, building customer insight is not a painless exercise. It simply cannot be reduced to creating a database to which only a few have access due to their hierarchical position in the firm and making some eye-catching reports. The dissemination of knowledge at all levels of the organization is a critical condition for achieving an organizational culture focused on learning. It is also the mechanism that allows knowledge to truly become a game changer that the customer can benefit from and ultimately improve business performance.

Finally, customer agility is about equipping the organization with enough technological capabilities to collect, store, organize, and analyze the vast amount of information that firms need about the customer. Tourism firms have been among the pioneers in generating customer information in large quantities and through multiple channels, both online and offline. Smart technologies play a crucial role in tourism firms when it comes to analyzing and exploiting that information and leveraging customer insights to inform the firm's decision-making process. Ultimately, technologies are key enablers for the firm to interpret changing customer demands and disseminate that knowledge to users in the organization responsible for making decisions. From the customer's perspective, technologies allow them to play a more proactive role in the design and configuration of the firm's value offering, either because they participate in product co-creation communities, or because they use any of the two-way communication channels available to them to create feedback loops between the firm and the customer. This more active participation of users and customers in value co-creation processes not only allows the tourism firm to be more responsive to changes in customers' demands, but also creates emotional and cognitive bonds with them that can have a positive impact on their evaluation of the quality of the products and services delivered by the firm.

When it comes to achieving greater customer agility, it should be noted that investments in technology are not as important as other factors. Owners and managers must keep in mind that technology does not play the central role in smart transformation, but rather technologies must be focused on providing timely support to gather information from customers, disseminate it, and manage the knowledge so generated across all levels of the organization to inform good decisions. In fact, tourism firms do not need large investments in state-of-the-art or highly sophisticated technologies to start providing customer agility. The three key customer data sourcing technologies (email, website, and database management technologies) can suffice if used effectively.

In the specific case of tourism SMEs, since their business model normally depends more on proximity to customers and on building personal relationships with them, they may be especially apt to detect changes in customer preferences. However, SMEs often lack the skills and resources to implement customer agility in the organization. Therefore, owners and managers must be aware of the limitations of their organization and implement strategies that allow them to continually make small organizational changes, one after another, to improve core business processes. Affordable and scalable new technologies, such as social media applications for customer service, can be of great help in this process.

Discussion Questions

- How are tourism firms doing in relation to organizational agility practices? What kind of practices and/or methodologies are firms implementing to achieve organizational agility?
- What barriers or impediments prevent the development of Agile practices in tourism firms? What can be done to overcome them?
- Are the owners and managers of tourism firms prepared to lead the transformation of their organizations following Agile principles?
- How should the transition from a conventional organizational model to another based on Agile principles be approached from the perspective

of business management? And from the operational and technological perspectives?

- What potential downsides and risks can the implementation of organizational agility entail for the tourism firm? How can these drawbacks be minimized?
- What are the technological implications of organizational agility for the tourism firm? How can they be anticipated and managed?
- How does the decentralization of decision-making affect tourism firms? In what cases is it positive (or negative) for the organization?

19 Data-Driven Agility

The Agile philosophy was originally introduced in 2001 in the software development industry (Beck et al., 2001). Since then, Agile principles, practices, and methods have continued to evolve as business needs for greater agility and responsiveness to changing customer demands have grown exponentially across industries. Firms are poised today to find new ways to develop self-learning capabilities and inform decision-making processes based on more comprehensive and accurate insights from customers. Not surprisingly, the application of Agile principles to the organizational context has developed strongly in recent years, in line with the unstoppable transition towards the smart firm. The combination of data processes with Agile methods is a great opportunity to make value creation in the smart firm more efficient and profitable and have a greater impact on performance.

Big Data is dramatically impacting business firms' use of information while increasing the challenges facing owners and managers. Recent advances in data science and data analytics techniques are further driving progress towards smarter organizations. As the very essence of the tourism business changes, so do the information and knowledge needs of tourism firms. This, in turn, is pushing changes in the way organizations apply Agile principles and practices (Larson & Chang, 2016). The challenges that smart firms must address are similar in nature to the founding principles of Agile. As such, Agile provides a framework to support smart business transformation by enabling a more dynamic, less formal, and more customer-centric way of working.

The Agile framework is well coordinated and aligned with the various stages of the data life cycle, which means it can be a suitable approach for working with Big Data as well as developing data analytics projects. In those cases, Agile practices provide the opportunity to spend little or no time establishing requirements up front and documenting them in detail before work

DOI: 10.1079/9781800620261.0019

begins. Instead, they emphasize the need to develop small projects in very short iterations to deliver rapid incremental improvements. This is consistent with the way prescriptive and predictive data analytics projects work, as they seek iterative discovery and continuous validation of behavioral patterns from data (Ambler & Lines, 2016). Noting these similarities between data-centric projects and Agile methods, the smart firm's use of Agile requires a number of adjustments to highlight the primacy of information and knowledge management over software development.

19.1 Agile in Big Data

The rise of Big Data in tourism and the new challenges it poses for the management of tourism firms explains why many organizations are inclined to implement Agile methods. The large influx of data generated by Big Data and the Internet of Things (IoT) requires new practices that allow the firm to manage data on distributed platforms (e.g., data warehouses, cloud, social media), keep applications up to date, and incorporate continuous testing solutions. This is generally an iterative process that, when in place, may allow the firm to proactively respond to changes in the business and customer environment.

Big Data and agility go hand in hand from an operational and strategic point of view. Big Data facilitates the firm's transition to agility by providing the insights and resources needed to make fast, timely decisions to deliver better customer value, while agility provides Big Data with the management practices and processes needed to extract value from the data more easily and consistently. In this way, owners and managers can leverage Big Data to identify common customer behavior patterns, personalize their product and service offerings, and set prices based on customer preferences and market trends. Ultimately, together Big Data and agility drive customer-centric management, while optimizing the use of organizational resources (Rialti et al., 2018).

The combination of organizational agility and Big Data further encourages integrated but decentralized decision-making, but only after the organization has become a self-learning organization and continuous interactions have been established between employees, customers, and the firm's stakeholders. For example, creating cross-functional teams that work in rapid cycles allows the firm to make incremental decisions, which is a big step forward towards organizational agility (Unhelkar, 2017). Self-learning organizations strengthen, update, and optimize Agile values and practices.

Notwithstanding its benefits for improving business performance and creating value, why haven't firms rushed then to implement Agile frameworks? No doubt business owners and managers have their own reasons. There are several obstacles that hinder Agile development in almost all functional areas of the business, which can be grouped into two categories: 1) lack of leadership; and 2) inadequate organizational culture (Joiner, 2019). Professionals working on the implementation of Agile practices also know very well that the lack of knowledge and experience in relation to Agile principles and methods explains why many organizations do not trust the Agile way of working and do not apply Agile practices to their data and transformation projects.

19.2 Agile Analytics

Big Data analytics initiatives are very different from any software development project on which Agile principles were originally built. For example, Agile project management approaches often fail to take into account many of the challenges of working with large amounts of data from which to extract actionable insights that inform decision-making. As a result, the methodologies used in software features development cannot be applied "as is" in Big Data analytics projects. The other side of the coin is that many teams working on data science and analytics application projects fail right when the projects are deployed to production. Very often, teams don't work collaboratively on short, continuous development and testing cycles, and when they deliver development products, they must wait to test them and get user feedback, thus making it difficult to manage resources and meet established time targets (Ullah, 2019).

Agile analytics is the application (or adaptation) of Agile practices and methods to data analytics projects with the aim of providing faster time to value, greater flexibility, and responsiveness. In practice, Agile analytics is a set of practices and methods that guide the organization in the construction of data warehouses and data analytics applications that respond to the unique and changing needs of the firm. Moreover, Agile analytics fosters collaborative relationships between data analysts, IT developers, and business users to ensure effective technical delivery by the execution team (Collier, 2012; Nagle et al., 2013).

Agile analytics focuses on developing iterative data discovery cycles that aim to extract value for the business faster by analyzing large amounts of data from multiple perspectives and dimensions, which is a more appropriate approach to the real circumstances in which tourism firms operate. The data to be analyzed can be structured and unstructured, as priority is given to the extraction of value and not to the process itself. Once the results are obtained, Agile analytics focuses on revealing new meanings and interpretations from data and finding new answers by avoiding bias. This implies important changes in the organizational and behavioral approach of the firm, in addition to the development of a creative mentality (Nagle et al., 2013). The Agile smart firm does not work according to a detailed work plan and a strict long-term schedule with detailed documentation, nor does it invest millions in data analytics projects. Instead, the firm establishes use cases that can deliver realistic short-term results for the business and can be specified as requirements for development. The tasks leading to the development of use cases are then assigned to cross-functional teams specialized in data analytics, IT, and other related functions. If some tasks require the intervention of experts from other teams, they can temporarily join the team. In this way, Agile analytics offers the ability to accommodate changes as they occur and rapidly scale projects as new requirements arise or impacts become apparent.

19.3 Agile Practices for Data-Driven Projects

Firms willing to move towards data-driven management need to improve their agility, which means having a good understanding of what an Agile framework is about and choosing which (Agile) practices and methodologies are best suited to implement in their data projects. Indeed, one major stumbling block for the adoption of an Agile framework in data-driven projects is the sometimes ill-informed choice of suitable tools and techniques that are valuable to teams. So, what Agile practices can be applied to data-driven projects to effectively improve the organization's business performance?

The fact is that this question is not easy to answer. There are many and varied Agile practices that can be applied in a wide variety of situations and contexts by teams seeking to extract value from data. Therefore, before deciding to apply Agile practices, it is always a good idea to first understand how Agile and data projects can fit together (Larson & Chang, 2016; Ullah, 2019). A thorough discussion on the Agile principles that are best suited to data projects and how they differ from traditional methods can be a good starting point for owners and managers. Some of the most important ones are presented below.

Incremental and iterative development

The Agile project development philosophy is embodied in iterative, short, incremental development cycles that are executed in an evolutionary style. Agile development cycles replace the traditional waterfall style by prioritizing the development of those requirements that the user values and continuously adapt development tasks as the customer or user provides feedback to the team. In each iteration, the team produces incremental improvements to the final product that are quickly made available to the customer/user for feedback. These iterations generally last from 2 to 4 weeks, although the final time depends on the nature of the tasks involved. At the end of each iteration an evaluation meeting is held to assess the new functionalities delivered by the team according to the requirements and the deviations from the work agreed upon in the kickoff meeting. Each new feature is only accepted after the customer or user has given their final approval.

Collaboration and self-management

Big Data and analytics projects are usually very complex. They require people with proficiency in multiple areas, from software and IT development to data science, and customer and industry expertise. Without the collaboration of people with different backgrounds it is almost impossible for data projects to be successful. Agile data projects enable mechanisms that promote collaboration among team members, as well as between team members and the customer/end user. Agile teams, unlike traditional project management teams, are independent and self-organizing to manage their tasks and make their own decisions. They decide the sequence in which the development tasks are carried out in each iteration and are responsible for the deliveries to the customer/user according to the committed requirements. Furthermore, Agile team members are discouraged from performing routine or repetitive tasks manually and, if there are any, attempts are made to replace them with automated tasks so as not to waste valuable time that should be spent developing quality features.

Interactions between individuals

Teams that work with formal and highly structured processes but have little experience are not as effective and fast as teams of experienced people who work together but without the burden of such formal processes. In the world of Big Data and analytics, it is essential to work quickly to create value, moving from data to knowledge and from knowledge to the interpretation of results in an agile way. The heaviest analytics processes (e.g., processes with transactional data) that used to take days to complete are now performed in just a few minutes using new Big Data technologies. Additionally, new professional profiles have emerged, such as the Chief Data Officer and data scientists, who are increasingly in charge of data processes. They put Agile principles into practice by applying machine learning algorithms for iterative pattern discovery and seek ongoing interaction with stakeholders for interpretation and guidance.

Early and continuous testing

For Agile teams to continuously deliver incremental data products to customers/users, they need to set up testing mechanisms in advance and test very frequently. Having automated test systems (e.g., analytics-driven testing) makes it easy for teams to test early in the development cycle and ensures that the data products created are reliable and accurate to customer specifications. On other occasions, Agile teams may use test-driven development whereby developers write a test after they have developed enough code to fail. After the first tests are completed, the code is updated, and new tests are rerun as needed until approved. If the code fails, the developers update the working code and test it again. Once the tests pass, the next step is to start a new development cycle to implement new features.

Speed of data analysis

As Big Data and data analytics projects are running at an ever-increasing speed, producing comprehensive documentation in data projects does not seem the best practice. A basic principle of Agile teams is that documentation is not essential, but communication is. As long as Agile teams meet the requirements of product development and these have been properly tested, not much literature is needed to explain things. Moreover, complete documentation not only does not help ensure the success of the project but may even contribute to its failure.

Documentation in Agile data projects is a secondary priority, justified only when the benefits to be gained from it outweigh the costs of preparing and keeping it up to date. Instead, the priority is to streamline data analytics processes as much as possible, ignoring documentation when it does not contribute to faster analysis. When for any reason firms decide to prepare documentation (i.e., when requested by stakeholders, to formalize contracts, to support communication with external stakeholders, for audit purposes), the content is only that which is strictly necessary, not vague or abstract, but concise and not detailed. Very often, Agile teams use the Agile Modeling method to identify what kind of documentation needs to be prepared in data projects, for example, contract templates, executive overviews, system documentation, user documentation, etc.

Response to change

In data project management, the traditional approach generally involves signing a contract between customer and provider with terms that tend to be as complete and detailed as possible. The scope of the contract includes the rights and obligations of the parties, the objective being that it is always strictly fulfilled, deviating as little as possible. Any change in the conditions established in the contract is considered negative and the parties must try to do everything possible to avoid it since it affects the time, resources, and cost of the project.

In Big Data and analytics projects, change is the rule, not the exception, as input data is constantly changing and is not structured or clean enough to be processed. Therefore, being prepared to respond to the changing conditions is a must in data projects. Unlike the traditional approach, the Agile approach attempts to eliminate the bureaucracy and resources required to draft a comprehensive contract between the parties. Instead, the parties must be prepared for changes and respond accordingly.

Collaboration and communication

Big Data and analytics projects consist of iterative discovery processes often carried out by data scientists who process the data and deliver results that must be verified and validated by stakeholders. This generally requires the use of multiple data sources and presentations in different formats using various tools to display graphs and figures. Timely and continuous collaboration between technicians and users/customers is thus essential to achieve the objectives pursued by data analytics projects.

However, while contracts may address expectations of collaboration between stakeholders at the beginning of a data project, it is difficult to gather all the necessary requirements, as they are often unknown, and it is difficult to anticipate all needs in advance. The way to solve this is to increase communication between stakeholders to continually approximate expectations and achieve a satisfactory delivery. The need to communicate effectively becomes a critical factor in developing an Agile data project. Agile communication encompasses all that occurs between team members, with customers and end users, as well as with operations staff and the firm's management.

Agile team members can choose from several modes of communication when working together, but the most effective communication is person-to-person and face-to-face, especially if shared modeling means are used (e.g., whiteboard, flip chart, paper, cards). It is crucial that communication between members of Agile teams is continuous and kept as simple as possible, and that members trust each other. If there is no mutual trust between team members, the information received and the people who provide it will not be trusted, and therefore the fundamental objective of communication will be lost.

Continuous Delivery

Continuous Delivery is a set of activities and workflow automation process used to develop a new functionality or feature, from ideation to delivery of value to the customer or end user. When firms use a Continuous Delivery approach, development-to-production time can be dramatically reduced, and data Agile teams can focus more on delivering quality data products. The process of Continuous Delivery consists of four stages: Continuous Exploration (CE); Continuous Integration (CI); Continuous Deployment (CD); and Release on Demand (this one intended for commercial software development firms).

CE lays the groundwork for what needs to be developed. Design thinking is generally used by teams to understand a market problem (or a customer need) and find a solution. CE takes the development requirements of the program's backlog and implements them. After implementing a new functionality or feature, the work is tested from start to finish before is validated in a staging environment. CD moves developments from the staging environment to a production environment, where they are verified and monitored to ensure that they work correctly. After testing in the production environment, the development team determines the right time to release them to customers/users. It must be noted that although these stages are described sequentially, a real-life project is not strictly linear. Rather, it is a learning cycle where team members work on all aspects in parallel, build a solution, test it, and iteratively learn from the feedback received.

Continuous review and improvement

Continuous improvement is a core Agile practice that makes teams more effective when they spend time regularly reviewing what works and what doesn't in a data project. Once the Agile team receives customer or user feedback, it takes immediate action to implement that feedback. The team uses the feedback received to learn and improve. Improvements can be introduced throughout the development process (i.e., changes in the duration of the iteration cycle), or can be made to the delivered product (i.e., changing data privacy requirements, data sources). Sprint retrospectives are a way to ensure the team takes time to review what they did well and where they need to improve. This encourages incremental improvement over time, thus becoming part of the DNA of the Agile team.

19.4 Agile Data-Driven Framework

As the number and scale of Agile data projects grows, there is an increasing need for owners and managers to understand the framework within which they will develop Agile projects. Agile frameworks seek to go beyond the usual linear approach based on a rigid, pre-established plan and timeline for data product development (e.g., waterfall development). Instead, they apply a less formal, adaptive approach focused on delivering a minimum viable product as quickly as possible that the organization can improve over successive iterative development cycles. This approach based on continuous iterations will keep teams working on small incremental improvements to the data product at regular time intervals, staying focused on responding to customer and user needs that arise throughout the project. Therefore, rather than waiting to have all the requirements developed and implemented before delivering the data product to the customer/user, Agile organizations seek to engage stakeholders early on so they can review the results and modify the requirements to meet their business goals.

Agile data-driven frameworks are customer-centric. Their purpose is to provide business owners and managers with the capabilities to identify common patterns of customer behavior and the insights needed to customize the organization's product and service offerings and pricing according to customer preferences. Besides a strong customer orientation, Agile frameworks pursue operational optimization, better risk management, and superior workforce utilization through Agile practices that can improve the performance of multidisciplinary teams working with data (Rialti et al., 2018). Figure 19.1 illustrates the workflow of an Agile data-driven framework that tourism firms can work on. This is not a rigid framework that organizations implement blindly, but rather a high-level approach that tourism firms can adapt to their own context and particular needs.

The workflow begins by stating the business objectives that the development of each data product pursues. It involves collecting functional and non-functional requirements from users and stakeholders and then translating them into work items that are noted down in the Product Backlog. Note that not all the requirements need to be collected at this stage. The aim of the Product Backlog is to manage all project work and it is filled in by the Product Owner (PO), who is also responsible for discussing each work item with the team to ensure they have the necessary capabilities to take on all required development tasks. Items in the Product Backlog that are selected for a development cycle are moved to the Iteration Backlog. This means that each work item goes

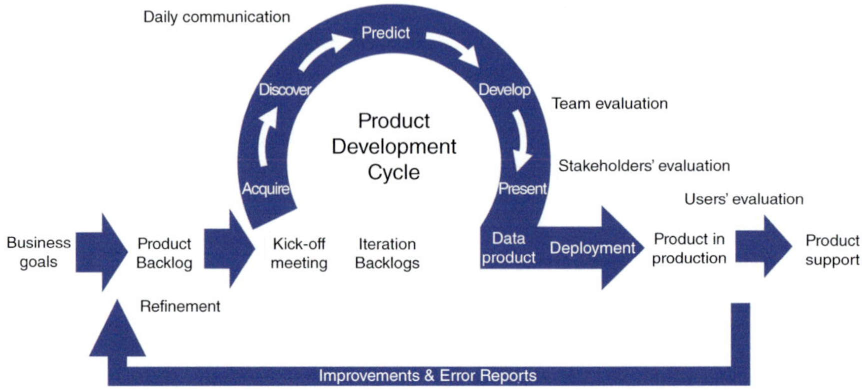

Fig. 19.1. Agile data-driven framework. Source: own elaboration based on Donker-Rostamy (2019).

through a data product development cycle comprising of acquisition, discovery, prediction, development, and presentation tasks. Once the work item meets all the conditions and acceptance criteria established in the Definition of Done (DoD) and is accepted by the PO, the product is considered ready for delivery to the user. At the end of each iteration, the work items are first moved to a stage environment to be presented to stakeholders and receive their feedback at evaluation meetings. Once the stakeholders are satisfied with the product, the operations team releases the product into the production environment. From this moment on the support team will provide reports on possible improvements and/or errors detected in the product to the development team so that they can make improvements in the next iterations.

Throughout the Agile work process, the status of the different project tasks is visible to everyone who is working on the product so that they are aware of the ins and outs of the tasks performed. As for the product users and stakeholders, they participate during the product development cycle and are aware of the progress of the work, offering feedback at the end of each iteration so that the development team delivers a high-quality and useful product. When the product does not deliver the result expected by the stakeholders, the product and/or the tasks are adapted as soon as possible to minimize deviations. Finally, it should not be forgotten that those firms that are willing to lead Agile data-driven projects must provide training, workshops, and meetings that make their teams improve their expertise in data-driven product development and accelerate the adoption of Agile practices in their daily work.

19.5 Challenges of Data-Driven Agility

When it comes to Agile data-driven project challenges, people skills and abilities rank high at the top of the list. The demand for data-savvy professionals has exploded in recent years in virtually every industry and especially in the data-intensive tourism sector. Today, data scientists are considered one of the most promising careers of the 21st century by the world's leading staffing companies, and universities are rushing to create training programs to educate qualified data specialists. Investing wisely in skills and technologies is one of the most pressing challenges for all organizations that want to adopt Agile practices successfully in their data projects.

The need for increasingly skilled people in data science and related technologies encompasses not only technical personnel, but also business leaders who need to be more knowledgeable in Big Data and analytics. This challenge is closely related to a major barrier noted by data project practitioners, which is a lack of sufficient commitment from leaders to undertake Agile projects. According to surveys, the lack of corporate sponsorship in the development of Agile analytics projects almost doubles in importance (35%) the construction of business cases (18%). This suggests that while the justification of data projects from a business viewpoint is already a major challenge, the actual implementation of a data project is even more so (Nagle et al., 2013).

There are challenges to data-driven agility at every stage of the data value life cycle. One of the most critical is the confidence that organizations have in the accuracy of their data. There is still a long way to go in this regard, as there are still many organizations that do not have much confidence in the accuracy of their data, nor do they have established processes that ensure compliance with data definitions and data standards between operating systems and technologies. Moreover, very few firms have their data in a single data repository or have implemented a master data management (MDM) system (or "master data warehouse") to create and maintain accurate and consistent master data sets, which is the way data-centric organizations have a single source of truth to improve business processes. This highlights the need for organizations to integrate the data they have and build a single truth of data across all business functions; even more so given the storm of new data they find daily.

Significant challenges also arise in the firm's ability to perform data analytics and report the results. Most firms are unable to find patterns and dependencies in multivariate analysis or perform predictive analysis on product or process performance. This amounts to saying that most firms rarely or never perform what can be considered standard data analysis. The same applies to the average time it takes an organization to develop a new or ad-hoc report/dashboard. Some surveys report that it would take most firms several days and that only a few could do it in less than 1 hour, which reflects the severe restrictions incurred by firms that work with data. Surveys also report that firms have significant difficulties when it comes to integrating new data sources, and that only a few can mine a new data set in less than an hour. All the above highlights the low maturity of data agility in organizations and the size of the challenge ahead until firms can unlock the full potential that data offers.

Data governance is one of the three main differences that distinguish firms that are capable of capturing value from data and those that are not. However, the number of firms that do not have data governance plans in place is still very high. Some surveys report that nearly 30% of firms' total time is spent on non-value-added tasks due to poor quality and availability (Petzold et al., 2020). These are reasons to believe that most firms do not have a data strategy in place and, therefore, are not treating their data as a corporate asset. This situation is especially alarming considering that data quality is one of the most common reasons for the failure of data initiatives. Organizations do not seem to be investing enough in the capabilities needed to achieve data quality maturity, and their behavior to maintain data quality are somewhat poor. Furthermore, firms should calculate the cost of working with bad or inaccurate data, which they rarely do. Consequently, organizations must rethink the way they manage their data quality policies and understand the implications of data governance policies.

Another key challenge of Agile data-driven projects is technology. Although less relevant than the above challenges, data technologies are generally not seen by owners and managers as a significant limiting factor for Agile data products. Only a relatively small number of firms consider the flexibility and speed of current database software (26% and 10%, respectively), and the rigidity of data warehouses and OLAP data models (12%) as a major barrier (Nagle et al., 2013). Agile data projects are greatly enhanced when the organization already has a pre-existing data warehouse that has been fed with data, otherwise building a data warehouse becomes a real challenge for the average organization. Agile is a value-based approach where customer/user knowledge and intelligence needs drive the development of data warehouse components and not the other way around. This prevents the firm's data warehouse from being "overbuilt" beyond its intended purposes (Collier, 2012). Owners and managers must note that although the data warehouse can help simplify data product development efforts, its implementation is not a prerequisite to perform data analytics and generate business intelligence for the organization.

Another technology-related challenge is the pervasive reliance organizations continue to have on spreadsheets for data management. The number of firms that still use spreadsheets as the only (or practically the only) tool for information processing and data analysis is still very high, even though spreadsheets are often riddled with errors and there are newer, less error-prone technologies available that are affordable. There really is a significant gap between the claims about big technology trends made by the industry's leading solution providers and the reality of how comfortable organizations are with them. This situation is further aggravated if we consider tourism SMEs with limited resources and lack of technical skills to tackle technology-based data projects. The responsibility to reverse this situation lies with each firm, but it is also necessary for technology providers to be aware of the challenges faced by tourism SMEs. Perhaps the technology industry should focus more on data tools and applications that are easy to use and for which users do not need highly specialized technical skills and knowledge. This could help firms reduce their dependence on spreadsheets and raise their level of technological maturity to meet the challenges of smartization. Recently, new start-ups (e.g., DataHero, DataCamp) have emerged aware of the opportunity to provide data management knowledge and tools in low-cost, on-demand models. Ultimately, it is important for organizations to implement a data governance program that enables them to meet data standards and maintain data quality, thus reducing reliance on spreadsheets as the primary tool for extracting value from data. Implementing an MDM solution that makes the organization share a common understanding and language of business data objects (e.g., customers, products, services) would be a big step forward in advancing Agile data project management.

Discussion Questions

- Are tourism firms ready to implement an Agile data-driven framework? What factors are for or against?
- What kind of Agile practices do you think would be easier (or more difficult) to implement in tourism firms? How could the difficulties be overcome?
- Can a tourism firm start implementing an Agile framework without having previous experience working with data? Does a firm need to reach a certain level of maturity in data management before starting to think about Agile practices?
- What benefits can a tourism firm expect from implementing an Agile data-driven framework? What threats can be expected?
- What Agile practices do you think best respond to the needs created by the smart transformation process of tourism firms?
- To what extent can Agile practices improve the work of people in tourism firms?
- Do you have any experience working with Agile practices, or know someone who has? Write down those aspects that can help to improve the performance of the firm.

20 Smart Leadership

Smart transformation is best managed when there are smart leaders who recognize the needs of the organization and are willing to guide it on the right direction. Leadership is a critical capability that smart firms need in order to address all the transformational changes required to build a data-driven culture in the organization. Furthermore, leadership is a critical competency to embrace Big Data throughout the organization and bring the benefits of data science to fruition (Unhelkar, 2017).

Paradoxically, as firms increase their reliance on data and algorithms for decision-making and the automation of routine, information-intensive tasks increases, so does the importance of interpersonal leadership skills between owners and managers. Smart firms need good leaders with a vision not only to improve some processes and make them work better, but to profoundly transform the organization around three key areas: customer experience, operational processes, and new business models (Kiron, 2017). What is at stake is not only how to manage data and implement data-driven technologies to get the most out of it, but to give the organization a new impetus that makes it stronger and more competitive. In the end, it is about making the tourism firm work in a more Agile way and be able to respond to the customer needs faster and precisely.

Today every smart leader understands that the world of technology and data is changing very rapidly, and they take very seriously how smart things are changing the way consumers and business behave. They are aware that Big Data, analytics, artificial intelligence (AI) and the Internet of Things (IoT) technologies are becoming ubiquitous and that a silent but powerful (Smart) Revolution is taking place that affects the management of organizations. Smart leaders have a clear mission: they work hard to make their organizations develop a smart mindset and be ready to respond to the disruptions caused by the use of smart technologies.

DOI: 10.1079/9781800620261.0020

The huge demand for change brought about by the Smart Revolution makes it inevitable for organizations to build a greater capacity for innovation and transformation, and this fundamentally affects leadership functions. In other words, it is simply not possible for the firm's transformation process to unfold and the firm to be reborn as a smart organization without renewed leadership. It is the job of leaders to understand how transformation affects existing organizations and how they can be threatened if they do not change, and become flexible, and do not reorganize towards increasingly digital and smart business models (Burchardt & Maisch, 2019; Earley, 2014).

20.1 Leaders and Smartization

In the last few years, business firms have gone from a lack of information as one of the main reasons for uncertainty to an overabundance of data that can now be quickly and affordably transformed into information and knowledge. Data offers organizations the potential to reduce risks in decision-making by enabling business leaders to access richer, more holistic insights into the challenges they face (Filatotchev & Nakajima, 2010). This means leaders will need to focus on Big Data and analytics skills and foster a data-driven decision-making culture with the right technologies to support it. In addition, the use of data by tourism firms strengthens organizational resilience, understood as the organization's ability to anticipate disruptions, adapt to disturbing events, and create lasting value in a turbulent environment (Kaivo-Oja et al., 2015).

The use of data and the possibility to create value for the organization is something that does not happen by chance or accident, nor is it something that can be easily achieved in a few weeks or months. Moreover, firms are unlikely to be more successful simply because they have access to good data. Becoming a smart organization requires well-defined leadership that shares a clear vision and goals and activates flexible talent management practices and an organizational culture that puts data at the center of the organization (Shamim et al., 2019). Smart leaders encourage the use of data and successfully guide the organization to extract value from Big Data and analytics, making it easier for data applications to consistently deliver their benefits to the organization. This requires smart organizational decisions and the commitment of visionary leaders who inspire a new data-centric culture by becoming role models in the use of data and with the skills and abilities to overcome the technical and organizational obstacles that arise when creating value from data (Brinch et al., 2021; Hume & West, 2020).

Smart firms' leaders view data as a key asset that all employees should have access to and know how to use. They believe that data and analytics can greatly enhance the organization's ability to innovate and respond to changing customer needs and preferences (Kiron, 2017) and ultimately view data as an asset key to a firm's competitive advantage. Smart leaders stand out because they emphasize the systematic use of data throughout the organization not only to address operational or tactical problems, or to create more compelling reports, but also to integrate data into decision-making processes that lead to the allocation of resources, put new products and services on the market, or promote changes in business models. To achieve their goals, smart leaders do not hesitate to build a flexible and decentralized organizational structure in which all its members collaborate and share their knowledge to extract value from data (Zeng & Glaister, 2018).

Smart leaders know how to connect the "big picture" questions that every business owner and manager asks, with the more specific, low-level questions that machines can answer, thereby reconciling machine-generated models with the nuances of human preferences (Andersen et al., 2018). Smart leaders are aware of the benefits of using data, but also know its limitations and adapt the role of data when formulating strategies and adjusting the expectations of data users in the organization. They never fail to communicate to the entire organization when data analytics reveals something insightful, because when the organization shares the insights generated by the data, it reinforces the importance and usefulness of the data and encourages people to pursue this line.

20.2 New Leadership Profiles

One of the factors on which the success of smart transformation depends is, in addition to technology, the type of leadership and the strategies deployed by the leaders in the organization (Bala, 2018). Becoming a smart firm requires significant financial, human, and technological resources, but especially leaders who possess a balanced mix of skills, attitudes, and behaviors that guide the organization towards "smartness".

Organizations must ensure that they have the right people in place to create a culture that makes the smart firm a reality. They need people who are operationally responsible for the firm's smart strategy and who can implement management practices that prioritize change initiatives and transformation projects aligned with the business strategy (Matt et al., 2015). This does not mean that leaders need to be experts in every single technology and practice that the firm implements, but they do need to be digitally fluent and have a strong understanding of the value of the investments needed to guide the firm's future growth.

The debate on who should lead the smart transformation of the firm is open in the academic and professional communities, and there is no consensus on who is the best candidate. In addition to the Chief Executive Officer (CEO) and the Chief Information Officer (CIO), who have traditionally played the leading role in the firm's IT transformation and innovation projects, new roles are emerging as a result of digitalization, among which are the Chief Digital Officer (CDO) and the Chief Transformation Officer (CTO). Some of the differences between CIOs and CDOs are that CDOs are not responsible for running IT within the firm, nor are they in charge of data (which is increasingly the responsibility of Chief Data Officers), rather they are change agents working in close collaboration with partners to create digital solutions that transform business models and processes. While CIOs have traditionally focused on back-office tasks and ensuring IT service levels, CDOs focus more on the customer and look at the applied side of business technologies. It remains to be seen what kind of interface is best between CIOs and CDOs in a context where both roles coexist, and whether the CIO's role will be neutral or have some leadership role in the transformation process (Brown & Brown, 2019).

However, it seems that at the moment these new roles are being created in many organizations with a limited time frame in mind, suggesting that organizations think that digital or smart transformation is going to take a limited amount of time and that at some point sooner rather than later these roles will no longer be needed. This belief may be even more widespread in SMEs, where there is often not as much variety in leadership roles as in large firms, and where it is common for the CEO or owner of the firm to be responsible for organizational transformation. This should not be a problem, after all it is about having a well-defined and clearly assigned role in the organization that exercises responsibility and leadership over the transformation process (Hönigsberg & Dinter, 2019). Once assigned a leadership role, the SME leaders will have to define the roles and responsibilities of the rest of the managers and employees of the organization to respond to the needs that arise from smart transformation.

Whoever ultimately takes the lead, it is critical that they focus on ensuring that the firm's technologies are used appropriately and remain aligned with the organization's business objectives, leveraging concrete smart strategies and fostering close collaboration between business and IT. Nevertheless, due to the expected long duration of the smart transformation processes, it is perhaps best that leadership is performed by a single role over time and that the responsibilities of the leader are agreed upon jointly with the firm's senior management from the outset, thus overcoming the usual resistance coming from the different areas of the organization. The above does not mean that in some firms the role of leader is not performed by a single role, but by several roles that share different responsibilities, for example, the CIO and the CDO.

20.3 Smart Leaders' Skills

Smart transformation has a strong impact on the leadership style of organizations. Taking the lead in firms that rely on data and knowledge to make decisions requires skills that differ substantially from conventional leadership skills focused on intuition and personal experience. Therefore, smart leaders need a new set of skills to tackle the deep transformation process that their organizations are about to undertake.

Smartization is not driven by technology. Deploying reliable and robust IT data systems is yet another component of the smart transformation process. It is not enough for leaders to be tech savvy, but they must also have extensive management skills to integrate IT processes and tools with business processes, and be able to recognize possible areas of synergy and friction between the two (Hanelt et al., 2021; Morabito, 2015). Just as important is that leaders focus their attention on multiple perspectives, such as customer centricity, systems that are flexible and easy to use, and information that is accessible and can be rapidly transformed into insights by the members of the organization. Without this mixed approach, it will be difficult for the organization to harness the business value of data centricity and smartness.

Smart leaders do not need to master all the technologies that are going to be needed, but they do need to be bilingual in both the "human" language and the language of machines, or put another way they must be able to communicate and work effectively with smart machines and with smart humans (Andersen et al., 2018). It is also vital that smart leaders recognize data as a critical organizational asset, thereby establishing a hierarchy of roles accordingly (e.g., data scientists, data analysts, project managers, business analysts) and locating decision-making according to the strategic nature of the data. All of these are essential skills for the leader to build a culture based on data, whose implementation in the smart organization will facilitate achieving an adequate return on investments in talent (Hupperz et al., 2021).

Communication and interpersonal skills

The smart transformation of tourism firms requires the active participation of all affected stakeholders. This makes communication skills paramount to ensure that they are all aware of the direction the organization will take and what future they can aspire to. Furthermore, smart leaders know that making people embrace change and share a vision is not easy. People will prefer the stability of the status quo to the uncertainty that comes with change. That is why it is very important that smart leaders stand out for their interpersonal and communication skills and know how to deal with resistance that may appear.

Engaging active followers to transform the organization implies convincing employees of all the good that awaits them if, working together, they manage to bring the vision into action. Initially, the leader's vision will be nothing more than an inspiring and well-crafted promise to attract the interest of others. But nice words are not enough – soon the leader will have to demonstrate the benefits of the vision and build trust with others, otherwise no one will be willing to work hard for years to come (Friedman, 2020).

Owners and managers must be aware that some ethical considerations may arise when engaging followers. For this reason, it is highly recommended that the leader presents herself not only as someone focused on the financial performance of the firm, but as an ethically responsible and authentic leader, communicative and inclusive, and concerned with the sustainability of the firm and the environment. Once trust among followers is achieved, the leader should be able to harness the creativity, knowledge, and collective wisdom of their co-workers to achieve her vision. Unfortunately, keeping this process on track is often hampered in the tourism firm due to the high proportion of seasonal and low-skilled workers, who pose an even greater challenge to the smart leader's abilities to orchestrate workforce motivation.

Not least important is that smart leaders have the skills to collaborate with external partners and suppliers. Being innovative and able to rapidly create new products and services when customer preferences change requires strong coordination skills both at the organizational level and with partners. This becomes even more apparent when the tourism firm is committed to open innovation as a model to foster collaboration with external actors seeking to provide the firm with more opportunities to exchange knowledge and deliver innovative responses. Team culture also plays an important role in maintaining knowledge exchanges and innovation performance in the smart tourism firm. Smart leaders will strive to shape this culture by acting as facilitators, mentors, and role models to others (Pesonen, 2020), mastering the double language of technology and business to provide the organization with valuable tools with which to increase the potential for knowledge exchange and collaboration. The construction of open and collaborative spaces that favor innovation will be greatly facilitated by the implementation of smart technologies, in the same way that these spaces will encourage the development of smart technologies, thus creating synergies between the two.

Transformational awareness

Given that the context of the tourism firm is complex and constantly changing, leaders must also develop transformational awareness in relation to the organization's processes and strategies (Hanelt et al., 2021). Smart leaders not only need to know how to sense and respond to changes within competitive dynamics, but they also need to be aware of the complex web of variables that influence the process of business transformation internally and how these are intertwined with the changes that occur in the market, the consumer, and technologies. This means that leaders must draw on the broadest and most diverse data sources and perspectives possible to fully assess the contextual conditions of the firm and incorporate complexity into their decisions. New tools and applications (e.g., Agile methodologies, design thinking) are constantly emerging to help leaders manage these new needs and support smart leadership.

Transformational awareness is closely related to the ability of smart leaders to accelerate organizational transformation strategies. This entails the ability to always be open to experimenting with new processes and conceiving innovative products and services based on available internal and external resources (Brown & Brown, 2019; Hanelt et al., 2021). The smart leader is constantly learning from customers and what is happening in the marketplace, driving continuous innovation and facilitating organizational adaptation, rather than spending her time planning and performing *ex ante* analysis that are of little or no use in a highly dynamic business ecosystem where opportunities and competitive advantages are fleeting.

20.4 Benefits of Smart Leadership

Having competent leaders is something that can provide significant benefits in the context of smart transformation and the development of a data-driven culture in the tourism firm. Below are some of the most significant benefits that the organization can gain in return (Fig. 20.1).

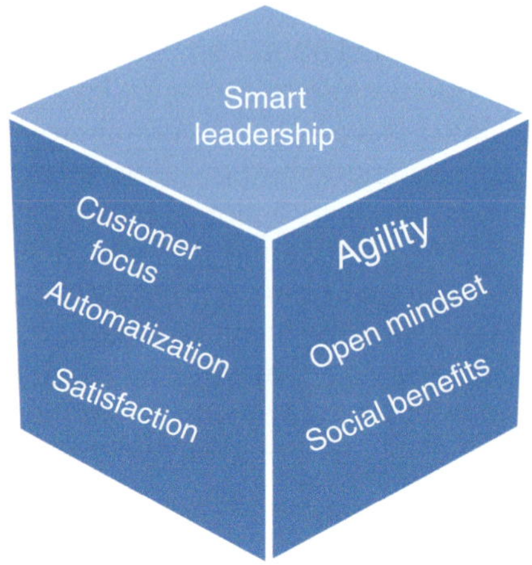

Fig. 20.1. Benefits of smart leadership. Source: own elaboration.

- **Increased customer focus**, which can translate into a better user experience and higher customer satisfaction as the smart organization consistently generates more insightful knowledge about customers.
- **Greater agility to deliver products and services**, through the development of an open mindset that leads the members of the firm to collaborate with partners and external actors and to engage in a culture of open innovation and continuous exchange of knowledge. By fostering more open and productive relationships, the firm will be in a better position to offer more innovative products and services and expand in the market.
- **Increased process optimization**, as smart leaders emphasize the need to embed new digital and data-driven technologies into key business processes and automate repetitive processes, allowing the organization to run in a more agile and decentralized mode.
- **Greater employee satisfaction**, since they are an important part of the leader's vision and should know what can be expected from their effort. Additionally, as employees are encouraged to participate in the process of defining and implementing technologies that meet their daily needs, they feel more motivated and prone to make their own decisions about how to organize their work and deliver benefits for the organization.
- **Greater focus on the social benefits obtained from the firm's actions**, since smart leaders have access to more information and knowledge of the environment and analyze more complex perspectives than conventional leaders. This in turn enables smart leaders to integrate broader perspectives into their decisions (e.g., insights, metrics, reports) and adopt more community- and environment-focused strategies and behaviors.

20.5 Smart Leaders and SMEs

Very often leaders in SMEs, especially those with a family structure, become a barrier for the organization to transform, especially when leaders are risk averse, which can end up affecting the organization's capacity for innovation and the development of a culture focused on data and knowledge. The personal qualities of leaders, their sociodemographic attributes, and even their educational level are factors that can influence risk appetite and affect the decision-making of owners and managers (Zapata et al., 2020).

In general, organizations with low risk-taking capacity are less likely to be innovative, with leaders often deciding what kinds of risks the organization can take (Pesonen, 2020). On many occasions, leadership in SMEs is highly centralized around the figure of the owner of the firm, and it is rare that the final decisions are made by other people or teams (Trenkle, 2019). This means that in SMEs, power is often highly concentrated in a single person who is the one who makes most of the important business decisions and has the decisive role in transformation decisions.

In addition, leadership in SMEs often suffers from a poor definition of responsibilities, a wrong vision of the future, or even an unwillingness to take risks, all of which hinder the potential competitive development of the firm. Very often SMEs remain focused on the idea of managing their daily operations instead of considering the organizational implications of transformation, so redefining leadership in the firm is not a priority for them. This may explain why many tourism SMEs continue to struggle with their digital transformation initiatives, indicating a lack of leadership in this key area (Hönigsberg & Dinter, 2019).

20.6 Agile Leaders

The leadership styles that have been working until now are no longer valid to tackle the challenges of the Smart Revolution. Simply put, leadership styles prior to the Smart Revolution were not designed to embrace data-driven technological breakthroughs, nor were they equipped to effectively manage the ultra-fast pace at which society and firms are transforming and technological change is occurring. This indicates that the landscape of leadership in business organizations is unequivocally changing forever.

In the highly uncertain and complex environment in which tourism firms operate today, it is quite difficult, if not impossible, for leaders to accurately predict potential future opportunities and threats. Gone is the idea of the "wise" leader on which organizations used to rely to devise the appropriate responses to challenges, and find solutions to all business problems. Nowadays, organizations struggle to learn around the clock and make change flow naturally and fast throughout the organization. It is thus not surprising that the role of the leader in business organizations is transforming to accommodate a more "agile" style and that "agile smart leaders" are emerging strongly.

Agile leaders apply the values and principles of Agile management to their own leadership style. They seek to mobilize the organization towards higher levels of organizational agility and orchestrate its resources for an effective transition in that direction. They are visionaries and create the right conditions so that the members of the organization are motivated and willing to seize the opportunities that come their way, sharing their knowledge and creativity along the way.

Agile leaders are transformational leaders, meaning they focus all their efforts on pursuing the crucial mission of driving profound change in the organization. They tap into the hopes and aspirations of organization members to gain their support and commitment to work together towards a high-flying vision (Gagel, 2017). As transformational leaders, they leverage followers to transcend their personal or departmental interests through inspiration, motivation, individual recognition, and intellectual stimulation, making their interests stay aligned with those of the leader. In a way, transformational leaders turn followers into disciples of change, so that change can permeate every corner of the organization.

Agile leaders do not lead by authority, but by example with open-mindedness, self-discipline, continuous self-evaluation, creativity, and capacity for innovation, thus being able to lead effectively in conditions that can change quickly and unpredictably. Agile leaders seek to eliminate bureaucratic constraints within the organization, and do not hesitate to tear down the silos that fuel internal conflict and foster subcultures that protect the interests of a few areas and departments of the firm. Instead of silos, they inspire people to become agents of change and strive to implement a culture of openness and creativity in which areas and departments openly share information and knowledge, while encouraging innovation and continuous experimentation (Attar & Abdul-Kareem, 2020; Forbes Coaches Council, 2018). Furthermore, they are aware of the importance of using broader and more diverse perspectives to address business problems and deliver solutions swiftly. This makes Agile leaders develop a keen eye to realize when something is not working and when the time comes to change things.

How can smart firms identify Agile leaders? Although no one has been a leader in a smart world until now, it is doubtful that the same criteria used in traditional organizations to recognize leaders (i.e., demonstrated leadership potential, proven track record) will continue to be useful for Agile leaders. HR departments will need a clear understanding of what Agile leadership is all about, which is key to unequivocally identifying Agile leaders, as well as promoting and scaling a true Agile organizational culture. Furthermore, owners and managers hiring Agile leaders should focus on recognizing Agile attributes, such as an innovation-oriented mindset, a passion for continuous learning, adaptability, and an entrepreneurial spirit. As can be seen, all these characteristics of Agile leaders are substantially different from the "wise" leader stereotype who used to rule conventional business organizations for decades by relying on their own knowledge, expertise, and the guts to make decisions based on instinct.

Discussion Questions

- What type of leadership style predominates in tourism firms? What factors reinforce this type of leadership?
- How do cultural differences and the size of the organization influence the type of leadership practiced in tourism firms?

- Who typically plays a leadership role in tourism firms? Is leadership exercised by one or several roles? What kind of roles are the most common and why?
- Are today's tourism firms leaders well prepared to face the challenges of smart transformation? Why? If not, what would it take for leaders to be well prepared?
- Are tourism firms sensitive to the emergence of new leadership roles as a result of the Smart Revolution? What new roles have consolidated?
- What skills of the tourism business leader are different from business leaders in other industries? What are the reasons for these peculiarities?
- What differences are there between an Agile leader and another who is not? Is the Agile leadership style applicable to tourism SMEs?

21 Open Innovation

Open innovation is a way of organizing innovation built on the idea that firms should use both internal and external ideas, knowledge, and technologies to improve their innovation activities (Bogers et al., 2019; Chesbrough et al., 2006). The notion of open innovation is built on the assumption that the best ideas for innovation are unlikely to come from a single firm, nor do the most talented people work for only one organization (Chesbrough, 2003). Implicit in open innovation is the idea that no firm can truly innovate in isolation, but rather must engage with a plethora of actors and partners to acquire ideas and resources from the external environment that ultimately help the firm stay ahead of competitors. This seemingly simple yet powerful idea that focuses on harnessing and enhancing the firm's internal innovation capabilities and exploring new business models has become an imperative for orchestrating innovation in smart tourism firms.

21.1 What Is Open Innovation?

The term open innovation was first enunciated by Professor Chesbrough in 2003, who proposed a paradigm shift in contrast to closed or in-house innovation, which is the one developed internally by the firm's R&D department. Since then, the concept has continued to evolve and has been redefined by Chesbrough and Bogers as "a distributed innovation process based on purposively managed knowledge flows across organizational boundaries, using pecuniary and non-pecuniary mechanisms in line with the organization's business model" (Chesbrough & Bogers, 2014).

According to the open innovation paradigm, the geographical footprint of innovation has changed dramatically, and the sources of ideas and knowledge have spread throughout the world. Moreover, the boundaries between firms,

DOI: 10.1079/9781800620261.0021

and between firms and their external environment, have become permeable to the point that firms can create value more effectively by integrating a broad variety of entities into their innovation process, such as suppliers, customers, experts, competitors, universities, consultants, etc. (Chesbrough, 2003). Hence, open innovation enriches the traditional innovation funnel, where the boundaries are typically closed, by bringing the external environment inside the funnel, so that innovations can even be exploited outside the firm (Fig. 21.1) (Pellizzoni et al., 2019). In other words, firms should not rely solely on their own internal research and ideas but also seek outside sources that can boost their innovation. This is known as the outside-in mode of open innovation and to embrace it firms must equip themselves with mechanisms to attract external actors and engage them in tasks that enrich the firm's value propositions. Many times these mechanisms involve the construction of platforms that connect the actors of innovation and integrate useful technologies to solve real problems. Other times firms rely on intermediate actors who, through their expertise and knowledge of the R&D system, foster connections between entities and help orchestrate the key resources that facilitate open innovation. Ultimately, the outside-in process of open innovation enriches the traditional innovation funnel by transcending organizational boundaries and making it permeable rather than closed. With open innovation, ideas, technologies, and solutions from the external environment can be brought into the funnel and the innovations thus developed can be exploited outside the firm.

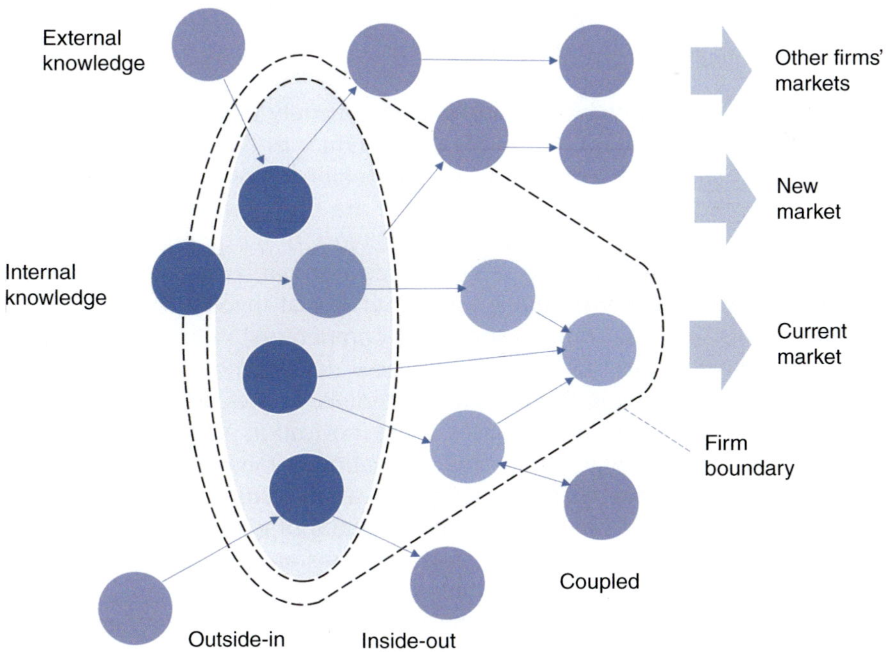

Fig. 21.1 The open innovation model. Source: own elaboration based on Chesbrough & Bogers (2014).

While the outside-in (or inbound) direction is perhaps the best known and most developed mode of open innovation, it is not the only one. There is also an inside-out (or outbound) mode and a coupled mode, which involves combined inputs and outputs of knowledge between innovation actors. The inside-out mode considers the firm's intellectual property (IP) as an enabler to access external ideas and allow others to use the firm's own ideas. This is possible because there are intellectual property rights (IPR) that grant their owners the right to exploit and share them with third parties under the established terms. Licensing a third party to exploit IPR in exchange for a royalty is a very common practice that encourages others to adopt the firm's own technology and know-how. This, in turn, can lead the firm to explore new business models that can bring new revenue streams and mitigate some of the risks assumed in technology development processes. Contrary to popular belief, the inside-out approach does not mean that the role of IPR loses its importance or is counterproductive – quite the opposite. It has been shown that collaboration and intellectual property rights are not substitutes but complementary to facilitate an open innovation process, and that firms become more collaborative after receiving a patent than before receiving it (Zobel et al., 2016).

The combination of the outside-in and the inside-out modes leads to the so-called coupled mode of open innovation. This mode requires that both processes occur simultaneously, which usually calls for some type of partnership agreement between firms, such as strategic alliances, joint ventures, or spin-offs, to jointly develop and commercialize innovations. Among the mechanisms used to implement a coupled open innovation process are the creation of communities of practice, the establishment of consortiums, and networks.

Today, open innovation includes a wide variety of different activities, from collaborations between industry and academia, to the development of open source software, crowdsourcing, or the relationships between corporations and start-ups (Bogers et al., 2019). There are firms that are even using open innovation within the more general framework of (corporate) innovation and entrepreneurship activities to drive change in organizational culture, such as when developing new products and services that involve multiple groups of stakeholders (e.g., customers, suppliers, competitors) with a view to increasing the efficiency and effectiveness of new development processes. In addition, firms are applying the exchange of ideas, knowledge, and technologies between various actors in all phases of the innovation process, including R&D, production, design, marketing, and innovation commercialization (Burchardt & Maisch, 2019). This has led to open innovation being considered an essential practice for any firm that relies on the potential of external ideas, knowledge, and technologies to create value. Tourism firms are increasingly motivated to integrate open innovation into their overall business strategy and combine external resources and skills with their internal capabilities, implementing specific processes that enhance learning and collaboration through networks and ecosystems.

21.2 Drivers of Open Innovation

Although open innovation emerges as a groundbreaking paradigm to which smart firms should start giving their full attention, the truth is that organizations have always relied on some form of external knowledge and ideas to innovate. What has changed is that today almost all firms agree that the best ideas and people are elsewhere, which is explained by the globalization of the business ecosystem, advances in education, and accelerated technological progress. Consequently, open innovation today is qualitatively and quantitatively very different from what it was in the pre-internet era. Knowledge and ideas no longer spring from a few leading innovation centers located in rich countries. Firms can now connect with communities large and small where cutting-edge ideas, knowledge, and technologies are being forged and shared, and new patterns of cross-functional collaboration are emerging. The exchange of knowledge and ideas has thus become tremendously efficient and having access to expertise and finding the right solution to a problem is no longer a matter of having substantial resources, but of knowing where to look at the right moment.

So, what are the main drivers that can accelerate the adoption of open innovation?

Since the 1990s, many global firms have significantly reduced their investment in research, leading to a sharp decline in internal R&D spending. An increasing number of corporate investors now take a short-term view of business, and there are plenty of shareholders who pressure top management to continually cut costs. As firm-driven innovation becomes rarer and technology cycles shorten, it becomes faster and more affordable to turn to external sources of R&D, leaving firms no other choice but to drive open innovation through suppliers, partners, consultants, universities, etc. This is an adverse trend that highlights the growing inability of firms to innovate and for which open innovation is only a "stopgap" solution (Bogers et al., 2019)

A second driver of open innovation has to do with digital technologies, which are significantly revamping the way firms organize innovation. Digital technologies are propelling the flow of vast amounts of information from one place to another and fostering open collaboration between firms to innovate. New ecosystems are emerging characterized by a coordinated collaboration where firms organize to improve their products and services and provide valuable solutions to those who need them. Given the growth that these ecosystems are having, everything seems to suggest that their economic and social impact will continue to increase and shape new forms of open collaboration in the near future (Brunswicker et al., 2015).

Open collaboration between firms opens up new opportunities to explore new business models and exploit emerging innovations. Those with the highest potential are related to the analysis, dissemination, and leverage of large amounts of data being generated by consumer behavior in smart ecosystems and platforms. Big Data thus becomes a key driver that, together with the technological infrastructure that sustains it (e.g., the Internet of Things (IoT)), further accentuates the need for participatory approaches among firms and fosters the emergence of open innovation systems (Burchardt & Maisch, 2019). Furthermore, the ubiquity of digital platforms requires the intertwining of multiple technologies whose level of sophistication and technical complexity means that they cannot be provided by a single vendor. Digital convergence thus becomes an essential commitment that every open digital ecosystem must put into practice. Both the IoT and smart destinations are good examples of this, as they require the orchestration of many types of partnerships and collaborations to be able to offer solutions. Through open innovation it is possible to achieve digital convergence more easily.

Closely related to the above is the role that social media are playing as creators of valuable knowledge assets for firms and as enablers of digital platforms for greater tourists engagement (Del Vecchio, Mele, et al., 2018a). Social media have become a source of intelligence for firms and a highly qualified provider of information on consumption patterns and consumer preferences in real time. The vast amount of data that is generated on social media about the experiences, opinions, and reviews of tourists is a rich source for open innovation and decision-making in tourism firms. However, despite the huge potential of social media for open innovation their use remains rare and firms have not yet reaped significant benefits from their use.

The combination of social media and smart technologies is radically changing firms' innovation strategies by enabling user empowerment and more participatory innovation processes. As more user knowledge and experiences are collected through social media and smart technologies, they become major players in the open innovation game plan (Brunswicker & Vanhaverbeke, 2015). As a result, the interest of firms in co-creating value with their users and customers is growing, as these too are willing to play an active role in the product and service development process. On the other hand, firms have begun to realize the vital role that users can play in the innovation process and have started to reorient their innovation strategies to harness co-creation opportunities (Jung & tom Dieck, 2017). There are good examples that highlight how users participate in the value co-creation process, from the Walt Disney theme park model, where employees and visitors engage in co-creating experiences, to the IKEA model where customers select, pick-up, transport, and even assemble the items they buy.

In the end, the idea of open innovation has become highly relevant to tourism firms, which have long been leaders in the adoption of innovative services (Abbate & Souca, 2013). Many tourism firms are used to creating value offerings with the collaboration of employees, suppliers, partners, and, especially, customers, who often become co-creators of their own tourism experiences and actively stimulate interactions to explore new ideas and solutions. Turning tourists and visitors from passive recipients of experiences into active participants in product and service development and co-creators of value it is expected to become an important source of competitive advantage for tourism firms in the coming years (Payne et al., 2008). The way in which tourists interact with firms to co-create experiences can be very varied. They can offer ideas, share content, and create their own personalized tourism experiences, with roles ranging from mere facilitators on the periphery of the firm's innovation process, to become involved as partners or having an active role contributing and exchanging knowledge with the firm. Be that as it may, tourism firms must provide users with co-production, participation, and personalization through appropriate mechanisms of joint exchange and interaction. This will pave the way for the tourist to become a key asset in the firm's innovation process, and a constant source of useful ideas to co-create opportunities in highly competitive markets (Yin et al., 2021).

Big Data and open innovation

Big Data becomes relevant for innovation when firms realize that external knowledge is a key source of the innovation process. Indeed, Big Data represents an emerging opportunity to improve the effectiveness of open innovation as it unfolds new discoveries and opportunities for entrepreneurship.

The open innovation paradigm can be a suitable approach to manage the large amount of data that is generated every day from different sources and in real time (Del Vecchio, Di Minin, et al., 2018). In fact, the relationship between Big Data and open innovation seems quite natural, as both deal with external sources of information and knowledge that can be highly impactful for business innovation (Del Vecchio, Mele, et al., 2018b). More specifically, Social Big Data, understood as the sum of all the data generated and shared through social media, is increasingly used by firms to acquire external knowledge that can be useful to develop new products, services, methods of distribution, etc. In light of its promising potential, new solutions are needed to make the most of available social data and turbocharge the innovation strategies of large and small firms.

From the perspective of open innovation, Social Big Data makes it easier for firms to create value together with customers and stakeholders by acquiring knowledge that can be used in the innovation process. Since Social Big Data is the result of very diverse and varied data that comes from customers and users with different levels of skills, knowledge, and experiences, open innovation can be a suitable approach to bring this knowledge to the different stages of the firm's innovation flow, including ideation, R&D, and commercialization (Del Vecchio, Di Minin, et al., 2018). In the ideation stage, Social Big Data can be a relevant source of knowledge about new trends, customer preferences, and acceptance of products and services, which if properly addressed by the firm, can help identify new customer needs and trends. Data generated through social media, when exploited using advanced analytical techniques, can be effective in predicting consumer behavior and confirming market trends, which in turn can support internal R&D and innovation processes, the development of new products and services, as well as improve business models and decision-making. No less important is that social media are an extraordinary promotion and marketing channel used by firms as a showcase for the commercialization of their products and services. Data generated from word of mouth and peer recommendations provides relevant insights into brand reach and perception, as well as user preferences, which can be very valuable for the commercialization and marketing strategies of firms.

Although Big Data challenges all industries, it acquires special relevance in knowledge-intensive industries such as tourism, where it can be a powerful tool to foster innovation and competitiveness and the basis for the development of smart firms. For tourism firms to develop open innovation strategies based on Big Data, it is important that they understand in advance how Big Data can be used to create outbound and inbound business opportunities, how Big Data can impact the absorptive capacity of firms, and how it can contribute to their competitive edge. Ultimately, tourism firms must understand all aspects related to creating value from Big Data (Del Vecchio, Di Minin, et al., 2018). There are many and varied ways to implement open innovation strategies using Big Data. Some may consist of customer communities, idea marketplaces, crowdsourcing, etc., which generate Big Data and allow firms to collect ideas and generate insights from groups of users. The use of sensors and devices in IoT ecosystems, in which a large number of potential contributors engage also represents a good opportunity to collect data and turn it into useful knowledge for firms.

Knowledge management

Competitiveness of tourism firms increasingly depends on the development, use, and exchange of knowledge-based assets from inside and outside the boundaries of the organization. Therefore, the development of strong collaboration and knowledge sharing networks and the application of modern knowledge management techniques become strategic resources to access the information that firms need to drive innovation. It is no coincidence that knowledge exchange plays a central role in the open innovation model, according to which firms identify, manage, exchange, and leverage knowledge effectively to improve their productivity and competitiveness (Laxamanan & Rahim, 2020). In short, firms should deliberately manage the inputs and outputs of knowledge to improve innovation processes and build on the results.

Some of the main knowledge management practices that can be put in place to implement the open innovation paradigm are described below.

- **Knowledge management in inbound open innovation:** Inbound open innovation activities involve managing streams of external information that complement, update, and improve the internal knowledge base available to the firm. Different knowledge management practices allow the firm to integrate internal and external knowledge by collecting, coding, storing, and transferring the knowledge acquired. These practices allow the employees and departments of the firm to explore their competitive environment in search of new ideas and knowledge, as well as to identify potential partners with whom to develop business opportunities to take to the market. It is usually a good idea to develop an internal database, more or less centralized, that contains all the information acquired and that can be reviewed by the members of the organization.
- **Knowledge management in outbound open innovation:** Outbound open innovation processes involve the transfer of information from the firm to

the external environment. Such an exchange makes sense when the benefits of transferring knowledge outweigh the potential gains of using it internally to develop new products and services. Knowledge management practices allow firms to decide whether to maintain or sell their knowledge, i.e., when a technology does not deliver the expected results and can be transferred to other firms without its competitive advantage being affected. Those firms that decide to transfer knowledge to other firms must identify in advance who their potential customers are and who they should contact internally. There are market knowledge management practices that help the firm to perform these tasks efficiently.

- **Knowledge management in coupled open innovation:** Coupled open innovation processes address the extensive use of knowledge by different firms, including information inflows and outflows. In this case, firms need knowledge management practices that combine both the acquisition and distribution of knowledge between organizations that can be heterogeneous, with different cultures, information systems, and strategies, which proves a rather challenging and complex task. In this context, it is important for the firm to be aware of the environmental conditions that make organizations engage in knowledge exchange practices, as well as the potential benefits they can obtain from outbound and inbound innovation flows. The use of smart technologies can accelerate the joint development of solutions through communication, collaboration, and participation of firms in iterative processes of generation and exchange of knowledge. At a later stage, the development of joint solutions will require the establishment of governance frameworks at the inter-organizational level to generate sustained positive effects on the innovation performance of firms.

The role of stakeholders

The way in which tourism firms carry out innovation is undergoing a major transformation. The involvement of stakeholders (e.g., customers, suppliers, platforms, authorities) has become central to innovation activities as the open innovation paradigm has taken root in tourism firms. These changes are accentuated by the prominent place that social media occupy in people's lives and in the activity of firms, which are replacing the traditional top-down approach to innovation with an instantaneous and flexible dialogue on equal terms aimed at co-creating value through innovation.

Social media along with smart technologies are radically changing innovation strategies, creating new user empowerment processes. By empowering users, firms recognize that users can make informed decisions when they have the information and tools to do so. Firms can greatly benefit from the knowledge and experiences of a large number of users thanks to digital tools that are ubiquitous and low cost, allowing users to actively participate in the open innovation workflow. Transparency is another key factor in achieving the right level of empowerment, which may lead users to prefer firms that innovate and provide them with the tools and capabilities to engage (Iglesias-Sánchez et al., 2019).

Therefore, for open innovation to work, flexible and two-way communication channels should be established with stakeholders. Social media have exceptional characteristics that encourage interactions between firms and people and are an ideal interface for identifying opportunities and promoting collaboration among stakeholders, which can lead to the development of highly productive products and services. However, despite the great potential to gather intelligence and become an important source of innovation, tourism firms have not yet sufficiently exploited the use of social media from an open innovation perspective. Interestingly, owners and managers seem to be well aware of the capabilities that these platforms can offer and that their products and services could be significantly improved by collaborating in one way or another with the stakeholders (Iglesias-Sánchez et al., 2019).

21.3 Challenges of Open Innovation

So far, we have seen that open innovation offers enormous potential to improve the knowledge capabilities of the tourism; however, it also poses many challenges that the organization cannot solve on its own and requires the collaboration and participation of other actors. Implementing the open innovation paradigm in the tourism firm is never a quick and easy process, not to mention that the first problems arise as soon as owners and managers try to determine what kind of practices constitute open innovation and what should be done first to improve the firm's innovation capabilities and competitiveness (Abbate & Souca, 2013). This raises difficult but key questions about how, when, and to what extent to open the boundaries of the firm, and how to find the right partners to start collaborating (Pellizzoni et al., 2019).

Open innovation usually implies an organizational cultural shock that, depending on the depth and intensity of the process, can result in a radical cultural change. The reasons are twofold: on the one hand cultural change is required so that the firm can harness all the benefits that open innovation promises; on the other hand, the application of open innovation practices in the firm implies changes that impact the culture of the organization (Burchardt & Maisch, 2019). Moreover, the impact that open innovation can have on the organizational culture depends on the degree of openness that the firm has when implementing the open innovation model, in such a way that the more open, the less impact on the culture. Culture is also affected by the frequency and duration of collaborations, so that when the firm regularly engages with external actors, the changes produced by the implementation of open innovation are smoother, regardless of the industry in which it operates (Bageac et al., 2020).

How organizations approach the process of implementing open innovation is another major challenge for the firm. It is not the same to address open innovation from the top down than from the bottom up. In general, the top-down approach is more desirable than the bottom-up approach, as this reinforces the development of open innovation practices within the organization. In those organizations where the introduction of open innovation paradigm is a novelty or involves time-consuming cultural changes, it is more effective to approach open innovation from the top down. In decentralized organizations, the successful implementation of open innovation will require strong leadership by top management, and a well-established culture based on the principles of autonomy and freedom in decision-making.

Once the firm has started to develop open innovation practices, it faces the need to create specific organizational structures that support the innovation workflow and extend achievements to all levels of the organization. These structures should focus on identifying knowledge opportunities that exist outside the firm and establishing explicit processes to acquire and transfer them within the organization. On other occasions, these structures should aim at creating connections with external agents and establishing collaborations with them that involve inflows and outflows of knowledge. In general, the units dedicated to open innovation in most of the organizations where they exist are small and characterized by their flexibility, responsiveness, and the ability to lead technical projects collaboratively (Bageac et al., 2020).

The possibilities of open innovation have been strongly boosted by digital technologies. New mechanisms for collaboration and interaction between firms are now possible supported by web-based software tools and digital services that foster the culture of openness and make open innovation practicable at all stages of the innovation process. These tools range from real-time conversation (e.g., Skype, Zoom), group writing applications (e.g., Google Drive, Teams), cloud document sharing (e.g., OneDrive, Dropbox), and advanced integrated workspaces (e.g., Slack). Most of the applications are provided as a freemium model.

Smart technologies are making possible the combination of open innovation practices and Big Data in the quest for firms to find new business opportunities and creating competitive advantages. In fact, Big Data and open innovation have their own challenges that become even more considerable and complex when mixed. This powerful combination poses significant challenges to the skills and knowledge of the workforce required to create and capture data that can be used for business innovation. Firms today must be able to collect, store, organize, process, and, above all, analyze, visualize, and interpret data, and they need the right people to accomplish all these complex tasks. The skills challenge is further amplified when data is used to perform open innovation processes that involve multiple actors. This is particularly relevant for SMEs for whom finding qualified personnel is not only difficult but also very expensive, so they will often need to turn to outsourcing practices.

The challenges that arise from the massive use of data for open innovation do not end here. Some of the most important ones have to do with the ability of firms to work with true and reliable data, both internal and external. Data for open innovation comes from multiple and often open sources that can be affected by various biases and inconsistencies. The consequences of unreliable, biased, or false data for the innovation process can be disastrous and lead to erroneous conclusions that can severely affect the performance of the firm in the marketplace. Owners and managers need to be aware of the importance of good data and spend time and resources to prepare their data mix well before bringing it into the innovation process, thus learning when to trust the results and when to criticize them.

Data privacy and security are also key challenges when firms work with Big Data and to which owners and managers must pay special attention if they are to be used in open innovation processes. This is especially relevant since open innovation involves multiple data sources from actors with diverse information and data management systems, some of which may contain weak processes that can limit reliable access to information and knowledge sharing. To overcome these risks that can arise throughout the firm data collection process, the open innovation paradigm encourages systems integration as a solution that can deliver even more competitive edge when it takes the form of ecosystems. However, systems integration is a major challenge that requires extensive cross-firm collaboration given the considerable heterogeneity of tools and management techniques associated with IT, knowledge management, and open innovation.

All the challenges above are further emphasized when it comes to SMEs that have limited resources to manage innovation processes or lack a structured approach to innovation. Often, owners and managers have few skills in or little knowledge about open innovation methods, so implementing an open innovation approach and coordinating collaboration with multiple actors becomes a challenging task for SMEs. In addition, the maturity period required for innovation initiatives performed by SMEs is usually much shorter than that of large firms, so they will tend to focus more on obtaining results that can be exploited in the short term than in sourcing the best possible innovations. Despite these particularities, SMEs generally have more agile and flexible decision-making processes in place and are capable of reacting to changes in their environment with greater determination, which makes them good candidates to adopt open innovation strategies (Torchia & Calabrò, 2019).

The benefits that SMEs can derive from open innovation practices are nonetheless significant. Collaborating with other SMEs or participating in clusters of firms can lead firms to break down traditional barriers and operate in the market as if they had a greater size and capabilities (i.e., better access to technological, financial, and human resources at low cost). Building partnerships of SMEs to work on open innovation also encourages member firms to innovate, allows them to share risks and sunk costs of failed projects and, ultimately, gives them the chance to benefit from the complementary capabilities and resources of others. However, none of this means that the firm's R&D department should cease to exist when the open innovation paradigm is implemented; quite the opposite, the in-house R&D department should acquire an even more decisive role under the open innovation paradigm. In fact, both open innovation and internal R&D are to be considered complementary. The real challenge is for the firm to develop absorptive capacity to recognize, acquire, and transfer knowledge and technology from external sources on the basis of its own expertise and know-how (Bogers et al., 2019).

Finally, open innovation is not without its drawbacks. Keep in mind that once a firm's knowledge assets are made available to other organizations to exploit, intellectual property can be difficult to protect and the benefits of innovation hard to capture (Abbate & Souca, 2013). Furthermore, the successful implementation of an open innovation strategy is based on the firm having a business model in place that is capable of retaining the value derived from innovation. Unless the firm has adequately addressed this important circumstance, open innovation can do more harm than good for the firm.

21.4 Agility and Open Innovation

The firms that best capitalize on open innovation are those that have the necessary organizational flexibility to restructure their processes, strategies, and business models in pursuit of the principles of the open innovation paradigm. This means the open innovation paradigm shares many principles and practices with the Agile framework context. Both aim to tear down corporate silos, strengthen open communication and collaboration between teams and people, and respond swiftly to changes in user/customer needs and preferences, and threats from competitors (Liao et al., 2019). User/customer engagement is continuous during the development of open innovation projects and takes place from the very beginning. Digital tools ensure that sharing and collaboration are much easier and that there is an open dialogue between the firm and the key stakeholders along the process.

From the moment the firm gathers complementary knowledge and technologies externally through inbound open innovation processes, it reduces the risks associated with experimentation, thus stimulating innovation, fostering communication with others, and providing greater flexibility in terms of the resources needed to innovate. Firms with greater external knowledge and skills are in a better position to make faster decisions and adapt their operating frameworks to changes in the environment and threats from competitors. In short, inbound open innovation processes positively influence organizational agility, while the latter improves inbound open innovation. The importance of outbound innovation is lower compared to inbound open innovation and the effects of agility on market capitalization are much less pronounced (Liao et al., 2019). Indeed, outbound open innovation is often the result of inbound open innovation. To harness the potential of both inbound open innovation and Agile principles, the firm must develop a culture and organizational set-up that addresses the following critical issues (Burchardt & Maisch, 2019):

- **Management must buy in:** To combine Agile principles and the open innovation paradigm and scale any mixed initiative across the organization, it is critical to gain the commitment and buy-in of the firm's top management. When top management understands both Agile and open innovation principles and shares the overall goals of the combined initiatives, it is possible to secure the necessary resources to drive open innovation and make critical decisions when needed.
- **Legal problems must be solved:** A common concern of top management when dealing with the open innovation paradigm is the protection of the resources at stake, especially those related to the protection of intellectual

property. Questions about how to protect certain data or information, how to establish collaboration agreements with other firms, and how to manage patents and other rights and obligations must be resolved beforehand.

- **Operational issues must be handled:** To act openly, with agility, and more focused on the needs of users/customers, it is vital to align the firm's operational processes in a different way. The firm will need continuous and deeper access to its customers and users to gain insight about their needs and preferences, it will have to collaborate with other actors with different goals and business models, and incentivize its workforce to share and collaborate, among other things.
- **Proactive change management to involve all stakeholders:** When the firm decides that it is going to implement the open innovation paradigm with Agile principles, all stakeholders must get involved. However, not everyone is ready to work in this way initially. Some won't know how to do it, others won't be able to, and there will surely be those who just don't want to do it. This is why the firm must perform proactive change management in order for all stakeholders to adapt to the new structure of values of the organization.
- **Decide when to be open and agile:** It is not always in the firm's interest to adopt an Agile and open innovation approach. Many times it can happen that a closed and traditional (waterfall) approach is more appropriate for what the firm needs. Sometimes making the transition to a new culture, new processes, or changing people's beliefs, habits and customs is simply an exaggerated effort that the firm cannot afford or does not want to make. Firms must be very aware of the pros and cons of implementing open innovation and Agile practices, and decide when is the best time to move forward.

A common way to address the issues above is to embed Agile principles into the hierarchical structures of the organization and progressively move step by step to achieve the desired goals. Owners and managers aiming to support smart organizational transformation will need to play a leadership role that understands, reflects, and promotes learning and creativity and promotes the development of key skills for everyone in the organization with the aim of creating competitive advantages for the firm. The drawbacks and limitations that the combination of open innovation and organizational agility entails for the tourism firm remains to be seen, as more research is needed on this particular topic.

21.5 Tips for Implementing Open Innovation

Although there are many tourism firms fully aware of the usefulness of the open innovation paradigm to drive the transition towards the smart organization, the problem is that firms usually do not have the capacity to put it into practice. Implementing an open innovation strategy in firms little or not used to opening their boundaries to the world, and given the number of factors affected, implies a high complexity. This situation becomes even harder when owners and manager must align the open innovation strategy with the firm's general business and systems strategy. Despite the obvious difficulties, owners and managers would better be convinced that the development of open innovation is essential to guarantee the continuity of tourism firms (Iglesias-Sánchez et al., 2019).

As owners and managers strive to implement open innovation practices, it is very important that they do so thinking about how they will create value (i.e., sourcing knowledge and technologies, establishing partnerships with other firms), and how they are going to capture that value, or put in another way, how they are going to integrate and transfer the new knowledge acquired and how they are going to convert it into new products and services that can be marketed. Unfortunately, there is no single formula that shows firms how to capture value, as every project and context is different (West & Bogers, 2013). However, it is paramount that firms choose an appropriate business model and the right smart technologies to support it. They must determine what knowledge they need to acquire to complement their internal capabilities and with what degree of openness they wish to operate. In the case of underutilized knowledge or technologies, the firm must also decide how much of that knowledge can be made available to other firms and thus generate additional revenue for the business. Ultimately, the firm must find a balance between open and closed innovation and work out how to develop both modes of innovation. Figure 21.2 shows some alternative forms of open innovation that are available to the tourism firm.

For tourism firms to be successful using external knowledge, they will need to realign the organization's processes and resources to accommodate that new knowledge. This will require developing different organizational practices, such as extensive delegation of responsibilities to people, lateral and vertical communication, and incentives to share knowledge (Bogers et al., 2019). Sometimes it will even be necessary to reshape the organizational culture itself to make it more participatory and collaborative. At times, firms may face the

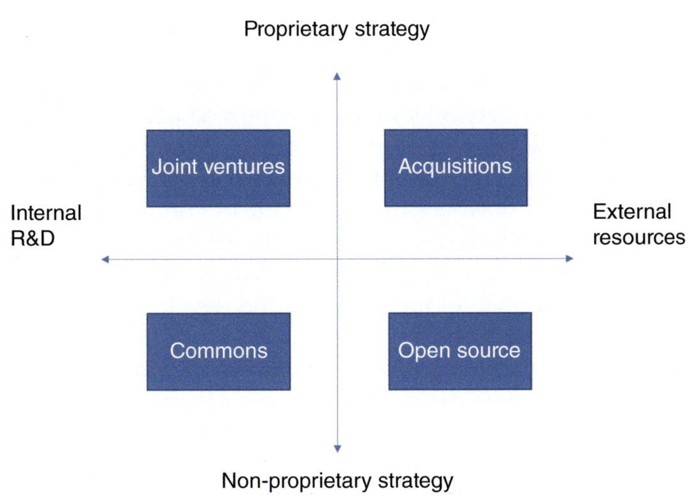

Fig. 21.2. Different forms of open innovation. Source: own elaboration based on Bogers et al. (2019).

challenge of not finding enough good partners to develop their open innovation initiatives, or perhaps the coordination work required is enormous given the interdependencies that exist between the actors involved. In these cases, it may be better to go it alone. Furthermore, owners and managers should not forget that open innovation has its limits. In situations when the available knowledge is out of date or there are no partners willing to risk a certain project, there is no way that open innovation can help. After all, for open innovation to be successful there must be a rich technology base that can be shared and complemented by the stakeholders.

Organizing for open innovation

There are multiple pathways to implement inbound (outside-in) open innovation that generally depend on the organizational agility of the firm. Very often firms choose between a team-centered model and an individual-centered model (Pellizzoni et al., 2019), although these are not the only viable approaches. The team-centered model is characterized by the presence of a strong managerial role with broad powers, and a clear and well-defined distribution of roles and responsibilities within the organization. Most surely top management commitment is guaranteed, and decisions are made top down, thus allowing open innovation to scale up quickly simply by adding more resources where needed.

The individual-centered model is a completely different approach. People work with a high degree of autonomy and are empowered to make their own decisions, which is reminiscent of Agile management principles and entrepreneurship dynamics. These principles are applied to managing the open innovation initiative and making it evolve through different iterative phases of development where open innovation is progressively learned and understood and gradually refined. The key to an effective implementation of the individual-centered model lies in granting each employee the appropriate level of autonomy to work, organize resources, and make decisions. Unlike the top-down approach, the individual model is bottom up, as employees draw on their formal and informal networks of relationships to accomplish innovation projects. While the individual approach is more affordable than the team-centric approach for the firm (i.e., there is no need to invest in dedicated roles, teams, and resources), it is however more difficult to control and to scale up as the commitment of top management is usually weaker and it is harder to secure the resources needed by open innovation. Nor is it easy to find the right people who can lead the launch of the initiative and develop a culture that permeates the rest of the organization.

There may be times when firms prefer to work with a mixed team and individual approach. In these cases, the firm's top management commits and participates actively during the first phases of open innovation and creates an ad-hoc structure in charge of managing the open innovation initiative. For their part, managers can decide what projects can be managed in a less structured way, engaging individuals with the ability to lead the tasks. In this way, it is possible to harness the benefits of both models without incurring large costs or causing, at least initially, large impacts on the organization (Pellizzoni et al., 2019).

Discussion Questions

- What impact does "openness" have on tourism firms? What factors influence the decision to adopt one level or another?
- How do tourism firms source smart technologies? What factors influence tourism firms' sourcing methods?
- How can tourism firms be classified according to their approach to open innovation?
- What kind of complementary external assets do tourism firms usually need to generate greater competitive advantages?
- What kind of culture is necessary to unfold open innovation in the tourism firm? What characteristics distinguish that culture?
- How and to what extent do tourism firms use social media to develop open innovation?
- Are tourism firms concerned about protecting their knowledge assets and making them available to other firms? What are the differences between large tourism firms and SMEs in this regard?

Part 4 What's Next?

Readers who have reached this part of the book have a broad perspective of what the Smart Revolution is all about, the technologies that are making it possible, and the management practices that are necessary to successfully tackle the smart transformation process. But where to start? What are the first steps an owner or manager should take, and in what direction, to build a smart organization? What's more, now that the reader understands the true magnitude of the challenge ahead for the tourism firm, what's next? Are such radical changes going to continue in the future? How can my organization be prepared for the ongoing challenge? Although the Smart Revolution has just begun and there is still not enough expertise and models of good practice to guide owners and managers on what works and what does not, in this part of the book the author ventures to suggest a framework of action that can be used to undertake the transition from a conventional firm to a smart one. Furthermore, those technological and non-technological trends that are very likely to shape the future capabilities of tourism firms are examined, along with their potential to create and sustain competitive advantage.

22 Smart Transformation Framework

Smart transformation has become a pervasive theme in the discourse of tourism business leaders. The slogans associated with transformation are continually repeated in all kinds of contexts and forums, almost always lacking depth and clarity. Today it is almost impossible to find a C-level executive from a large or even medium-sized tourism firm that has not publicly committed to the need for a digital or smart transformation (Trenkle, 2019). However, it cannot be said for sure that owners and managers really understand what a smart tourism business means. Different stakeholders usually have very different, if not inconsistent, views on what it means to be a smart hotel or a smart tour operator. This makes it extremely difficult to implement truly smart business models. Some business leaders consider that "smartness" serves to improve the brand image and reputation of a business, others that it is related to the optimization of operations and cost reduction, and there are still some who think that it is a source of competitive advantages to attract customers (Mehraliyev et al., 2020).

The continual advancement of smart technologies and the transformational practices that make it possible to implement and extract value from them bring an untold number of challenges for firms of all sizes. Transformation is a complex process in which not only technologies and technology specialists are involved, but everything that the organization is and does comes into play. Dealing with smart transformation ultimately involves significant risks that are highly likely to end up resulting in failure for the organization. For this reason, business owners and managers must fully understand the nature of smart transformation and decide to tackle it early and with the help of tools that can guide them through this complicated landscape. One such tool available to owners and managers is the Smart Transformation Framework (STF).

Today, there are some STFs available for business leaders to analyze, diagnose, and establish a strategy to address smart transformation if by STFs we also mean digital transformation frameworks that are the precursors of STFs. These

 Managing the Smart Revolution in Tourism Firms: Innovation and Value Creation in the Era of Data (F.J. Navarro Meneses)
DOI: 10.1079/9781800620261.0022

frameworks generally provide a blueprint or systematic plan that includes standardized dimensions, action areas, and work tasks to help a firm become a "smarter" organization. Choosing an STF to work across the smart transformation process is essential for the organization because this way it will have a clearer and more precise roadmap for how to address the technological and organizational disruption the organization will face, and how to adapt the structure, operations, technologies, culture, etc. to the changes to come.

An STF is not going to protect the organization from disruption, nor is its mission to encapsulate it to prevent change and its negative effects at all costs. It's quite the opposite. The mission of an STF is to showcase the options that the organization can take to tackle change and start a transformation process, trying not to leave anything important behind and avoiding making the mistakes that others have already made.

22.1 Components of the Smart Transformation Framework

Smart transformation is a journey that involves a complex environment of tasks, resources, and capabilities, but one that can also create considerable opportunities for tourism firms that commit to change. When a firm embarks on smart transformation, it is seeking to become a "wiser" and more agile organization capable of responding to the changing needs of its customers and the environment through a superior ability to make faster and more accurate decisions. However, smart transformation initiatives can become quite a chaotic and risky enterprise if the organization does not have a roadmap that organizes the process and clearly sets forth the strategies to be pursued. Fortunately, firms have STFs at their disposal as a methodological tool that can assist them in this process of profound change, giving them greater confidence and certainty.

An STF is a structured model made up of management dimensions, areas of action, and monitoring processes that guide the organization's smart transformation efforts in an orderly and systematic manner. An STF is essential for tourism firms to work effectively in each phase of the transformation process and find answers to the challenges posed by the dynamics of change. It also empowers members of the organization to assess the status quo, understand the current state and capabilities of the organization, and identify where there are gaps and what are the key areas to start with to achieve the desired goal. Generally, an STF is the result of the analytical synthesis of the best practices developed to date in the field of digital/smart transformation, drafted by interdisciplinary teams made up of different stakeholders (i.e., members of the academy, specialists from the technology industry, business firms, consultants, government representatives, etc.). By distilling all this valuable insights from experience into a few key practices, the STF may illuminate the way forward for business leaders to drive smart transformation.

Based on the review of the literature carried out by several authors (Bumann & Peter, 2019; Teichert, 2019), six key dimensions or core action areas of an STF can be identified: strategy, organization, culture, technology, customers, and people (Fig. 22.1).

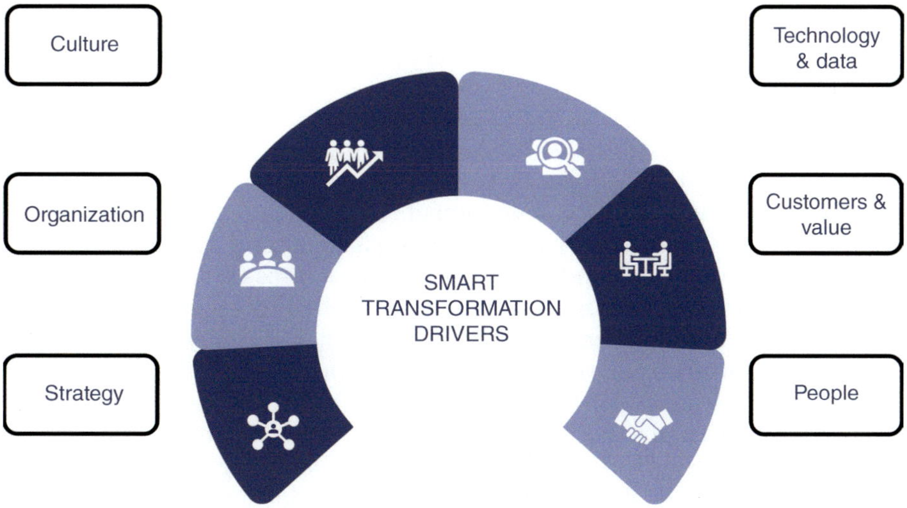

Fig. 22.1. Smart transformation drivers. Source: own elaboration.

Strategy

Formulating a smart transformation strategy is key for the tourism firm to achieve a successful transformation. However, experience shows that there is no consensus among academics, firms, vendors, and consultants on what is the best way to articulate this type of strategy or what relevance it should have for the organization. Opinions are varied and divided in this respect, between those who consider that the transformation strategy is just another IT strategy that must be subject to (and aligned with) the general business strategy of the firm; those who consider that transformation requires a strategy that merges the IT strategy and the business strategy; and there are still those who argue that smart transformation is so important that it requires a specific top-level strategy separate from other functional strategies of the firm and that coordinates the drivers of smart transformation.

As organizations approach the early stages of smart transformation, it's common for them to pay more attention to specific technologies (particularly those focused on operations) and be less concerned with developing an organization-wide strategy. These organizations are often more reluctant to invest in relevant skills for their employees. However, as firms mature in the transformation process, they become more collaborative and more likely to accept risks. They also feel better equipped to craft a transformation strategy and are even willing to have a dedicated person or team to lead transformation initiatives.

The transformation strategy should be well documented and communicated at all levels of the organization so that employees understand and internalize it. Moreover, the firm should ensure that the smart transformation strategy has committed enough resources to be implemented and that it is updated regularly to adapt to changes in the environment, and trends, and the lessons learned along the way. Finally, it is key that the strategy has the support and recognition of the firm's leaders, not only at the time of its conception, but especially when it is time to implement the planned actions.

Organization

Because everything in the environment surrounding the tourism firm is changing faster than ever, the organizational model must also transform to support an ever-changing market and operating context. If firms continue to work according to traditional standards, they will have a hard time adapting to the Smart Revolution and will end up dying. Consequently, the way in which organizations are governed and work become key factors. Rigid processes and hierarchical structures are no longer useful to deal with the complexity around the firm, instead firms need to become more flexible and open. Organizations undertaking smart transformation must foster organizational agility and collaboration within and outside the firm. Firms should adopt open innovation practices that extend to all areas or units that produce innovation within the organization, and speed up development cycles and shorten time to market for products and services. It is no longer enough for firms to think of innovation as something they should do from time to time or when they see that everyone else is doing it or the market changes, but that innovation is a continuous process based on short development cycles (which will give the firm a greater capacity to adapt).

Tourism firms need to adopt an approach based on collaboration with third parties and partnership in networks and ecosystems through which they can build win–win relationships with stakeholders and leverage the capabilities of others. This also means that many firms will be inclined to implement Agile working methods and lean operations that, away from traditional hierarchies and overly formal processes, allow them to rapidly iterate the processes of designing, refining, and delivering products and services to customers. All these changes mean that employees will need to adapt accordingly, and new roles will have to emerge to take full responsibility for the smart transformation process. Many organizations will create new roles to meet this need, such as the Chief Digital Officer (CDO) or the Chief Transformation Officer (CTO), although in other cases, it will still be the Chief Information Office (CIO) who is at the forefront of the transformation. There will be organizations in which the roles (CDO/CTO and CIO) coexist and it is necessary to maintain a balance between them.

Culture

Culture is a key factor in achieving a successful smart transformation. Smart firms have the capacity for rapid experimentation, invest in talent, and work with Agile practices focused on extracting value from data. These types of firms develop a culture that encourages freedom to experiment and offer their employees the chance to continuously test new ideas, thus increasing the chances of having an impact on the firm's performance. That's why, in addition to highly valuing hard skills (e.g., analytical and mathematical methods, computer science, etc.), they will also value soft skills, such as the ability to lead change initiatives, manage teams, perform tasks autonomously, make responsible decisions, creativity, etc.

Culture development implies a strong commitment on the part of the leadership team, which must always support the change strategy in the organization and communicate a clear idea of the target culture in line with the vision and purpose of the organization. Leaders must lead by example and show through their own behavior what changes need to be made and why they are necessary. This highlights how important it is for a firm that seeks to transform to have leaders who envision the future, know how to communicate their vision to others, and inspire and motivate people to pursue a purpose, being aware of the risks that all this entails.

Technology and data

The use and development of technologies is another key action area within the STF. Smart firms have a strong focus on the implementation and use of technologies that makes them different from the rest. While for many organizations the technology dimension boils down to having a flashy front-end and a well-designed website, or doing a "facelift" here and there from time to time to feel up to date and on trend, for smart firms it means making an extra effort that includes the complete transformation of their technological capabilities. These organizations operate beyond "off-the-shelf" technology solutions that have a broad installed base, and instead explore and exploit a mix of emerging technologies such as Big Data, analytics, artificial intelligence (AI), and the Internet of Things (IoT). They also take advantage of modern architectures such as the cloud and application programming interfaces (APIs) to add cutting-edge technology capabilities in key areas of the business at very low cost. These implementations are often accompanied by the introduction of Agile work methods to gain flexibility and speed in responding to customer needs. In this way, smart firms break ties with legacy systems that hamper their innovation efforts and are a drag on their agility.

Additionally, with the growing threats of cybercrime, firms are increasingly concerned with IT security and protecting access to key assets and resources. This involves not only the personnel in charge of IT tasks, but all the employees of the organization, who must understand what is at stake and adhere to established security rules to ensure continuity of business operations and access to key business resources, including data. On the other hand, data is a key asset for smart tourism firms and their ability to use data to inform business planning and decision-making is critical. Exploiting data is particularly effective to improve revenue management and use dynamic prices, as well as to understand the type of products and services that customers demand, to identify new markets and potential customer segments, etc.

Customer and value

Firms must leverage the new capabilities offered by smart technologies to make more effective and efficient use of their data, especially customer data. When data is used properly, organizations can tailor their products and services based on the data generated by their customers. Data not only affects the customer orientation of firms, but it can also be used to measure performance so that the organization can act accordingly. If something goes wrong, the firm can more easily trace the causes and provide solutions. That's why smart firms care about acquiring new capabilities and competencies that allow them to collect more data, turn it into insights, and use it for things like predicting customer behavior, delivering personalized products and services, and offering better customer experiences.

Increasingly, tourism customers expect firms not only to be able to respond to their needs when they need it and where they need it, but also to be able to remember them and anticipate their future demands. This enables customers to engage with firms across multiple touchpoints both digital and physical, leading firms to bridge the digital and physical worlds to offer hybrid interaction channels that provide flexibility and convenience to the customers. This poses significant challenges for tourism organizations as they must ensure that the customer experience is consistent across all touch points and that communication is designed to support a multi-channel strategy. Furthermore, since organizations are starting to involve customers in all phases of the product and service innovation process, customers increasingly play a relevant role as co-creators of the products and services that they later consume. Ultimately, the tourism firm must define its own value creation model, focusing on building a value network that will generate revenue (i.e., B2C, B2B, online sales, etc.). All the above are areas in which smart technologies are going to play a crucial role.

People

In addition to the many technological resources that are necessary to achieve a successful smart transformation, it is crucial that tourism firms find ways to equip their employees with the necessary skills and competencies to compete in a smarter world. Developing people skills involves attracting more talent, enhancing expertise, implementing new technologies, and understanding and exploiting the opportunities that technology can offer (OECD, 2020). Smart organizations will have to develop and implement appropriate training schemes to expand the competencies and skills that their employees will need. This involves taking stock of the capabilities of the workforce and the resources available to the firm and assess whether there is room for upskilling. To do this, firms must actively seek collaboration with universities, incubators, and other institutions to gain access to the critical talent they need. Furthermore, organizations will need to create a workplace that is attractive and flexible for their future workers. This means offering flexible working hours, digital workspaces, mobile work, etc. that increase employee motivation and lower infrastructure costs, and attract digitally savvy millennials.

22.2 Getting Started with Smart Transformation

Smart transformation can be approached through the sequence of phases contained in the STF (Fig. 22.2), each of which contains a set of key tasks to advance the change process. These phases are usually independent of the industry (although some industries may develop its own STF) and the size of the organization, and encompass generic topics that are applicable to most organizations. Using an STF helps organization members lay the groundwork for how they can work on smart transformation, as well as create an Agile way of thinking and acting. The STF also helps stakeholders better align and engage in the key activities of each phase. Figure 22.2 shows the five phases that can guide a tourism firm to start the smart transformation process. Once these five phases are completed, the firm must begin the implementation of the actions that have resulted from the STF.

Understand the problem

Smart transformation must start with a collective brainstorming effort where everyone in the organization understands the problem they are facing and is aware of the challenges ahead. Sometimes this is not an easy task because, although most tourism firms are used to sensing the effects that occur in their business (i.e., customers stop buying, customers are not loyal), they often don't know the reasons.

In this first phase, the firm must assess its readiness and maturity to tackle the smart transformation process. To do this, hypotheses must be formulated about what is happening in the firm, the depth of the disruption that is taking place around it, and what may happen in the future. This phase should also include a collective reflection through which the firm's ambitions are made explicit, and a shared vision of its future is established. These hypotheses will have to be verified in the next phase once more information is gathered.

Collective creativity techniques can be used, such as visioning and brainstorming, in order to extract expert knowledge from the leaders and members of the organization. Workshops can be held with external experts from the tourism and technology industry to help create a holistic view of the transformation process. Once all this knowledge has been gathered, it is convenient to establish a research plan that lays the foundations for aggregating all the information that will have to be analyzed before defining the strategies and concrete actions of transformation.

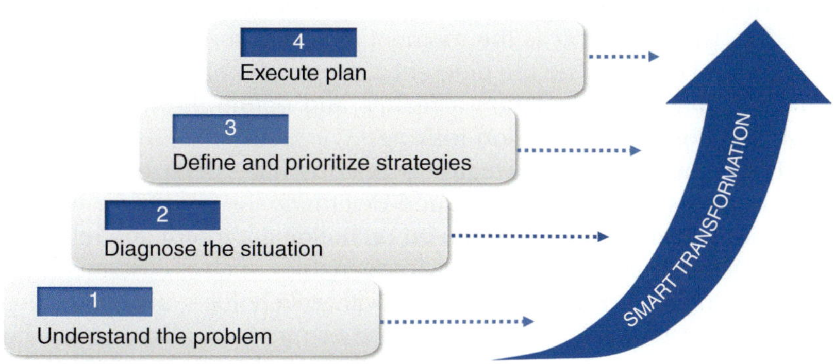

Fig. 22.2. Phases of the Smart Transformation Framework. Source: own elaboration.

It is also important that the work plan to be followed by the members of the organization is established and that the people who are going to participate in the work team are designated, in addition to deciding which are the main work milestones, the calendar, the decision-makers, etc. It is also highly recommended that at this stage a transformation control and monitoring unit is appointed to oversee the progress of the transformation process. This unit may be formed by a representation of the organization's leaders with the ability to decide on the resources and the course of action to be taken by the organization.

Diagnose the situation

From all the information and expert knowledge gathered through the research work in the previous phase, the firm will be able to better understand where it comes from and where it can go. With this valuable information in its hands, it is time for the firm to make the most objective diagnosis possible about its current situation and its level of digital and smartness maturity, reimagining what its future could look like and how it could capture value from it. The work to be carried out in this phase includes an exhaustive evaluation of the organization's capabilities and the analysis of the functions that contribute to the creation of competitive advantages for the business. Through this diagnosis, the firm will obtain a clear picture of the change initiatives it is currently carrying out and its status. Furthermore, the diagnosis will allow the firm to define its opportunity space; in other words, determining what are the competitive advantages that the firm has and which ones it could focus on in the future to untap its potential, and identifying the existing gaps that it will have to fill in order to achieve its transformation ambitions.

Define and prioritize strategies

Now is the time to make sense of the diagnosis and decide "what to transform", i.e., decide on the business model, the organizational structure, people, processes, value offerings, customer engagement model, etc. The firm must build its own strategic vision and develop a realistic roadmap that can be put into practice. Therefore, this is the moment in which the firm's teams must reflect broadly and critically on the different strategic alternatives available to the firm to make smart transformation a reality, and define the new futures that will serve as a context for decision-making.

Once the different alternative scenarios have been established, the firm must make its strategies explicit and determine the priority for their execution. Strategies should not only be focused on building a robust and reliable IT framework, but must pay attention to multiple variables, such as customer focus, exploring and exploiting data, creating value with stakeholders, and facilitating fast and efficient decision-making. Without these elements in place, it will be difficult for the strategies to have any relevant effect on the firm's smart transformation.

Transformation strategies should cover a whole set of functional (e.g., human resources, ICT, finance), operational (e.g., processes, sales, marketing, products), and corporate strategies (e.g., business models, organizational structure, leadership). All these strategies must be integrated into a road map that shows the way forward towards the organization's vision.

The firm must draft business cases that outline the way in which the required investments are going to be made and the benefits that are expected to be obtained from each strategic line. Early investments are unlikely to generate returns in excess of the cost of capital, while shareholders and the board of directors are likely to expect immediate results. Therefore, the ideal is to devise short-term change initiatives that can generate quick results and, in the process, serve to finance the next long-term initiatives. Existing and emerging technologies that may have a greater impact on the firm should be identified, as well as the future technological architecture of the firm and how smart technologies will be used. This will require the firm to have a solid execution and performance measurement framework.

Execute plan

In this phase, the firm establishes the deployment plan for the smart transformation strategies. This includes a change management program that ensures that the changes introduced by smart transformation can be integrated into the organization, and that employees are fully aware of what is going to change, why it is going to change, and what can be expected from change. As the organization begins to experience change and the benefits begin to materialize, its management and leadership model will need to adapt. That is why it is important that the firm has a 360-degree vision of the transformation process and is prepared for continuous learning.

The firm should lay the groundwork for a new governance model that supports the transition from the current situation to the future. This new model should contemplate how the firm's organizational framework will evolve and the skills and abilities that future employees must have, as well as the way in which employees will collaborate inside and outside the organization. It is also key that the firm establishes an objective operating framework that states how the processes are going to be organized and how the functions and responsibilities are going to be distributed within the organization. Finally, the firm will need to measure the value and impact of smart transformation initiatives and use the results to refine the change implementation processes.

Culture change among owners, managers, and employees will be a key indicator of the success of the smart transformation process. Without an innovative and data-centric culture, smart transformation will not bear fruit and will be doomed to failure. It will require strong leadership to direct the implementation of transformation strategies and empower the people who work in the organization to contribute, innovate and become more creative and productive. In short, a governance model will have to lay the foundations for a culture that supports high performance by people.

22.3 The Role of Platforms

Digital platforms play a very important role in tourism today due to their innovative business model and their proven ability to concentrate supply and demand in a single place. Platforms business models are characterized by a high level of productivity that is the result of the innovative way in which multiple stakeholders have been engaged to co-create content, collect and exploit market data, and share value among users and customers (OECD, 2017a). In just a few years, platforms have experienced strong growth to the point that many tourism firms around the world now seek their place within these ecosystems, having to adapt their business models in accordance with the requirements that govern them.

Digital platforms such as Tripadvisor, Airbnb, and Booking.com are key to the daily operations of many tourism firms, but they are also key enablers of digital and smart transformation. These platforms provide firms with technical and business functionalities that go beyond online transaction processing, as they offer value-added functionalities related to data access, data analysis, and market intelligence that allow SMEs to learn more about their business and customers (L. Li et al., 2018). In this way, through the platforms tourism SMEs have an easier time overcoming many of the barriers in the effective use and adoption of digital and smart technologies and are encouraged to enter the culture of data and acquire new resources and analytical capabilities.

Furthermore, as digital platforms grow and evolve, platform providers are updating the services they offer and refining the terms of business to adjust them to the changing needs of customers and users. Platforms encourage firms to innovate to meet market and customer demands and to follow the rules set by the platform providers in order to maintain competitive advantages over their competitors (Jiao & Chen, 2019). Together, all these functions and capabilities provided by the platforms allow many tourism SMEs to increase their skills, grow their businesses in the digital world, and expand their markets. Additionally, some platforms offer management-oriented services, such as learning sessions, webinars, conferences, meeting events, etc., which are instrumental in helping tourism firms join open innovation ecosystems and prepare for digital and smart transformation. Without many of the services and support these platforms provide, it is likely that many owners and managers would not have cared about digital transformation and would still be far from understanding the true dimension of the Smart Revolution.

22.4 Smart Transformation in SMEs

SMEs are a large part of the value creation units and the backbone of most economies in the world. However, more than two-thirds of SMEs are low-tech, which means that a very high proportion of them struggle daily to make progress in digital/smart transformation and still have years of work ahead to reach a level that allows them to be competitive in the new smart context (Hönigsberg & Dinter, 2019). Provided that smart transformation can be considered a radical innovation itself, since it implies a profound change in processes, in the firm's value proposition, in the relationships with customers and suppliers, and even in the business model, it still offers many opportunities to tourism SMEs. The other side of the coin is that smart transformation also imposes important obligations: having adequate financial resources to face change, having qualified personnel to lead and manage the process, overcoming resistance and organizational challenges, reformulating legacy technologies and the operations model, ensuring data access and reliability, etc. Many of these are also common obstacles to innovation.

There are many owners and managers who think that SMEs will be the most negatively affected by smart transformation due to the complexity of the process and the amount of resources it requires (Beliveau et al., 2018). There is no doubt that these factors are major obstacles, but it is also clear that SMEs are among the main beneficiaries of smart transformation. Very often SMEs are in a better position to deliver smart transformation faster than large firms and can more easily develop new smart IT frameworks from scratch. Owners and managers must be aware that SMEs and their organizational characteristics differentiate them from other types of firms, which makes smart transformation a distinctive process due to the following factors (Trenkle, 2019):

1. **Leadership:** Which is usually exercised by the owner of the firm who is the one that makes most of the important management decisions that affect the business.
2. **Organization:** Which is usually of a hierarchical type, with the owner and managers located at the top of the pyramid and accumulating various roles. The rest of the staff is at the base and is focused on executing the operational tasks of the business and supporting decision-making.
3. **Process design:** With a low degree of work distribution, low openness to external collaboration, and a high dependence on "universal" technologies that are more focused on supporting business operations and with little orientation towards the creation and management of knowledge.
4. **Sales and marketing:** Mainly focused on regional and/or local customer segments, who are offered specialized services in which personal and human relationships prevail.
5. **Innovative capacity:** Usually without a formal R&D department with the right people and resources; scarce and very short-range R&D activities; limited financial resources; and a more reactive than proactive purpose.

Operational aspects and especially those that have to do with people (i.e., organizational structure, leadership) have a higher relevance in SMEs when compared to process frameworks and well-defined formal responsibilities, which predominate in large firms. The human component is key in the conception of smart transformation in SMEs and helps to better frame the context in which SMEs develop. Owners and managers, assisted by small management teams, are generally the ones who take responsibility for transformation actions. In SMEs there is normally no structure of C-level responsibilities as in large firms with perfectly defined roles and responsibilities. Hence, it is the owner or managers who must find a way to personally acquire the knowledge to improve their skills and understand the terms of the Smart Revolution. Below are some of the most relevant obstacles that are specific to SMEs and that every owner and manager should take into account when tackling the smart transformation.

Economic and financial

Financing smart transformation activities is one of the hardest parts for SMEs, especially as their financial resources are often very limited. The volume of investments required for smart transformation is significant, both in monetary terms and in terms of people and time. Therefore, when dealing with smart transformation, it is always desirable to have a solid financial structure. However, it is quite common to see that banks are reluctant to lend money to SMEs for ambitious technology projects when there is no business case that can fully guarantee a level of performance. This problem is exacerbated when the ROI of smart technologies is often not clear, either because the potential of some technologies is not always evident, or due to the difficulties faced by SMEs when measuring or anticipating results. Most of the time, owners and managers have no choice but to seek other financing options to raise sufficient resources, and this includes, in addition to the firm's own funds, applying for public aid programs (e.g., digitalization programs, economic development programs, loan guarantees, R&D funds, etc.) and other less conventional private capital formulas such as venture capital and crowdfunding.

Another key economic/financial issue is that owners and managers seem to be very clear that when the products and services offered by the firm have a low value, it is not worth betting on smart technologies that are complex and expensive. This contrasts with the pressure they constantly have to keep margins stable (if not growing) and take action immediately when margins deteriorate for whatever reason. Smart transformation can offer owners and managers significant opportunities to increase efficiencies and exploit new revenue streams, so it would certainly be worthwhile to evaluate different financing options.

Organizational and cultural

Smart transformation requires SMEs to reconsider their organizational structure and the principles and values on which the organizational culture is based. This usually drives SMEs to rethink the structure of the organization, redefine roles and responsibilities, and ensure that they have the appropriate employees and skills to successfully address the transformation process. Obstacles often appear as with leaders reluctant to make deep changes at the organizational level; organizations that oppose or are resistant to change; employees who fear losing their jobs or worsening their working conditions and do not want to get involved or committed; lack of understanding of the principles of the Smart Revolution and its implications for business, etc.

Recent studies suggest that the best way to assign responsibilities at the forefront of digital transformation is to share them between the CDO and CIO (Singh & Hess, 2020). However, in SMEs there is not usually this diversity of roles, and most often it is the owner or manager of the firm who becomes the person in charge of the smart transformation process. This is not necessarily a problem, as having a single person in charge is generally considered a success factor (Hönigsberg & Dinter, 2019). The fact that the firm has a person responsible for the transformation is not a guarantee of success either – it all depends on how this person exercises responsibility with vision, commitment, and leadership. Furthermore, SMEs often assess how the digital/smart transformation process will integrate within the firm's existing structures, and they will rarely create an autonomous entity to manage the transformation, except when the transformation initiative becomes very ambitious. SMEs are also more limited than large firms when it comes to acquiring skills and talent, and they tend to focus more on developing their own staff and attracting young talent with digital skills.

Competencies and resources

SMEs face the usual problem of lack of technical knowledge when dealing with transformation processes. In general, SMEs struggle to find qualified employees with the required technical competences that they need to incorporate into their teams and that have the necessary knowledge and skills. In addition to this lack of qualified employees, there is also the barrier of high effort involved in coordinating the tasks and activities associated with the transformation to ensure that the entire organization pursues the same objectives.

Legal

In many developed and developing countries bureaucracy and restrictive laws and regulations are a major obstacle to implementing smart transformation. Among the legislation that most affects the development of smartization is that related to the privacy of personal data, data security, and consumer protection. In other countries, however, legal barriers are less relevant.

Technical

The pro-technological ambition of the owners and managers of the firm to adopt the latest smart technologies is crucial to anticipate the intensity in the implementation of smart transformation and the benefits for the organization. This personal characteristic is often the result of the knowledge and previous training of the owner/manager, the entrepreneurial and dynamic character, personal experience, existing skills in the workforce (and in the firm's networks), and the financial situation of the firm. Some owners and managers consider that smart technologies are facilitators of the strategic objectives of the firm and by implementing them, new business opportunities can arise. These individuals actively seek out new technological advances and constantly scan the technology landscape for new opportunities. Other owners and managers perceive smart technologies more as necessary tools to support business operations, and as such these technologies are important to remain competitive. Still there are other owners and managers who choose to stay as far away from the technology as possible, and consider it a no-go field that they prefer to leave in the hands of second-level IT technicians. Last but not least, SMEs often have quite different ICT systems in place, and their data is stored in different silos that often do not communicate with each other, which is a major obstacle to smart transformation. Technological immaturity is common in SMEs and this is reflected in the great uncertainty about the reliability of systems and data, as well as in the weak IT infrastructure and the interoperability issues that arise between systems and applications.

Implementation process

SMEs generally lack a methodical approach when it comes to implementing smart transformation. Often this is due to the very novelty of the smartization phenomenon and the fact that there is not yet enough experience and knowledge that can be crystallized into a commonly accepted methodological approach or, if it exists, it is not easily adapted to the specific needs of SMEs. On other occasions, the reasons are that many owners and managers are unaware of the existence of STF approaches and that the number of best practices is still small.

Discussion Questions

- What methodological approaches could be more valuable for tourism firms to successfully address smart transformation? What elements should these approaches include consider to be adopted by tourism firms?
- What factors, other than financial constraints, may have a larger influence when prioritizing smart transformation strategies?
- What factors does the tourism firm depend on to develop an organizational culture that accelerates smart transformation?
- What factors differentiate smart transformation in SMEs and in large firms?
- Do you know of any good practice from an SME that shows how smart technologies have been successfully adopted? Describe the fundamental aspects that characterize it.
- What role do you think platforms such as Tripadvisor, Airbnb or Booking.com will have in the future as tourism firms become smarter?
- Which of the specificities discussed in this chapter do you think are shaping most the transformation process of tourism SMEs?

23 The Future of Smartization

The advance towards smartization is accelerating in the tourism industry and affects all the actors in the ecosystem, especially tourism firms. Milestones unthinkable a few years ago have been reached that are completely transforming tourism as we know it, generating new oceans of opportunities that will be exploited by a new type of organization with capabilities radically different from today's and whose business models will be built on smart technologies. Eventually, the big question is whether and to what extent the smartization of tourism firms will be good for consumers and firms themselves, and whether it will mark the beginning of an era of wealth and prosperity for all, or what is to come is a painful adjustment process with unpredictable costs in the business, economic, and social spheres for which firms and society are not prepared. In other words, where is the Smart Revolution taking us and what are the implications for the future of tourism?

Many critics of smartization fear the effects of the rate at which job obsolescence will occur due to the mass adoption of smart technologies. They argue that while the impact of the Industrial Revolution was fully felt after two centuries and that of the digital revolution took three or four decades, it may not be more than a decade or two before the world feels the full effects of the Smart Revolution. There are many who fear that the magnitude of the jobs that will be displaced is such (up to 47% of all occupations according to some studies), that the social structure in most countries will not be able to withstand this rapid and profound restructuring of the labor market (Makridakis, 2017). Even if governments were able to launch massive retraining programmes, it would not be easy to avoid major social disruption, as the new jobs will require skills that are not readily available. This concerns labor groups such as taxi drivers and bus drivers, who would surely be among the first to be replaced by autonomous vehicles, but the range of jobs at stake is much broader.

DOI: 10.1079/9781800620261.0023

The changes brought by the Smart Revolution will predictably extend to consumer buying habits, which presumably will change so drastically that they will eventually eliminate the need for thousands of jobs. Entire industries have already begun to experience these effects, such as the banking sector, where customer service offices have been closing for a few years and thousands of employees are being laid off or retired; or in retail, where neighborhood stores are closing while the volume of electronic commerce increases. This profound social and economic transformation not only affects less qualified professional profiles, but also highly qualified professions, such as financial advisors, medical specialists, or those related to high-level computing tasks, which could be at risk of being automated using algorithms made available for some time now.

Yet, despite this somewhat daunting perspective, it is perhaps fair to say that the Smart Revolution is not going to be the end of the world as we know it. It is true that the world is going to change a lot and in a short time, but we should not underestimate the ability of human beings and societies to adapt to change. In the end, it has been this ability to adapt that has brought humanity to its greatest levels of wealth. Although it is foreseeable that the vast majority of jobs that involve physical work or that have to do with data processing will disappear (78%), as they are automated or replaced by robots, many others will continue to be difficult to replace, such as those that involve some intellectual expertise (18%) or managing others (9%) (Chui et al., 2016). On the other hand, jobs related to soft skills will become even more important and will grow in the coming years.

The next question we could ask ourselves is: will the new jobs created offset the jobs lost as a result of the Smart Revolution? It is not surprising that there are answers for all tastes. There are those who argue that labor supply is fixed, and that people will compete with the rest of the factors of production (including robots) for a place in the chain. Others think that human desires and needs are infinite, and that entrepreneurs will always devise ways to harness them and create new opportunities. There is nothing wrong with finding reasons to worry, but perhaps we should not waste too much time in sterile debates that lead nowhere, such as whether the jobs emerging from the Smart Revolution will also be automatable, and focus instead on how best to prepare ourselves to face a social, economic, and business context that is going to be radically different and that will create sustainable opportunities for all.

Let us think that the Smart Revolution will surely increase the number of new firms that will exploit Big Data, analytics, and artificial intelligence (AI) technologies, with the support of new tools born from crowdfunding and venture capital that will improve their chances of success. The innovative ideas that are going to shape the Smart Revolution will come from anywhere and will not require large and expensive R&D laboratories to be devised, nor will they require large amounts of funding to be developed or commercialized (Makridakis, 2017). Today, some countries have stepped forward, given the growing importance of innovation and entrepreneurship to their economies, and continually encourage their innovators to start new businesses (e.g., Israel, China, France, United Kingdom, New Zealand). These efforts will lead to a greater openness of innovation processes in firms and will emphasize cooperation between knowledge and innovation centers around the world to accelerate technological innovation. Possibly this new spiral will result in a higher level of competition worldwide, which in turn will favor a more equitable distribution of employment and wealth among nations.

Tourism is going to experience many changes in the coming years and owners and managers must be prepared to know how to manage them. Maybe that's why this book caught your attention. It remains to be seen how the emergence of smart tourism firms will affect tourism value offerings and to what extent the authenticity of the tourist experience will be transformed. For example, in the context of smart tourism augmented reality is being used as an efficient tool to present information and add value to the tourist experience, although the true impact that the new experiences powered by smart technologies will have on authenticity and value to consumers is still unknown. Owners and managers should not forget the role that the attitudes of residents will play in places where smart tourism activities take place. Smart tourism seeks to improve the tourism experience of travelers, but also affects the quality of life of residents (Santos-Júnior et al., 2020), so there will be residents who appreciate it and others who don't. For this reason, it will be necessary for all the actors in the ecosystem to seek combinations that allow a balance to be achieved between smart technologies and the variables that affect the quality of life in local communities, such as accessibility, social inclusion, human and social capital, innovation, sustainability, entrepreneurship, etc.

The following sections examine some of the trends that will have the greatest impact on the future development of smart capabilities in tourism firms, according to the author's own predictions. These include access to the combined capabilities of Big Data, analytics, and AI (BDAI) "as-a-Service", which will bring tourism SMEs closer to realizing the promises offered by the Smart Revolution; the concern for the ethical aspects that emerge from the smartization process and that threaten the balance between firms, people, and society; the development of superior computational capacities through cognitive technologies and quantum computing, which will raise everyone's raw intelligence and bring machines closer to human reasoning (in addition to accelerating the replacement of jobs by machines in a large number of tasks); and the advances towards the connected and autonomous vehicle, which will radically transform the way we move around and experience tourism.

23.1 BDAI "as-a-Service"

The combination of Big Data, analytics, and AI (BDAI) will generate great business opportunities and competitive advantages for tourism firms in the future. Through the integration and interoperability of these technologies, tourism firms will transform their operations and business model and, in the medium and long term, obtain significant gains in efficiency and productivity (Samara et al., 2020). The inclusion of BDAI in the culture of people will better prepare them to open new markets and explore (and exploit) new niches in existing markets, will allow the creation of new value offerings focused on the experiential component and personalization, and, ultimately, will contribute significantly to boosting tourism growth.

With the proliferation of available data and computing capabilities increasing at an exponential rate, BDAI technologies are gaining momentum. The rise of the "as-a-Service" trend is contributing significantly to this growth, with more firms accessing IT infrastructure and applications as a service though the cloud. BDAI-as-a-Service (BDAIaaS) thus represents a source of great opportunities to accelerate the incorporation of tourism firms to the benefits of Smart Revolution, especially tourism SMEs.

How it works

BDAIaaS is an approach to provide tourism SMEs with access to the infrastructure, hardware, and software necessary for Big Data, analytics, and AI capabilities. Through this model of access and use of IT, SMEs could increase the pace of adoption of smart technologies, instead of keeping it slow. BDAIaaS offers an integrated service that merges different capabilities and technologies that are very complex by nature, making it easy for users to work with them by combining:

- **Platform as a Service (PaaS),** so that access to the infrastructure, operating system, and databases is managed and scalable depending on the specific needs of the firm at any moment.
- **Software as a Service (SaaS),** so that machine learning, analytical models, and visualization can be carried out in a public (or private) cloud and the firm would pay only for the services consumed.

Unlike the license fee model where customers would pay high license fees up front to install BDAI on-premises (on their own servers), the "as-a-Service" model offers customers access to BDAI capabilities on demand. In other words, instead of offering BDAI as a product subject to rapid technological obsolescence, BDAI is delivered as a pay-as-you-go service. In this way, firms do not need to make large investments in installations and updates of "packaged" solutions from time to time, thus drastically reducing the time it would normally take an installation (weeks or months). By storing all data and applications in a central hub, firms choosing BDAIaaS could avoid the high price of hard drive space and the need to purchase and maintain their own hardware. BDAI delivery time would also be much shorter, and the usual problems associated with heavy installations would be avoided. In addition, since firms might pay per user, it would be possible to choose which employees would have access to BDAI capabilities rather than buying in bulk for the entire organization, thus making IT purchasing more flexible.

Benefits

BDAIaaS offers opportunities for data scientists to rapidly process and analyze terabytes of data from widely dispersed sources and varied formats (e.g., text, images, etc.), making possible the flexible use of the capabilities provided by solution providers according to the specific needs of the firm. Firms could have access to endless advanced BDAI capabilities that would always be up to date, through a pay-per-use model in which the firm would be the one to decide which options to "activate" and which ones to "deactivate" according to their needs. BDAIaaS providers would be interested in providing their customers with a wide variety of options that meet their needs, each at a higher price.

Since cloud computing is at the heart of BDAIaaS, tourism firms would have at their fingertips all the shared processing capacity they needed at any given time with the click of a button. Similarly, when firms no longer needed that capacity, they could simply "turn off" the cloud service and stop incurring unnecessary costs. Firms could take advantage of BDAIaaS to minimize the costs of infrastructure and implementation of the different tools, in addition to quickly extracting meaning from complex data sets created from statistical analyses and models (Larson & Chang, 2016). In short, tourism firms would substantially improve the time to value generation. Today research continues on what is the best option to deliver BDAIaaS solutions and how to make them mainstream for firms. Still, the "aaS" model is expected to grow strongly in the coming years, moving beyond the cloud to enter our daily lives through mobile devices, thus creating a rich universe of new services for consumers and businesses.

23.2 Ethics

The ethical considerations that arise from the smartization process often go unnoticed by the owners and managers of tourism firms, as well as by the ICT industry. However, ethical behavior and decision-making in smart contexts will play an increasingly important role in the future for all stakeholders in the tourism ecosystem, since they will provide new perspectives to understand both smartization and the emerging organizational and social impacts.

The development and use of smart technologies depend on the processing of large amounts of heterogeneous data (Big Data) to drive innovation and support competitive advantage. However, smart technologies may raise ethical issues that can undermine the credibility of data-driven smart tourism firms. The importance of ethics is going to be more evident as smart transformation extends to an ever-increasing number of actors and becomes deeper and irreversible for all tourism organizations. How are fundamental values such as privacy, security, equality, transparency, and autonomy going to be affected in a smarter world dominated by Big Data? Should some data be a public good? Will consumers be bound to disclose personal data to some extent to make society safer and more efficient? (Christen et al., 2019; Yarali et al., 2020). Whatever the answers may be, owners and managers will need to decide how ethical smart practices (e.g., ethical Big Data, ethical analytics, ethical AI) are implemented within the organization, and how they are to be communicated inside and outside the organization.

On the other hand, the ethical behavior of the leaders and people of tourism organizations in relation to smart technologies has been little studied, even though there are still many decisions to be made throughout the smartization process. Most of the research on innovation and technology in tourism has focused on the effects they have on innovators, and what factors affect their adoption among users and consumers, and how they can become a source of competitive advantage. However, innovations such as smart technologies have much broader societal and individual impacts that need to be considered. An example is Airbnb, which although it originally emerged as a platform with a well-defined socio-innovative approach to sharing, today has considerable negative impacts on destinations, small firms, and people.

Consequently, ethics must be seen as a complementary approach that helps owners and managers to glimpse multifaceted perspectives of the smartization phenomenon, more so as they are currently poorly studied and have a notorious impact on both the way smart transformation is carried out and the results that are pursued. This does not mean that ethics should be used as a safeguard against the way smart transformation is taking place, but rather as an additional dimension that can help tourism firms strike a better balance between the elements that drive change and sustain performance over time.

Ethics and employees

Firms must ensure that their business objectives do not conflict with the moral views of their employees, although this can sometimes be contradictory from the perspective of financial objectives (Vial, 2019). As the capabilities offered by smart technologies increase, firms must seek to align business model decisions with the values and principles of employees. To illustrate the relevance of this topic, take the case of Google and see how the company changed its well-known motto "Don't be evil" in the 2018 Code of Ethics revision to its equally popular statement: "And remember ... don't be evil and if you see something that you think isn't right – speak up!" (Carbone, 2018). Despite this powerful claim, the tech company has come under fire many times in recent years by US regulators and its broad community of critics, to the point that former US Treasury Secretary Steven Mnuchin himself urged the Justice Department to review the power that companies like Google have over the US economy.

Ethics and performance

As value creation networks in tourism become more complex and involve more varied actors, the possibility for the firm to maintain full control over its level of performance becomes a challenge. In this context, ethics plays a key role as it can guide the design and use of smart technologies to achieve short-term objectives, while making the firm sustain its level of performance in the long term (Vial, 2019). For example, some tourism firms may try to increase their proximity to the customer and personalize their experiences by anticipating latent preferences that are not always made explicit through primary and secondary data sources. Although these practices may be profitable for the firm's performance in the short term, they have sometimes been shown to be undesirable, not because they are illegal, but because some stakeholders consider them morally reprehensible. This is the case with Facebook, which is continually called out for its unethical practices and has led some prominent US business corporations (e.g., Microsoft, Verizon, Coca-Cola, Unilever, Ford, Starbucks) and civil rights organizations (e.g., the National Hispanic Media Coalition, the Anti-Defamation League, Free Press, etc.) to join boycotts on the use of the popular social network, hence the call for people not to use the social network owing to its bad ethical practices (Elgan, 2020).

Tourism firms must remain vigilant to these risks, as customers and other stakeholders are bound to become critical if certain moral values are violated. In addition, it is important for firms to demonstrate that they are using data responsibly and are not only focused on the business case. To assess ethical risks, tourism firms can start by taking the customer's point of view and asking themselves: Would the customer agree to disclose their data if they knew exactly what was being done with it? What are the possible benefits that customers would be willing to accept to provide their data (Christen et al., 2019)? In a context in which smart technologies represent an increasing part of the value proposition of tourism firms, the ethical bases in force until now (i.e., corporate social responsibility) should be reviewed on the basis of smartization, incorporating new areas of interest such as ethical performance, data governance, etc. By doing so, tourism firms will obtain a more complete and richer understanding of the phenomenon of smartization that will surely allow them to address the process with greater success. It should not be forgotten that consumer trust and acceptance are pre-requisites for the successful implementation of smart technologies. This will not only require tourism firms to inform consumers in a transparent and understandable way about how data is collected and used, but firms must also offer them freedom of choice depending on the service.

Ethics and stakeholders

As value creation networks increase in complexity, tourism firms must satisfy the multiple, sometimes contradictory, demands of value co-creators. This issue becomes even more apparent in digital ecosystems such as platforms that, by definition, rely on multiple parties and for which data is at the core of their competitive value proposition. Firms must balance the demands of multiple parties, without compromising the firm's performance or its ability to sustain its competitive advantages over time. At times this may require the firm to redefine perceptions of what is considered right and wrong. For example, for a platform owner, an ethical challenge may be how to ensure that one party's demands are not met at the expense of others (i.e., if the platform grants access to the data to one party, then the other party should not perceive it as a breach of security and privacy).

23.3 Cognitive Computing

Cognitive computing technologies are fundamentally different from all other forms of computing used so far. Cognitive systems continually learn from their interaction with data (structured and unstructured), contexts, and people, and thus improve their learning and reasoning capabilities over time. They may be considered the third phase of the evolution of AI, which goes from traditional AI through artificial general intelligence to cognitive systems. Similar to how human learning evolves from birth to adulthood, a cognitive system learns and becomes more intelligent over time from the information it gains each time it interacts with its environment and the experience it accumulates.

A cognitive system is different from automated systems, which are those that can sense certain parameters of the environment and perform some action according to the data they detect. For example, the light turns on automatically when we enter a smart room. This is automation. If the light changes its color depending on the moment of time, or the weather, this is cognitive. Conventional automated systems are not able to perceive or communicate human emotions – basically what they do is process "if-then-what" conditions. Instead, cognitive systems mimic some aspects of human thought: they can learn, reason in some way, and suggest responses, incorporating emotional elements into their interaction with humans. While automated systems are preconfigured and coded by humans and follow rigid rules, cognitive systems can define their own rules through continuous learning and experience (Pramanik et al., 2018).

Cognition is also different from AI. The former emphasizes the learning process and recalls any facts that it has previously learned. AI is the ability of a machine to understand and decide what action to take and when among the given options, but it needs a set of well-defined rules codified by a human being. In other words, AI is based on the human being's perception and anticipation of possible complex scenarios and their ability to codify them in a system.

This paradigm shift from rule-based computing to autonomous learning and reasoning is changing computing forever (Hamm & Kelly, 2013). With the advent of Big Data, cognitive computing promises to unlock a vast amount of knowledge now hidden under the rich and huge amount of data that tourism firms generate. Therefore, the old dilemma about whether man or machine will prevail loses relevance to give way to a new era in which machines would not replace humans but would join them to expand their capabilities and help them make decisions by leveraging Big Data analytics (Castaldi et al., 2018; Gudivada et al., 2019). However, deploying these capabilities requires a major shift in the way business organizations approach problems, use technologies, and operate.

How it works

Cognitive computing is based on computer systems that mimic the human brain; that is, they have the ability to process natural language, learn from experience, interact with humans, and make decisions based on what they learn. Cognitive computing systems are learning systems that use a wide range of principles and techniques from cognitive science, neuroscience, data science, nanotechnology, machine learning, and cloud computing (Gudivada et al., 2019; Noor, 2015). Cognitive systems incorporate integrated data analytics and automated management that enable iterative interaction with the outside world to capture and analyze data, reason to hypothesize, learn from experience, and interact with humans naturally to achieve specific goals. Throughout a learning process, cognitive computing systems improve over time and produce their own knowledge, reducing errors and improving the quality of analysis and predictions.

Cognitive systems are not programmed by default but can improve themselves by learning through incremental interactions and training based on previous experiences and data sets. Therefore, unlike conventional programmable computers, cognitive computing is not limited to deterministic constraints, but has a dynamic essence by continuously detecting and learning from the environment and improving its decision-making capabilities autonomously (Pramanik et al., 2018). By learning from past errors and successes, cognitive systems enable humans to discover new relationships and behaviors that would otherwise go unnoticed in a large volume of data. This unique combination of analysis, problem solving, and communication with humans in a natural way creates a new way of interacting between humans and machines, turning machines into allies to increase human reasoning ability and support decision making.

Applications

Cognitive computing started to gain attention in 2011 when IBM's Watson computer played two champions of the TV game show *Jeopardy* and won. Watson had access to 200 million pages of structured and unstructured information stored on four terabytes of disk and was able to respond directly to questions posed in natural language. Today's new cognitive products span applications from cognitive cyber-physical systems (with built-in intelligence) to mechatronic components (combination of mechanics, electronics, and computation) that can monitor their own state and are capable of self-configuration, self-protection, self-optimization, and self-repair, as well as communicating with other cognitive products (Castaldi et al., 2018). The number of cognitive solutions in the market has not stopped growing, such as Watson from IBM, Azure from Microsoft, Deep Mind, Enterra Solutions, to name a few. Meanwhile, Google, Amazon, and Apple are also working on solutions focused on specific applications.

Cognitive computing technologies are game changers since they carry out some of the functions similar to human cognition (including learning, understanding, planning, solving problems, etc.) and deliver a variety of improvements in business functions, including production processes, logistics, financial management, waste management, and more. Cognitive systems can be key drivers of automation in knowledge management activities and in providing greater intelligence to the firm's products and services (Lytras & Visvizi, 2021; Noor, 2015). By applying new cognitive capabilities to existing automation, core business processes could not only run faster, but could emulate human judgment. This would help firms become more efficient and agile in their business, as well as provide meaningful information for decision makers to solve problems for which they previously had neither the skills nor the necessary resources.

Emerging cognitive systems are incorporating greater abilities to recognize behavioral patterns in Big Data, improvements in natural language processing and complex communication, increased self-learning, and other capabilities that used to be uniquely human. This opens new frontiers for distributed cognitive sensors (e.g., OrCam, Neurocam), robotic applications, and large socio-technical cognitive systems (e.g., smart cities, cognitive infrastructures, etc.). Chatbots are an example of these systems, which today allow human–machine interaction with a high level of fluidity and dynamism. From the creation of Eliza, the first chatbot in 1966, to the present with Alexa (2015), Cortana (2015), and Woebot (2017), there are thousands of chatbots on the market with applications in fields like marketing, support systems, health care, entertainment, education, and cultural heritage, which are becoming less robotic and more intelligent (Adamopoulou & Moussiades, 2020).

The Internet of Things (IoT) is another breeding ground for the use of cognitive computing. If the IoT is not intelligent and interactive, its capacity is very limited, so Cognitive IoT (CIoT) represents a great opportunity to add more advanced functionalities and autonomous behavior. The CIoT can dynamically interact with connected "things", learn from the environment, generate meaning, make decisions, and transmit them to humans based on the domain in which the IoT is applied. This augmented IoT through cognitive capabilities would mean a giant leap towards the full potential of the IoT and would elevate "things" to a higher level of intelligence and interaction. The recent development of specialized processors for cognitive computing, coupled with advances in Big Data tools and deep learning, are driving new and transformative applications in many industries. IBM's TrueNorth and Intel's Loihi chips have been designed to emulate the functions of a human brain through thousands of neurons that form synapses interconnected through circuitry. Tests have shown that they can be used for speech recognition and pattern identification.

In the app realm, Google has released the Cloud Natural Language API (https://cloud.google.com/natural-language/), part of the larger Cloud Machine Learning API family, which provides developers with natural language understanding technologies, including sentiment analysis and of entities, content classification, and parsing. Google's rival app Microsoft Azure Cognitive Services (https://azure.microsoft.com/es-es/services/cognitive-services) aims to make AI available to all developers and data scientists through an API, which could easily add AI capabilities and accelerate advanced decision-making across a large number of applications.

Challenges

Neuromorphic systems, which are modeled after the human brain, are among the cutting-edge cognitive computing technologies that have the potential to be more generalizable. These neuromorphic systems, based on the principles of neuroscience, are often pitted against computation-driven machine learning on the road to artificial general intelligence (also known as strong AI). The former intend to take advantage of neuroscience to achieve general intelligence from the development of processing models similar to the human brain. The latter aims to solve practical tasks leaving aside most of the principles of neuroscience in favor of brute force optimization (a problem-solving technique that involves listing all possible solutions and checking which is correct) and the use of a large volume of data (Deng et al., 2021). Nonetheless, with the help of Big Data, high-performance processors, and algorithms based on advanced artificial neural networks, the machine learning pathway has so far achieved better results than neuromorphic computing, especially in terms of accuracy. This is not meant to be a dilemma in which firms must choose between the two technologies. The real challenge lies in rethinking what the advantages of the human brain are and pointing out what should be the objectives to be achieved by neuromorphic computing to bridge the "gap" between neuromorphic computing and machine learning.

New hardware, programming languages, and applications will need to be developed in the coming years to drive cognitive computing forward. The new hardware will include neuromorphic machine technologies to process sensory data (e.g., images and sounds), and respond to changes in the data in ways not specifically programmed. Research is being carried out in the development of new neuromorphic chips that can overcome physical limitations and considerably reduce the power requirements of traditional processors. In 2017, the microprocessor manufacturer Intel developed Loihi, a neuromorphic chip with 128 cores and 130,000 neurons, which integrates memory, computing, and communication and works in parallel. In 2021, Intel introduced its second-generation neuromorphic chip Loihi 2 (Fig. 23.1), together with Lava, an open source software framework for developing neuro-inspired applications. According to the commercial information provided by Intel, Loihi enables accelerated learning in unstructured environments that require autonomous operation and continuous learning, with low power consumption and high performance and capacity (Intel, 2022). Additionally, Intel Labs, the research arm of Intel Corporation, has established the Intel Neuromorphic Research Community (INRC) as a collaborative research environment that brings together academic, government, and industry research teams from around the world with a view to overcoming the challenges of neuromorphic computing.

Fig. 23.1. Loihi 2 neuromorphic chip. Source: Walden Kirsch/Intel Corporation.

The confluence of cognitive technologies with Big Data, analytics, the IoT, and cloud computing will greatly expand the number of applications and the impact of cognitive computing. In the course of the coming years, a new generation of cognitive devices and systems will be developed that will be the result of the fusion of cognitive technologies with new tools and devices that will allow users to easily interact in a kind of continuous and dynamic "conversation" with technologies and will impact on the operational processes and decision-making systems of firms. Perhaps this will make it possible for humans and machines to work together in the future to extend human capabilities, especially those associated with knowledge, finding relevant patterns in large dynamic data, and making optimal decisions; and discover a new generation of cognitive products.

23.4 Quantum Computing

The idea of quantum computing arose from the reflection made by scientists about the fundamental limits of computing. They surmised that if technology continued to comply with Moore's Law (which states that the number of transistors in a microprocessor doubles approximately every 2 years), then the shrinking size of the circuits within a silicon chip would eventually reach a point where individual elements would not exceed the size of a few atoms. Consequently, the physical laws that would govern the properties and behavior of circuits on a subatomic scale would no longer be the classical ones but rather quantum mechanics. This gave rise to the possibility of creating a computer based on the principles of quantum physics (Rao et al., 2015). Feynman was one of the first to try to answer this question in 1982 by creating an abstract model showing how a quantum system could be used to do calculations. The model explained how a machine could have the ability to perform quantum physics experiments inside a quantum mechanical computer. Later, in 1985, Deutsch proposed the possibility that a quantum computer could be general purpose and published a theoretical paper that went on to show how a quantum computer could have capabilities far beyond those of a traditional classical computer. After Deutsch published that article, the quest began to find practical applications for such a machine.

In recent years, there has been relentless progress in both quantum hardware and quantum algorithm development that has brought quantum computing much closer to reality. As an emerging paradigm, quantum computing offers the potential to provide a significant computational advantage over conventional classical computing by exploiting the principles of quantum mechanics. The world's leading technology companies such as IBM, Google, Microsoft, and Intel, as well as some ambitious start-ups like Rigetti Computing and IonQ, are racing to develop the first large-scale universal quantum computer that could solve many computationally complex and intractable problems in areas like data science, finance, drug design, etc. (Gill et al., 2020).

How it works

Unlike classical computers, which store and process information as a series of bits that can take only a binary value ("0" or "1") and are then manipulated through Boolean logic gates arranged in succession to produce a result, the quantum computer encodes information into quantum bits (or qubits). According to the principles of quantum mechanics, qubits can take the values "0", "1", or any ratio of "0" and "1" in the superposition of both states, with a given probability of being a "0" and a given probability of being a "1" (i.e., 1/3 of "0" and 2/3 of "1"). The quantum computer operates on qubits by executing a series of quantum gates on a single qubit or a pair of qubits that are reversible. Therefore, quantum computers have access to an exponentially large computational space, where "n" qubits can be in a superposition state of 2n possible outcomes at any given moment in time. This allows quantum computers to tackle problems where computational complexity is the main bottleneck for classical machines, i.e., quantum computers can tackle problems with very large data sets with only a small number of qubits. The second key property of quantum computing is entanglement. Unlike classical bits, whose value can be set independently of other bits, qubits can have entangled states. In an entangled state, the properties of the qubits are bound together despite their physical separation. This means that measuring one qubit alters the properties of the other qubits that are in the same entangled state. Entanglement is a fundamental propriety that can exploited for dense coding and quantum simulation of correlated systems (Gill et al., 2020). Quantum computing will exploit the properties of both superposition and entanglement in such a way that the probability of the desired outcomes will increase while the probabilities of all other outcomes will decrease. Together, superposition and entanglement create computational power that can solve problems exponentially faster than classic computers.

Applications

In the past few years, global research and development has focused on building quantum computers that support increasingly complex industrial applications. Some of the most important programs worldwide include the United States National Quantum Initiative to promote quantum research and development and increase the country's economic and national security; the Quantum Technologies Flagship initiative of the European Union, launched in 2018 and with an estimated budget of €1 billion; the UK National Quantum Technologies Program (NQTP) with £1 billion for collaboration between industry, academia, and government; the Center for Quantum Computation & Communication Technology in Australia; and the National Laboratory for Quantum Information Science that China is building and that has a US$10 billion budget.

Big tech industry players are not far behind. Computer giants such as IBM, Microsoft, Alibaba, and Google are at the forefront of quantum computing developments, and new start-ups such as D-Wave Systems, Xanadu, Quantum Circuits, and Rigetti Computing compete to be the first to launch a scalable industrial computer. Ever since Google declared in 2019 that it had achieved quantum supremacy with Sycamore (a processor that could perform a calculation in 200 seconds that would take the world's fastest supercomputers 10,000 years to solve), there have been spectacular new announcements every so often. In December 2020, Chinese scientists at the University of Science and Technology in Hefei announced that they had built a quantum computer that could perform some calculations nearly 100 trillion times faster than the world's most advanced supercomputer. IBM has also announced that it has broken the 100-qubit barrier with its Eagle processor, and Microsoft has developed an open source programming language, Q#, that can develop and run quantum algorithms.

According to forecasts by the IT market intelligence company IDC, customer spending on quantum computing will grow from US$412 million in 2020 to US$8.6 billion in 2027. This represents a compound annual growth rate (CAGR) in 6 years of 50.9%. IDC anticipates that these large investments will see current limited quantum computing capabilities progressively overtaken by a new generation of more powerful quantum computing solutions, leading to new use cases and market segments adopting quantum computing as a source of competitive advantage (IDC, 2021). Applications of quantum computing are many and varied (e.g., modeling of commercial and passenger traffic, weather forecasting, blockchain, cryptocurrencies, etc.) and the number continues to increase every day. It is most likely that quantum computing will become a great ally of tourism firms in the future.

Machine Learning

Quantum computing promises to speed up machine learning algorithms to analyze classical data. Although it has not yet been fully demonstrated whether quantum machine learning can provide superior computational efficiency compared to classical methods, the results obtained in quantum principal component analysis and quantum neural networks look promising (Gill et al., 2020). Efficient searching and sorting of large data sets have become high-priority tasks for many large firms, including those in the tourism industry. Other already widespread machine learning applications such as voice, image, and handwriting recognition have become challenging tasks for traditional computers in terms of speed and accuracy. Here is where quantum computing could be of great help, improving pattern recognition and processing these complex problems in a time span that would take traditional computers hundreds of years (Jha, 2021).

However, classical database software is no longer sufficient to run quantum algorithms (e.g., Grover's algorithm). Furthermore, the use of AI and reinforcement learning could provide more computational power to manage, for example, the data generated by IoT devices. Nonetheless, it will still be necessary to develop new software that does the work of classical databases in the quantum world and to develop large machines with capacities of millions of qubits. On the other hand, quantum computers consume less energy than classical ones and, therefore, processing high data-intensive problems using quantum machine learning algorithms promises to see reduced energy costs.

Robotics

Robots use high graphics processing power (GPU) to solve computational tasks that are very data-intensive, such as vision, movement, optimal control, etc. and where quantum computing could help perform calculations at considerable speed. Quantum computing may improve the ability of robots to sense their environment and, by using cloud-based quantum computing, solve highly complex problems. In addition, the main kinematics issues associated with the mechanical movement of robots could be solved with quantum neural networks that recognize the moments of friction and inertia of the joints. Other typical issues of robot operation, such as identifying the reasons for the inconsistency between expected and observed behavior, could be solved using quantum algorithms. Quantum computing could also help reduce the complexity implicit in AI-based robotics by using quantum random walks to speed up response time and accuracy, rather than using information deduced from graph search.

Quantum cloud

Creating a safe and efficient environment for quantum computing in the cloud is an area that has great potential. A quantum cloud computer is a quantum computer that can be accessed over a network. Top tech companies like IBM (IBM Quantum), Google (GCP), Microsoft (Azure), and Amazon (Braket) have launched initiatives that combine quantum computers with cloud computing and that do not need to have a physical quantum computer installed. In this way, users have the opportunity to access quantum computers in the cloud to solve complex problems that require powerful computing (Soeparno & Perbangsa, 2021). The different quantum cloud computing services on offer today provide different architectures and performance levels. As competition continues to intensify among the big tech players, quantum cloud computing services will continue to offer specifications that promise ever better performance and faster runtimes.

Social networks

Social networks handle a huge volume of data every day as their use by millions of users continues to grow. Soon the amount of data in social networks will be so large that classic computing capabilities will no longer be able to process it easily and quickly in an acceptable time. Processing Social Big Data with relational databases where objects are semantically linked through multiple relationships is a major challenge. Mining such a relational database often requires enormous computing power in terms of hardware and software to deliver reasonably accurate and timely results. Therefore, to handle all this data and extract value from it, it will be necessary to have a large computing capacity that is also fast and efficient. Instead of the limitations presented by classical computing, quantum computing offers the ability to perform complex computation with social network data in an easy and efficient way, taking advantage, for example, of a graph theoretic representation of social network attributes to model sophisticated data structures and their interactions (Rao et al., 2015). Similar use cases include the modeling of different types of networks, such as telecommunications networks, traffic and transportation networks, tourist mobility networks, etc.

Quantum simulators

Although our capacity to predict complex systems has increased over time, state-of-the-art predictions still need new capabilities, especially in the field of social science applications. However, these developments have been restricted by the computational power that is currently available. Simulation models need to continuously calculate and recalculate visitor flow, optimal mobility routes, air traffic, prices and distribution, etc. Until now, simulations have used conventional computing tools and techniques, but with quantum computing it could be possible to solve these problems more easily and in less time in a controlled environment using small-scale "quantum simulators" of 50-100 qubits. Furthermore, quantum computing would allow for a higher degree of sophistication and levels of complexity in simulated systems, which could have a significant impact on simulation applications for science, and decision-making, including decisions related to tourism firms and destinations.

Quantum cryptography

Quantum cryptography is another application with great potential for the future of quantum computing. Quantum cryptography differs from traditional cryptographic systems in that the key element of its security model is based on the laws of quantum mechanics rather than classical mathematics. Since copying data encoded in a quantum state is not possible, the chances of being attacked by a cybercriminal are reduced. Furthermore, compared to traditional cryptography, quantum cryptography increases the probability of intrusion detection and improves performance. However, although with quantum computing firms could protect themselves better, it also adds the risk of being able to crack many of the conventional cryptographic systems used today, such as RSA, which uses the factorization of integers and is valid both for encrypting and for signing digitally. For example, by harnessing the power of quantum superposition with Shor's algorithm, it is possible to factor very large numbers in a matter of seconds and crack data encrypted in this way.

The rapid advance of quantum computing has also opened up the possibility of blockchain decryption using the Grover and Shor algorithms. These algorithms pose a threat to both public-key cryptography and the hash functions used by the blockchain, forcing the blockchain to be redesigned to resist quantum attacks, thus creating what are known as post-quantum or quantum-proof cryptosystems (Fernandez-Carames & Fraga-Lamas, 2020). National organizations dedicated to cybersecurity, such as the US National Security Agency (NSA), have warned about the impact that quantum computing could have on IT products and recommends increasing the level of security through new approaches such as elliptic curve cryptography (ECC). On the other hand, since post-quantum schemes require high execution times, in addition to storage and computing resources that consume more energy, further research will be necessary to optimize cryptosystems that maximize computational and energy efficiency. Post-Quantum Bitcoin is an experimental example of a post-quantum digital signature scheme used with the main Bitcoin blockchain.

Challenges

Although advances in quantum computing look promising and the tech industry is getting closer to "quantum supremacy" (solving a problem in a quantum computer that is intractable in a classical machine), there are numerous challenges ahead that will likely take years to resolve. One of the most important is how to build a large-scale, fault-tolerant universal quantum computer that can unleash the full potential of quantum computing in real-world applications. Key milestones have been achieved in this regard. For example, in late 2021, IBM introduced Eagle, its most powerful quantum system with a 127-qubit processor, and the first to break the 100-qubit barrier. The system contained new chip cooling and packaging technologies, including a new cryogenic platform to keep temperatures low and make the system more stable. These technologies will serve as the building blocks for IBM's next two quantum systems: the 433-qubit Osprey, scheduled for launch in 2022, and the 1121-qubit Condor, expected in 2023 (Scannell, 2021).

Developing a large-scale quantum computer has significant challenges, including the development of quantum hardware that reduces the effects of decoherence. Decoherence refers to the interactions that a qubit has with its environment and that can cause disturbances (or incoherent states) that collapse the superposition and lead to errors in the quantum information. For example, small variations in temperature or electric or magnetic fields can cause quantum information to degrade in the computer. Before any quantum computer can solve a complex problem, the industry will have to find a way to keep decoherence and other sources of error at acceptable levels; meanwhile the field of quantum error correction will remain one of the most active research areas in the field of quantum computing. Tests conducted with the NISQ (Noisy Intermediate-Scale Quantum) algorithm, specifically designed for quantum processors, seem promising to address these technical challenges.

Finally, it is noteworthy that quantum computing does not yet have its own high-level programming language. What this means is that the algorithms are processed by building quantum circuits to which available quantum gates or operations are systematically applied to find the desired solution. An example of industry efforts to address this issue is a visual programming tool provided by IBM for beginners to code and learn how to use a quantum computer.

23.5 Connected and Autonomous Vehicles

Connected and autonomous vehicles (CAV) have the potential to disrupt transport-related industries, including tourism, by transforming urban spaces, modes of transport use, tourism employment, and the overall visitor economy. The arrival of CAVs on the market will generate a new world of business and social opportunities that, however, are not free from challenges and threats (Cohen & Hopkins, 2019). CAVs are much more than "autonomous taxis" and "constructive policy dialogue"; they raise concrete questions about how to design the cities in which we want to live and how tourist experiences will be designed in the future. At the epicenter of CAV innovation are urban environments, whose spatial morphology will have to gradually transform along with tourism activities to accommodate CAVs. CAVs also have the potential to shift current mobility practices, as well as impact the way public administrations raise revenue through vehicle taxes, the livelihoods of transport drivers, and, in general, the viability of transportation systems as we know them (Bissell, 2018).

CAV navigation will gradually relegate the driver to a "hands-off", "feet-off" role (with or without the ability to quickly regain control of the vehicle) to reach a fully automated mode in its most advanced stage. According to the most optimistic forecasts, CAVs could start to hit the mass market as early as 2025, first in parts of Asia, Europe, and the US, and later becoming the leading mode of road transport globally in the 2040s. This means that all transport-related industries will gradually transform, and the tourism sector is not going to be an exception. This potential for transformative change will create tremendous opportunities for new and existing players in the automotive industry and beyond. In fact, the world's leading vehicle manufacturers have been developing automated driving capabilities for years, and new entrants in the motor industry, like Google, Apple, and Uber, now compete to develop commercially viable advanced technologies for fully equipped automated vehicles. Still, CAVs need to overcome numerous challenges related to security, cost, and customer perceptions to achieve full implementation.

Applications

Tourists will be among the first consumer groups to experience the CAV revolution. The role of tourists within urban mobility schemes is key to CAV innovation, which focuses on overcoming unsustainable practices and optimizing routes and flows of people according to environmental conditions. For example, CAVs have been tested as autonomous passenger shuttles to and from airports throughout Europe (e.g., Gatwick, Heathrow, Amsterdam, Brussels, Paris, Frankfurt, etc.), drastically reducing travel times and saving tons of carbon emissions (Cohen & Hopkins, 2019). New start-ups dedicated to the development of innovative mobility solutions are also emerging, such as CAVs specially designed for spaces that receive a large volume of visitors and workers, such as airports.

Many plans are underway around the world to continue expanding the use of CAVs in tourism environments, mainly as sustainable transport solutions that help overcome congestion and pollution in daily and daytime urban commutes. This ranges from airport shuttles and transfers, to taxis, car rentals, and vehicles used for guided city tours. CAVs are expected to provide "last mile solutions" that facilitate multimodality, such as moving urban tourists between transport stations (e.g., airport, train, bus, and maritime stations). By totally or partially removing the driver, the cost of transport could be significantly reduced, although at first it is likely that due to the novelty and the reduced supply and demand of the market, there will be an overprice in CAV mobility, as has happened in the electric vehicle market.

Benefits

The benefits described below are those expected to anticipate the increase in consumer demand for CAVs, as well as to justify the public R&D funding programs that need to be carried out to boost these innovative technologies.

1. First, there is the discourse of security as one of the main factors in favor of automated mobility. According to this argument, CAVs could eliminate up to 90% of driving accidents by reducing driver risks and driving errors. This is especially relevant in tourist contexts, in which due to ignorance of driving rules and the road environment, fatigue, and the playful nature of tourism transport, driving accidents with tourists occur.
2. Second, CAVs will help reduce congestion because fewer accidents are expected, but also because fewer CAVs will be needed to meet mobility demands compared to human-driven cars (Kellerman, 2018). Nonetheless, this argument is debatable because it will largely depend on the property schemes that prevail in the market and on whether the CAV will replace the individual ownership model with one of shared ownership. In any case, it seems that due to the connected nature of these vehicles (vehicle-to-vehicle and vehicle-to-infrastructure), traffic flows could be optimized, leading to a reduction in mobility congestion.
3. Third, CAVs can deliver environmental benefits, particularly in terms of fuel savings and reduced carbon emissions. This is mainly due to the electric traction of these vehicles and the elimination of unsustainable driving practices in favor of so-called "green driving", such as sudden braking or driving at high speeds.
4. Fourth, CAVs promise to improve accessibility for non-drivers, whether they are elderly, low-income, or disabled, which will contribute to greater social equity in urban settings. While CAVs are likely to hit the market in the world's richest and most developed cities, in the long term and as their use becomes more popular, CAVs are expected to become a solution that favors the integration of mobility of the most disadvantaged groups.

5. Finally, due to the non-polluting nature and low environmental impact of CAVs, they can be used in urban spaces where vehicles driven by people have already been prohibited due to their high impact. Likewise, the elimination of obstacles to driving and the transformation of the cabins, in which passengers will be able to read, eat, or access audiovisual content, will improve the experience of the tourist, who will be able to use CAVs as an innovative tool to create new tourism experiences.

Achieving some or all of these benefits will depend not only on the degree of intelligence and automation achieved by CAV technologies, but also on fundamental questions related to the governance of the new mobility framework, such as vehicle ownership models (i.e., shared versus private); the ratio of CAVs versus non-CAVs on the roads; the technical specifications of performance, autonomy, and consumption; and the public regulations that will govern the use of CAV (Cohen & Hopkins, 2019).

Challenges

CAVs will transform tourism in the coming years and will contribute to reconfiguring the urban space, creating new challenges that will put the automobile industry, consumers, and those responsible for urban planning to the test. It is possible that by improving the travel experience and reducing travel costs, there will be an increase in the induced demand for mobility by residents and tourists. Higher levels of CAV use by urban tourists could negatively affect efforts made to encourage responsible use of public transport by visitors, potentially leading to "overtourism" in the form of hordes of small CAVs congesting urban tourist spaces. In a scenario like this, there would foreseeably be a shortage of urban land available for parking, which would call into question the planning of city centers. Better travel conditions could make people want to travel longer distances for work, entertainment, or for business or leisure trips, which could encourage people to move their places of residence to more distant places, taking advantage of lower property costs. This could lead to urban sprawl and further increase car dependency, with tourists perhaps turning to private vehicles and public or shared transport declining.

New urban tourism destinations could emerge as CAVs grow in popularity, either as new attractions that were previously difficult to access, or secondary cities now emerging as new destinations due to more affordable transport connectivity. The new mobility facilitated by the CAV could redefine what is the "commercial district" or the "center" of cities, by connecting urban spaces in ways never seen before. In short, CAVs could affect the flows and experiences of urban and interurban mobility. It has been argued that the location of the hotels would be less important among the selection criteria of tourists since it would no longer be so important that the hotels need to be connected by public transport (Bainbridge, 2018). The same would happen with other types of tourist facilities, such as tourist accommodation, bars, or restaurants, which in many cases could be easily reached through CAVs. It could also happen that travelers decide to spend the night in their CAV while it drives them to their final destination, instead of spending the night in a hotel. This idea of CAVs as "hotels on the move" would affect the behavioral patterns of travelers, and hotels and tourist accommodation.

The arrival of CAVs on the market is likely to threaten many tourism-related jobs in the future, notably those of professional drivers such as taxi and bus drivers, and those travel businesses that rely on human drivers, although it is likely that the replacement of professional human drivers would be gradual and in stages. There are also many questions about which business models will prevail in the CAV market. With the shared (on-demand) CAV model, up to ten privately owned vehicles could be replaced by each shared CAV, and trips would surely be more affordable and attractive to tourists. However, CAV users would be forced to spend time with strangers in the confined space of the cabin, which in some cultural contexts may be considered unacceptable due to the expectation of having to interact. On the other hand, the environmental benefits of CAVs will depend on the shared/private ownership models and the type of propulsion with which these vehicles are equipped. Although CAVs for urban use are most likely to be electric vehicles, it should not be forgotten that some electric vehicles have higher overall emissions than combustion engine vehicles, and that the true emissions reduction of electric vehicles depends on the production processes and fuels for electrification.

Finally, certain ethical, security, and privacy concerns have also been identified in the development and use of CAVs. Ethical questions arise, for example, when discussing the degree of protection that passengers traveling in a CAV should have and whether CAVs should be programmed to protect passengers above all else, even when the safety of those outside may be threatened. The recent death of some pedestrians due to accidents caused by CAVs in the USA has drawn the attention of the media and opened a public debate about the limits of the automation of these vehicles. Not to mention the possible malicious use that terrorists or people with murderous intentions could make of CAVs, which could be directed against the crowds in busy urban areas. Finally, there is also the issue of the privacy of the personal data recorded during the trips made in the CAVs, as well as the data related to the places visited, the routes taken, the travel times, the stops made, etc., which that could be used to direct advertising or marketing actions to the occupants.

Discussion Questions

- Is it a realistic scenario to think that machines will replace human beings in most of the tasks that involve human reasoning? Should we humans be worried?
- What are the main ethical problems that currently affect the tourism firm in relation to smart technologies?
- Can ethics contribute to increasing the firm's performance? How?
- What main advantages (disadvantages) would adopting the "as-a-Service" model have for the tourism firm?

- What potential practical benefits could quantum computing bring to the tourism firm? What factors will its adoption depend on?
- What are the main impacts that the implementation of the CAV could have on tourism firms?
- In what ways could CAVs affect the quality of the experience perceived by tourists?

References

Abbate, T., & Souca, M.L. (2013). Open innovation and online intermediaries: A review of theory and its implications for tourism. *Proceedings of the International Conference Marketing – From Information to Decision, 6*, 9–22. http://search.ebscohost.com/login.aspx?direct=true&db=bth&AN=93425296&site=ehost-live (accessed 23 June 2022).

Accenture. (2017). The future of food: New realities for the industry. In *FMI. The Voice of Food Retail*. https://www.fmi.org/docs/default-source/midwinter-2018/kurt-salmon-and-fmi-future-of-food-report-(2).pdf?sfvrsn=c35c796e_0 (accessed 12 April 2022).

Accenture. (2020). *Busting the Myths of Agile: How the pandemic accelerated the adoption of the intelligent operating model* (August). https://www.accenture.com/_acnmedia/PDF-132/Accenture-COVID-19-Busting-MythsAgile-transformation.pdf (accessed 23 June 2022).

Adamopoulou, E., & Moussiades, L. (2020). Chatbots: History, technology, and applications. *Machine Learning with Applications*, 2(July), 100006. https://doi.org/10.1016/j.mlwa.2020.100006

Adukaite, A., Reimann, A.M., Marchiori, E., & Cantoni, L. (2013). Hotel mobile apps. The case of 4 and 5 star hotels in European German-speaking countries. In *Information and Communication Technologies in Tourism 2014* (pp. 45–57). Springer.

Aguilar, E., Remeseiro, B., Bolaños, M., & Radeva, P. (2018). Grab, pay, and eat: Semantic food detection for smart restaurants. *IEEE Transactions on Multimedia*, 20(12), 3266–3275.

Akkaya, B., & Tabak, A. (2020). The link between organizational agility and leadership: A research in science parks. *Academy of Strategic Management Journal*, 19(1), 1–17.

Alahmari, A., & Duncan, B. (2020). Cybersecurity risk management in small and medium-sized enterprises: A systematic review of recent evidence. *2020 International Conference on Cyber Situational Awareness, Data Analytics and Assessment, Cyber SA 2020*, June. https://doi.org/10.1109/CyberSA49311.2020.9139638

Alberts, C., Dorofee, A., Stevens, J., & Woody, C. (2003). Introduction to the OCTAVE Approach. In *Networked Systems Survivability Program*. Carnegie Mellon Sofware Engineering Institute. https://www.fmi.org/docs/default-source/midwinter-2018/kurt-salmon-and-fmi-future-of-food-report-(2).pdf (accessed 30 July 2022).

Ambler, S.W., & Lines, M. (2016). The disciplined agile process decision framework. In *International Conference on Software Quality* (pp. 3–14). https://doi.org/10.1007/978-3-319-27033-3_1

Andersen, E., Johnson, J. C., Kolbjørnsrud, V., & Sannes, R. (2018). The data-driven organization: Intelligence at SCALE. In A. Sasson (Ed.), *At the Forefront, Looking Ahead* (pp. 23–42). Scandinavian University Press.

Antoniadis, I., Spinthiropoulos, K., & Kontsas, S. (2020). Blockchain applications in tourism and tourism marketing: A short review. In A. Kavoura, E. Kefallonitis, & P. Theodoridis (Eds.), *Strategic Innovative Marketing and Tourism* (pp. 375–384). Springer International Publishing. https://doi.org/10.1007/978-3-030-36126-6_41

Ashton, K. (2009). That "internet of things" thing. *RFID Journal*, 22(7), 97–114.

Atlam, H.F., Walters, R.J., & Wills, G.B. (2018). Fog computing and the Internet of Things: A review. *Big Data and Cognitive Computing*, 2(2), 10.

Attar, M., & Abdul-Kareem, A. (2020). The role of agile leadership in organisational agility. In B. Akkaya (Ed.), *Agile Business Leadership Methods for Industry 4.0*. Emerald Publishing Limited. https://doi.org/10.1108/978-1-80043-380-920201011

Attaran, M. (2017). The Internet of Things: Limitless opportunities for business and society. *Journal of Strategic Innovation and Sustainability*, 12(1), 11.

AWS. (2022). What is a data lake? https://aws.amazon.com/es/big-data/what-is-a-data-lake/ (accessed 23 June 2022).

Azuma, R.T. (1997). A survey of augmented reality. *Presence: Teleoperators & Virtual Environments*, 6(4), 355–385.

Baculard, L.-P., Colombani, L., Flam, V., Lancry, O., & Spaulding, E. (2017). *Orchestrating a Successful Digital Transformation*. Bain & Company. https://www.bain.com/contentassets/dd440ca288d34c16ba8cd3ab6ef69a04/bain_brief_orchestrating_a_successful_digital_transformation.pdf (accessed 30 July 2022).

Bae, S.J., Lee, H., Suh, E.-K., & Suh, K.-S. (2017). Shared experience in pretrip and experience sharing in posttrip: A survey of Airbnb users. *Information & Management*, 54(6), 714–727.

Bageac, D., Reynaud, E., Fortun-Auad, S., & Thi, N.T.V. (2020). Organizational aspects of open innovation. *Institut de Socio-Économie Des Entreprises et Des Organisations (Écully, Rhône). Recherches en Sciences de Gestion*, 138, 41–68. https://www.proquest.com/scholarly-journals/organizational-aspects-open-innovation/docview/2444524837/se-2?accountid=14475 (accessed 23 June 2022).

Bainbridge, A. (2018). *Autonomous Vehicles & Auto-Tours. What is an auto-tour and how will autonomous vehcicles impact tours, attractions & cities?* http://www.destinationcto.com/docs/AutoTour.pdf (accessed 15 April 2022).

Bala, M. (2018). Digital transformation: Review of concept, digital framework, and challenges. In V.P. Gupta and D. Bansal (Eds.), *Theoretical and Empirical Development in Management and IT* (pp. 133–152). Swaranjali Publications.

Barkan, M., & Tapliashvili, N. (2018). Cryptocurrency use in medical tourism. *Scientific and Practical Cyber Security Journal (SPCSJ)*, 2(4), 104–110.

Bashkirova, O. (2016). *Knowledge-Based Business-Model (The Concept of Smart Organization)*. https://papers.ssrn.com/sol3/papers.cfm?abstract_id=2842509 (accessed 30 July 2022).

Bazzaz Abkenar, S., Haghi Kashani, M., Mahdipour, E., & Jameii, S.M. (2021). Big Data analytics meets social media: A systematic review of techniques, open issues, and future directions. *Telematics and Informatics*, 57(March), 101517. https://doi.org/10.1016/j.tele.2020.101517

Beck, K., Beedle, M., Van Bennekum, A., Cockburn, A., Cunningham, W., Fowler, M., Grenning, J., Highsmith, J., Hunt, A., & Jeffries, R. (2001). *Manifesto for Agile Software Development*. https://agilemanifesto.org (accessed 30 July 2022).

Beliveau, M., De Santa-Eulalia, L. A., Mosconi, E., & Cadieux, N. (2018). How can SME embark in a digital transformation in the context of the 4th Industrial Revolution? *2nd International Symposium on Supply Chain*, 4(0), 37.

Berger, R. (2015). *The Digital Transformation of Industry*. Study commissioned by the Federation of German Industries (BDI), Munich. http://www.Rolandberger.Com/Publications/Publication_pdf/Roland_berger_Digital_transformation_of _industry_20150315.Pdf.

Berndtsson, M., Forsberg, D., Stein, D., & Svahn, T. (2018). Becoming a data-driven organisation. *ECIS 2018 Proceedings*, 1–9.

Berntsson Svensson, R., Feldt, R., & Torkar, R. (2019). The unfulfilled potential of data-driven decision making in agile software development. *International Conference on Agile Software Development*, 69–85.

Berntsson Svensson, R., & Taghavianfar, M. (2020). Toward becoming a data-driven organization: Challenges and benefits. In F. Dalpiaz, J. Zdravkovic, & P. Loucopoulos (Eds.), *International Conference on Research Challenges in Information Science* (pp. 3–19). Springer International Publishing.

Bissell, D. (2018). Automation interrupted: How autonomous vehicle accidents transform the material politics of automation. *Political Geography*, 65, 57–66. https://doi.org/https://doi.org/10.1016/j.polgeo.2018.05.003

Boes, K., Buhalis, D., & Inversini, A. (2015). Conceptualising smart tourism destination dimensions. In *Information and Communication Technologies in Tourism 2015* (pp. 391–403). Springer.

Bogers, M., Chesbrough, H., Heaton, S., & Teece, D.J. (2019). Strategic management of open innovation: A dynamic capabilities perspective. *California Management Review*, 62(1), 77–94. https://doi.org/10.1177/0008125619885150

Brandão, A., Mamede, H.S., & Gonçalves, R. (2018). Systematic review of the literature, research on blockchain technology as support to the trust model proposed applied to smart places. In Á. Rocha, H. Adeli, L.P. Reis, & S. Costanzo (Eds.), *World Conference on Information Systems and Technologies* (pp. 1163–1174). Springer International Publishing.

Brinch, M., Gunasekaran, A., & Fosso Wamba, S. (2021). Firm-level capabilities towards big data value creation. *Journal of Business Research*, 131(July), 539–548. https://doi.org/10.1016/j.jbusres.2020.07.036e

Brown, N., & Brown, I. (2019). From digital business strategy to digital transformation – How: A systematic literature review. *Proceedings of the South African Institute of Computer Scientists and Information Technologists* 2019, 1–8.

Brunswicker, S., Bertino, E., & Matei, S. (2015). Big Data for open digital innovation – A research roadmap. *Big Data Research*, 2(2), 53–58. https://doi.org/10.1016/j.bdr.2015.01.008

Brunswicker, S., & Vanhaverbeke, W. (2015). Open innovation in small and medium-sized enterprises (SMEs): External knowledge sourcing strategies and internal organizational facilitators. *Journal of Small Business Management*, 53(4), 1241–1263.

Bumann, J., & Peter, M. (2019). Action fields of digital transformation – A review and comparative analysis of digital transformation maturity models and frameworks. *Digitalisierung und andere Innovationsformen im Management*, 2(November), 13–40.

Buonincontri, P., & Micera, R. (2016). The experience co-creation in smart tourism destinations: A multiple case analysis of European destinations. *Information Technology & Tourism*, 16(3), 285–315.

Burchardt, C., & Maisch, B. (2019). Digitalization needs a cultural change – Examples of applying Agility and Open Innovation to drive the digital transformation. *Procedia CIRP*, 84, 112–117. https://doi.org/10.1016/j.procir.2019.05.009

Busulwa, R., Evans, N., Oh, A., & Kang, M. (2020). *Hospitality Management and Digital Transformation: Balancing efficiency, agility and guest experience in the era of disruption*. Routledge.

Caddeo, F., & Pinna, A. (2021). Opportunities and challenges of blockchain-oriented systems in the tourism industry. *2021 IEEE/ACM 4th International Workshop on Emerging Trends in Software Engineering for Blockchain (WETSEB)*, 9–16.

Calvaresi, D., Leis, M., Dubovitskaya, A., Schegg, R., & Schumacher, M. (2019). Trust in tourism via blockchain technology: Results from a systematic review. *Information and Communication Technologies in Tourism* 2019, 304–317.

Çapar, H. (2020). Using cryptocurrencies and transactions in medical tourism. *Journal of Economic and Administrative Sciences.* https://doi.org/10.1108/JEAS-07-2019-0080e

Caralli, R.A., Stevens, J.F., Young, L.R., & Wilson, W.R. (2007). *Introducing Octave Allegro: Improving the information security risk assessment process*. Carnegie-Mellon University Software Engineering Institute.

Carbone, C. (2018). Google revises code of conduct, removes "Don't be evil". *New York Post*, 22 May. https://nypost.com/2018/05/22/google-revises-code-of-conduct-removes-dont-be-evil/ (accessed 23 June 2022).

Castaldi, L., Iscaro, V., Maresca, P., & Mazzoni, C. (2019). Digital transformation and internationalization: A complexity approach. Paper delivered at conference, *Digital Transformation and Internationalization of Firms: Prospects, Challenges and Future Agenda*, University of Rome LUMSA, Palermo, 7–8 February.

Castellani, D., Rullani, E., & Zanfei, A. (2017). Districts, multinationals and global/digital networks. *Economia e Politica Industriale*, 44(4), 429–447.

Centobelli, P., & Ndou, V. (2019). Managing customer knowledge through the use of Big Data analytics in tourism research. *Current Issues in Tourism*, 22(15), 1862–1882. https://doi.org/10.1080/13683500.2018.1564739

Cepeda, J., & Arias-Pérez, J. (2019). Information technology capabilities and organizational agility: The mediating effects of open innovation capabilities. *Multinational Business Review*, 27(2), 198–216. https://doi.org/10.1108/MBR-11-2017-0088

Cetina, C., Giner, P., Fons, J., & Pelechano, V. (2013). Prototyping dynamic software product lines to evaluate run-time reconfigurations. *Science of Computer Programming*, 78(12), 2399–2413.

Chen, C.L.P., & Zhang, C.-Y. (2014). Data-intensive applications, challenges, techniques and technologies: A survey on Big Data. *Information Sciences*, 275, 314–347.

Chen, F., Deng, P., Wan, J., Zhang, D., Vasilakos, A.V, & Rong, X. (2015). Data mining for the internet of things: Literature review and challenges. *International Journal of Distributed Sensor Networks*, 11(8), 431047.

Chen, Z., Alfred, R., & Eboy, O.V. (2021). Modeling tourism using spatial analysis based on social media big data: A review. *Computational Science and Technology: 7th ICCST 2020,* Pattaya, Thailand, 29–30 August, 2020, 724, 437.

Chesbrough, H. (2003). *Open Innovation: new imperative for creating and profiting from technology*. Harvard Business Press.

Chesbrough, H., & Bogers, M. (2014). Explicating open innovation: Clarifying an emerging paradigm for understanding innovation. In *New Frontiers in Open Innovation* (pp. 3–28). Oxford University Press.

Chesbrough, H., Vanhaverbeke, W., & West, J. (2006). *Open Innovation: Researching a new paradigm*. Oxford University Press on Demand.

Chesbrough, H., Vanhaverbeke, W., & West, J. (2014). *New Frontiers in Open Innovation*. Oxford University Press.

Chessell, M., Scheepers, F., Nguyen, N., van Kessel, R., & van der Starre, R. (2014). Governing and managing big data for analytics and decision makers. *IBM Redguides for Business Leaders*. IBM Redbooks.

Chessell, M., Scheepers, F., Strelchuk, M., van der Starre, R., Dobrin, S., & Hernandez, D. (2018). *The Journey Continues: From data lake to data-driven organization*. IBM Redbooks.

Chiheb, F., Boumahdi, F., & Bouarfa, H. (2019). A new model for integrating Big Data into phases of decision-making process. *Procedia Computer Science*, 151(2018), 636–642. https://doi.org/10.1016/j.procs.2019.04.085e

Christen, M., Blumer, H., Hauser, C., & Huppenbauer, M. (2019). The ethics of Big Data applications in the consumer sector. *Applied Data Science*, 161–180. https://doi.org/10.1007/978-3-030-11821-1_10

Chuang, C.-M. (2020). A current travel model: smart tour on mobile guide application services. *Current Issues in Tourism*, 23(18), 2333–2352.

Chui, M., Manyika, J., & Miremadi, M. (2016). Where machines could replace humans – and where they can't (yet). *McKinsey Quarterly* (July 8). https://www.mckinsey.com/business-functions/mckinsey-digital/our-insights/where-machines-could-replace-humans-and-where-they-cant-yet (accessed 30 July 2022).

Chung, N., Han, H., & Koo, C. (2015). Adoption of travel information in user-generated content on social media: The moderating effect of social presence. *Behaviour & Information Technology*, 34(9), 902–919.

Chung, N., & Koo, C. (2015). The use of social media in travel information search. *Telematics and Informatics*, 32(2), 215–229.

Cloudera. (2022). *Apache Hadoop Ecosytem*. https://es.cloudera.com/products/open-source/apache-hadoop.html (accessed 4 October 2022).

Cohen, S.A., & Hopkins, D. (2019). Autonomous vehicles and the future of urban tourism. *Annals of Tourism Research*, 74, 33–42. https://doi.org/https://doi.org/10.1016/j.annals.2018.10.009

Collier, K. (2012). *Agile Analytics: A value-driven approach to business intelligence and data warehousing*. Addison-Wesley.

D'Aveni, R.A., Dagnino, G.B., & Smith, K.G. (2010). The age of temporary advantage. *Strategic Management Journal*, 31(13), 1371–1385.

Dahlström, P., Desmet, D., & Singer, M. (2017). The seven decisions that matter in a digital transformation: A CEO's guide to reinvention. *McKinsey Digital*. https://www.mckinsey.com/business-functions/mckinsey-digital/our-insights/the-seven-decisions-that-matter-in-a-digital-transformation (accessed 30 July 2022).

Damnjanović, V., Lončarić, D., & Dlačić, J. (2020). Teaching case study: Digital marketing strategy of Accor Hotels: Shaping the future of hospitality. *Tourism and Hospitality Management*, 26(1), 233–244.

Davenport, T.H., & Ronanki, R. (2018). Artificial intelligence for the real world. *Harvard Business Review*, 96(1), 108–116.

De Saulles, M. (2016). The Internet of Things and business. In *The Internet of Things and Business*. Routledge https://doi.org/10.4324/9781315537849

Del Giudice, M. (2016). Discovering the Internet of Things (IoT) within the business process management: A literature review on technological revitalization. *Business Process Management Journal*, 22(2), 263–270. https://doi.org/10.1108/BPMJ-12-2015-0173

Del Vecchio, P., Di Minin, A., Petruzzelli, A. M., Panniello, U., & Pirri, S. (2018). Big Data for open innovation in SMEs and large corporations: Trends, opportunities, and challenges. *Creativity and Innovation Management*, 27(1), 6–22. https://doi.org/10.1111/caim.12224

Del Vecchio, P., Mele, G., Ndou, V., & Secundo, G. (2018a). Creating value from Social Big Data: Implications for smart tourism destinations. *Information Processing and Management*, 54(5), 847–860. https://doi.org/10.1016/j.ipm.2017.10.006

Del Vecchio, P., Mele, G., Ndou, V., & Secundo, G. (2018b). Open innovation and social big data for sustainability: Evidence from the tourism industry. *Sustainability (Switzerland)*, 10(9). https://doi.org/10.3390/su10093215

Deng, L., Tang, H., & Roy, K. (2021). Understanding and bridging the gap between neuromorphic computing and machine learning. In *Frontiers in Computational Neuroscience*. Frontiers Media SA. https://doi.org/10.3389/978-2-88966-742-0

Donker-Rostamy, M. (2019). *Agile Facilitates Data-Driven Solutions*. Open University of the Netherlands.

Dorcic, J., Komsic, J., & Markovic, S. (2019). Mobile technologies and applications towards smart tourism – state of the art. *Tourism Review*, 74(1), 82–103. https://doi.org/10.1108/TR-07-2017-0121

Dredge, D., Phi, G., Mahadevan, R., Meehan, E., & Popescu, E.S. (2019). *Digitalisation in Tourism: In-depth analysis of challenges and opportunities*. Executive Agency for Small and Medium-Sized Enterprises (EASME), European Commission. https://ec.europa.eu/docsroom/documents/33163/attachments/1/translations/en/renditions/native (accessed 30 July 2022).

Dreyfus, H.L. (1972). *What Computers Can't Do: The limits of artificial intelligence*. Harper & Row.

Duan, Y., Edwards, J.S., & Dwivedi, Y.K. (2019). Artificial intelligence for decision making in the era of Big Data – Evolution, challenges and research agenda. *International Journal of Information Management*, 48, 63–71. https://doi.org/10.1016/j.ijinfomgt.2019.01.021

Earley, S. (2014). Agile analytics in the age of big data. *IT Professional*, 16(4), 18–20.

Elgan, M. (2020). Why it's unethical for businesses to use Facebook. *Insider Pro*, 30 September. https://www.idginsiderpro.com/article/3584055/why-its-unethical-for-businesses-to-use-facebook.html (accessed 23 June 2022).

Facts and Factors. (2022). *Online Travel Booking Market*. https://www.fnfresearch.com/online-travel-booking-market (accessed 23 June 2022).

Fang, H. (2015). Managing data lakes in big data era: What's a data lake and why has it became popular in data management ecosystem. *2015 IEEE International Conference on Cyber Technology in Automation, Control, and Intelligent Systems (CYBER)*, 820–824.

Feng, J., Yang, Y., Shen, H., & Cai, Z. (2017). Development of lighting control system for smart hotel rooms. *International Journal of Performability Engineering*, 13(6), 913.

Fernandez-Carames, T.M., & Fraga-Lamas, P. (2020). Towards post-quantum blockchain: A review on blockchain cryptography resistant to quantum computing attacks. *IEEE Access*, 8, 21091–21116. arXiv:2004.07917. https://doi.org/10.1109/ACCESS.2020.2968985

Filatotchev, I., & Nakajima, C. (2010). Internal and external corporate governance: An interface between an organization and its environment. *British Journal of Management*, 21(3), 591–606.

Fleisch, E., Weinberger, M., & Wortmann, F. (2015). Business models and the Internet of Things. In *Interoperability and Open-Source Solutions for the Internet of Things* (pp. 6–10). Springer.

Fletcher, G., Groth, P., & Sequeda, J. (2020). Knowledge scientists: Unlocking the data-driven organization. https://doi.org/10.48550/arXiv.2004.07917

Forbes Business Insights. (2021). *Cyber Security Market, 2021–2028*. https://www.fortunebusinessinsights.com/industry-reports/cyber-security-market-101165 (accessed 23 June 2022).

Forbes Coaches Council. (2018). *What Does it Mean to Be an Agile Leader?* Forbes. https://www.forbes.com/sites/forbescoachescouncil/2018/06/29/what-does-it-mean-to-be-an-agile-leader/?sh=7fb4cd54db4f (accessed 23 June 2022).

Forman, E.H., & Selly, M.A. (2001). Introduction: management decision-making today. In *Decision by Objectives: How to convince others that you are right* (pp. 1–14). World Scientific. https://doi.org/10.1142/9789812810694_0001

Fortune Business Insights. (2020). *The Global Big Data Technology Market Size*. https://www.fortunebusinessinsights.com/industry-reports/big-data-technology-market-100144 (accessed 23 June 2022).

Friedman, H.H. (2020). *Organizational Agility, Visionary Leadership in the Age of VUCA*. https://doi.org/10.2139/ssrn.3728372

FTC. (2015). *The Internet of Things: Privacy and Security in a Connected World*. Federal Trade Commission.

Gagel, G. (2017). The intersection of organizational agility and transformational leadership: A literature review. *Academy of Management Proceedings*, 2017(1), 10895. https://doi.org/10.5465/ambpp.2017.10895abstract

Gaurav, J., & Kongar, E. (2021). Value creation via accelerated digital transformation. *IEEE Engineering Management Review*, 49(2), 63–72. https://doi.org/10.1109/EMR.2021.3054813

GeeksforGeeks. (2021). *Hadoop Ecosystem*. https://www.geeksforgeeks.org/hadoop-ecosystem/ (accessed 23 June 2022).

Gestrin, M.V., & Staudt, J. (2018). *The Digital Economy, Multinational Enterprises and International Investment Policy*. https://t4.oecd.org/investment/the-digital-economy-multinational-enterprises-and-international-investment-policy.htm (accessed 26 July 2022).

Ghemawat, S., Gobioff, H., & Leung, S.-T. (2003). The Google file system. *Proceedings of the Nineteenth ACM Symposium on Operating Systems Principles*, 29–43.

Ghorbani, A., Danaei, A., Zargar, S. M., & Hematian, H. (2019). Designing of smart tourism organization (STO) for tourism management: A case study of tourism organizations of South Khorasan province, Iran. *Heliyon*, 5(6), e01850. https://doi.org/10.1016/j.heliyon.2019.e01850

Ghule, V., & Sakhare, S. (2017). Smart organization. *2017 IEEE 7th International Advance Computing Conference (IACC)*, 826–830.

Gill, S.S., Kumar, A., Singh, H., Singh, M., Kaur, K., Usman, M., & Buyya, R. (2020). Quantum computing: A taxonomy, systematic review and future directions. *Journal of Sotware: Practice and Experience*, 52(1), 66–114. https://doi.org/10.1002/spe.3039

Gillet, F. (2018). *Boost CX Quality by Using IoT in Customer Journeys*. https://www.forrester.com/report/Boost-CX-Quality-By-Using-IoT-In-Customer-Journeys/RES137069 (accessed 23 June 2022).

Grecuccio, J., Giusto, E., Fiori, F., & Rebaudengo, M. (2020). Combining blockchain and IoT: Food-chain traceability and beyond. *Energies*, 13(15), 3820.

Greineder, M.J., Blohm, I., & Leicht, N. (2020). Conceptualizing the agile work organization: A systematic literature review, framework and research agenda. *BLED 2020 Proceedings*, *25*.

Gretzel, U., Sigala, M., Xiang, Z., & Koo, C. (2015a). Smart tourism: Foundations and developments. *Electronic Markets*, 25(3), 179–188.

Gretzel, U., Werthner, H., Koo, C., & Lamsfus, C. (2015b). Conceptual foundations for understanding smart tourism ecosystems. *Computers in Human Behavior*, 50, 558–563.

Gudivada, V.N., Pankanti, S., Seetharaman, G., & Zhang, Y. (2019). Cognitive computing systems: Their potential and the future. *Computer*, 52(5), 13–18.

Guttentag, D.A. (2010). Virtual reality: Applications and implications for tourism. *Tourism Management*, 31(5), 637–651.

Hamm, S., & Kelly, J. (2013). *Smart Machines: IBM's Watson and the era of cognitive computing*. Columbia University Press. https://doi.org/10.7312/columbia/9780231168564.001.0001

Hanelt, A., Bohnsack, R., Marz, D., & Antunes Marante, C. (2021). A systematic review of the literature on digital transformation: Insights and implications for strategy and organizational change. *Journal of Management Studies*, 58(5), 1159–1197.

Harari, Y.N. (2016). *Homo Deus: Breve historia del mañana*. Debate.

Henriette, E., Feki, M., & Boughzala, I. (2015). The shape of digital transformation: A systematic literature review. *MCIS 2015 Proceedings*, 10, 431–443.

Hoffman, D.L., & Novak, T. (2015). *Emergent experience and the connected consumer in the smart home assemblage and the Internet of Things*. https://doi.org/10.2139/ssrn.2648786

Högberg, K. (2021). Strategic responses to digital disruption – an exploratory study of digital transformation in hospitality. *AMICS 2021 Proceedings*, 10. https://aisel.aisnet.org/amcis2021/org_transform/org_transform/10/ (accessed 4 October 2022).

Hönigsberg, S., & Dinter, B. (2019). Toward a method to foster the digital transformation in SME networks. *ICIS 2019 Proceedings*, 9. https://aisel.aisnet.org/icis2019/business_models/business_models/19 (accessed 30 July 2022).

Hume, E., & West, A. (2020). Becoming a data-driven decision making organization. *The CPA Journal*, 90(4), 32–35.

Hupperz, M., Gür, I., Möller, F., & Otto, B. (2021). What is a data-driven organization? *AMCIS 2021 Proceedings*, 6. https://aisel.aisnet.org/amcis2021/strategic_is/strategic_is/6 (accessed 30 July 2022).

IATA. (2022). Air passenger numbers to recover in 2024. https://www.iata.org/en/pressroom/2022-releases/2022-03-01-01/ (accessed 23 June 2022).

IBM. (2022). What is data science? https://www.ibm.com/cloud/learn/data-science-introduction (accessed 23 June 2022).

IDC. (2021). IDC forecasts worldwide quantum computing market to grow to $8.6 billion in 2027. 29 November. https://www.idc.com/getdoc.jsp?containerId=prUS48414121 (accessed 23 June 2022).

Iglesias-Sánchez, P.P., Correia, M.B., & Jambrino-Maldonado, C. (2019). Challenges of open innovation in the tourism sector. *Tourism Planning and Development*, 16(1), 22–42. https://doi.org/10.1080/21568316.2017.1393773

Information Threat Group. (2017). *Insider Threat: Spotlight report*. http://crowdresearchpartners.com/wp-content/uploads/2017/07/Insider-Threat-Report-2016.pdf (accessed 30 July 2022).

Intel. (2022). Neuromorphic and probabilistic computing. Beyond today's AI. https://www.intel.es/content/www/es/es/research/neuromorphic-computing.html (accessed 23 June 2022).

Irdeto. (2019). *New 2019 Global Survey: IoT-focused cyberattacks are the new normal*. https://resources.irdeto.com/global-connected-industries-cybersecurity-survey/new-2019-global-survey-iot-focused-cyberattacks-are-the-new-normal (accessed 23 June 2022).

Ivanov, S., & Webster, C. (2019). Conceptual framework of the use of robots, artificial intelligence and service automation in travel, tourism, and hospitality companies. In *Robots, Artificial Intelligence, and Service Automation in Travel, Tourism and Hospitality* (pp. 7–37). Emerald Publishing Limited. https://doi.org/10.1108/978-1-78756-687-320191001

Jandail, S., & Ratan, R. (2014). A proposed novel approach for sentiment analysis and opinion mining. *International Journal of UbiComp (IJU)*, 5(1/2), 1–10. https://doi.org/10.5121/iju.2014.5201.

Jankel, N. (2015). AI vs human intelligence: Why computers will never create disruptive innovations. *Huffington Post.* https://www.huffpost.com/entry/ai-vs-human-intelligence-_b_6741814 (accessed 23 June 2022).

Javornik, A. (2016). Augmented reality: Research agenda for studying the impact of its media characteristics on consumer behaviour. *Journal of Retailing and Consumer Services*, 30, 252–261.

Jesse, N. (2018). Internet of Things and Big Data: The disruption of the value chain and the rise of new software ecosystems. *AI and Society*, 33(2), 229–239. https://doi.org/10.1007/s00146-018-0807-y

Jha, N. (2021). Short review on quantum computing and it future trends. *International Journal of Research in Engineering and Science (IJRES)*, 9(7), 71–75.

Jiao, E.X., & Chen, J.L. (2019). Tourism forecasting: A review of methodological developments over the last decade. *Tourism Economics*, 25(3), 469–492. https://doi.org/10.1177/1354816618812588

Jingen Liang, L., & Elliot, S. (2021). A systematic review of augmented reality tourism research: What is now and what is next? *Tourism and Hospitality Research*, 21(1), 15–30.

Jöhnk, J., Ollig, P., Oesterle, S., & Riedel, L.-N. (2020). The complexity of digital transformation – Conceptualizing multiple concurrent initiatives. *Proceedings of the 15th International*

Conference on Wirtschaftsinformatik (WI), 1051–1066. https://doi.org/10.30844/wi_2020_j8-joehnk
Joiner, B. (2019). Leadership agility for organizational agility. *Journal of Creating Value*, 5(2), 139–149. https://doi.org/10.1177/2394964319868321
Joy, B. (2000). Why the future doesn't need us. *Wired* (1 April). https://www.wired.com/2000/04/joy-2 (accessed 30 July 2022).
Juc, N., & Misrahi, Ti. (2021). *Travel & Tourism Economic Impact 2021*. https://wttc.org/Portals/0/Documents/Reports/2021/Global%20Economic%20Impact%20and%20Trends%202021.pdf?ver=2021-07-01-114957-177 (accessed 23 June 2022).
Jung, T.H., & tom Dieck, M.C. (2017). Augmented reality, virtual reality and 3D printing for the co-creation of value for the visitor experience at cultural heritage places. *Journal of Place Management and Development*, 10(2), 140–151. https://doi.org/10.1108/JPMD-07-2016-0045
Kaivo-Oja, J., Virtanen, P., Jalonen, H., & Stenvall, J. (2015). The effects of the Internet of Things and Big Data to organizations and their knowledge management practices. *International Conference on Knowledge Management in Organizations*, 495–513.
Kaplan, A., & Haenlein, M. (2019). Siri, Siri, in my hand: Who's the fairest in the land? On the interpretations, illustrations, and implications of artificial intelligence. *Business Horizons*, 62(1), 15–25.
Kaspersky Lab. (2018). *Damage Control: The cost of security breaches (IT security risks special report)*. https://media.kaspersky.com/ pdf/it-risks-survey-report-cost-of-security-breaches.pdf (accessed 30 July 2022).
Kauffmann, E., Peral, J., Gil, D., Ferrández, A., Sellers, R., & Mora, H. (2019). A framework for Big Data analytics in commercial social networks: A case study on sentiment analysis and fake review detection for marketing decision-making. *Industrial Marketing Management*, 90(August), 523–537. https://doi.org/10.1016/j.indmarman.2019.08.003
Kellerman, A. (2018). *Automated and Autonomous Spatial Mobilities*. Edward Elgar Publishing.
Kelly, E. (2015). Introduction: Business ecosystems come of age. *Deloitte Insights* (16 April), 3–16. https://www2.deloitte.com/us/en/insights/focus/business-trends/2015/business-ecosystems-come-of-age-business-trends.html (accessed 30 July 2022).
Kemp, R. (2014). Legal aspects of managing Big Data. *Computer Law and Security Review*, 30(5), 482–491. https://doi.org/10.1016/j.clsr.2014.07.006
Kiron, D. (2017). Lessons from becoming a data-driven organization. *MIT Sloan Management Review*, 58(2). https://sloanreview.mit.edu/case-study/lessons-from-becoming-a-data-driven-organization (accessed 30 July 2022).
Kounavis, C.D., Kasimati, A.E., & Zamani, E.D. (2012). Enhancing the tourism experience through mobile augmented reality: Challenges and prospects. *International Journal of Engineering Business Management*, 4(1), 1–6. https://doi.org/10.5772/51644
KPMG. (2015). *FEEL FREE Cyber Security Dashboard: Monitor, analyse and take control of cyber security*. KPMG Advisory NV. https://silo.tips/download/cyber-security-dashboard-monitor-analyse-and-take-control-of-cyber-security (accessed 30 July 2022).
Kung, L., Kung, H.-J., Jones-Farmer, A., & Wang, Y. (2015). *Managing Big Data for Firm Performance: A configurational approach. AMCIS 2015.* https://doi.org/10.13140/RG.2.1.4847.9843
Kurzweil, R. (2005). *The Singularity Is Near: When humans transcend biology*. Penguin.
Kurzweil, R., Richter, R., Kurzweil, R., & Schneider, M.L. (1990). *The Age of Intelligent Machines* (Vol. 580). MIT Press.
Larson, D., & Chang, V. (2016). A review and future direction of agile, business intelligence, analytics and data science. *International Journal of Information Management*, 36(5), 700–710.
LaValle, S., Lesser, E., Shockley, R., Hopkins, M.S., & Kruschwitz, N. (2011). Big Data, analytics and the path from insights to value. *MIT Sloan Management Review*, 52(2), 21–32.

Law, R., Chan, I.C.C., & Wang, L. (2018). A comprehensive review of mobile technology use in hospitality and tourism. *Journal of Hospitality Marketing & Management*, 27(6), 626–648.
Laxamanan, G., & Rahim, R.E.A. (2020). A systematic literature study from 2013 to 2018: The role of knowledge in open innovation. *International Review of Management and Marketing*, 10(2), 84–95. https://doi.org/10.32479/irmm.9337
Lee, I. (2019). The Internet of Things for enterprises: An ecosystem, architecture, and IoT service business model. *Internet of Things*, 7, 100078.
Lee, I. (2020). Internet of Things (IoT) cybersecurity: Literature review and IoT cyber risk management. *Future Internet*, 12(9). https://doi.org/10.3390/FI12090157
Leybourn, E. (2013). *Directing the Agile Organisation: A lean approach to business management*. IT Governance Ltd.
Li, J., Xu, L., Tang, L., Wang, S., & Li, L. (2018a). Big Data in tourism research: A literature review. *Tourism Management*, 68, 301–323.
Li, L., Su, F., Zhang, W., & Mao, J.Y. (2018b). Digital transformation by SME entrepreneurs: A capability perspective. *Information Systems Journal*, 28(6), 1129–1157. https://doi.org/10.1111/isj.12153
Li, M., Yin, D., Qiu, H., & Bai, B. (2021). A systematic review of AI technology-based service encounters: Implications for hospitality and tourism operations. *International Journal of Hospitality Management*, 95(May), 102930. https://doi.org/10.1016/j.ijhm.2021.102930
Li, X., & Law, R. (2020). Network analysis of Big Data research in tourism. *Tourism Management Perspectives*, 33, 100608. https://doi.org/10.1016/j.tmp.2019.100608
Li, Y., Hu, C., Huang, C., & Duan, L. (2017). The concept of smart tourism in the context of tourism information services. *Tourism Management*, 58, 293–300.
Liao, S., Liu, Z., & Ma, C. (2019). Direct and configurational paths of open innovation and organisational agility to business model innovation in SMEs. *Technology Analysis and Strategic Management*, 31(10), 1213–1228. https://doi.org/10.1080/09537325.2019.1601693
Liebana-Cabanillas, F., Carvajal-Trujillo, E., Villarejo-Ramos, Á.F., & Higueras-Castillo, E. (2020). Antecedents of the intention to use NFC mobile applications in tourism. *Journal of Hospitality and Tourism Technology,* 11(2), 369–383.
Lu, Y., Papagiannidis, S., & Alamanos, E. (2018). Internet of Things: A systematic review of the business literature from the user and organisational perspectives. *Technological Forecasting and Social Change*, 136, 285–297.
Lv, H., Shi, S., & Gursoy, D. (2021). A look back and a leap forward: A review and synthesis of big data and artificial intelligence literature in hospitality and tourism. *Journal of Hospitality Marketing & Management*, 31, 145–175.
Lytras, M.D., & Visvizi, A. (2021). *Artificial Intelligence and Cognitive Computing: Methods, technologies, systems, applications and policy making*. Multidisciplinary Digital Publishing Institute.
Ma, M., Wang, P., & Chu, C.-H. (2013). Data management for Internet of Things: Challenges, approaches and opportunities. *2013 IEEE International Conference on Green Computing and Communications and IEEE Internet of Things and IEEE Cyber, Physical and Social Computing*, 1144–1151.
Maalem Lahcen, R.A., Caulkins, B., Mohapatra, R., & Kumar, M. (2020). Review and insight on the behavioral aspects of cybersecurity. *Cybersecurity*, 3(1). https://doi.org/10.1186/s42400-020-00050-w
Madera, J.M., Dawson, M., Guchait, P., & Belarmino, A.M. (2017). Strategic human resources management research in hospitality and tourism: A review of current literature and suggestions for the future. *International Journal of Contemporary Hospitality Management*, 29(1), 48–67.
Mahraz, I., Benabbou, L., & Berrado, A. (2019). A systematic literature review of digital transformation. In *Proceedings of the International Conference on Industrial Engineering and Operations Management*. IEOM Society International.

Makridakis, S. (2017). The forthcoming artificial intelligence (AI) revolution: Its impact on society and firms. *Futures*, 90, 46–60.

Mang, C.F., Piper, L. A., & Brown, N.R. (2016). The incidence of smartphone usage among tourists. *International Journal of Tourism Research*, 18(6), 591–601.

Mariani, M.M., Buhalis, D., Czakon, W., & Vitouladiti, O. (2016). *Tourism Management, Marketing, and Development: Performance, strategies, and sustainability*. Springer.

Mariani, M., Baggio, R., Fuchs, M., & Höepken, W. (2018), Business intelligence and Big Data in hospitality and tourism: A systematic literature review. *International Journal of Contemporary Hospitality Management*, 30(12), 3514-3554. https://doi.org/10.1108/IJCHM-07-2017-0461

Markoff, J. (2016). *Machines of Loving Grace: The quest for common ground between humans and robots*. HarperCollins Publishers.

Matt, C., Hess, T., & Benlian, A. (2015). Digital transformation strategies. *Business & Information Systems Engineering*, 57(5), 339–343.

McAfee, A., Brynjolfsson, E., Davenport, T.H., Patil, D. J., & Barton, D. (2012). Big Data: The management revolution. *Harvard Business Review*, 90(10), 60–68.

McCarthy, J. (2007). What is artificial intelligence? http://jmc.stanford.edu/artificial-intelligence/what-is-ai/index.html (accessed 21 March 2022).

Mehraliyev, F., Chan, I.C.C., Choi, Y., Koseoglu, M.A., & Law, R. (2020). A state-of-the-art review of smart tourism research. *Journal of Travel & Tourism Marketing*, 37(1), 78–91.

Mihet, R., & Philippon, T. (2018). *The Economics of Big Data*. Mimeo.

Mirarab, A., Mirtaheri, S.L., & Asghari, S.A. (2019). Value creation with big data analytics for enterprises: A survey. *Telkomnika (Telecommunication Computing Electronics and Control)*, 17(6), 2790–2802. https://doi.org/10.12928/TELKOMNIKA.v17i6.11962

Miraz, M.H., Ali, M., Excell, P.S., & Picking, R. (2015). A review on Internet of Things (IoT), Internet of Everything (IoE) and Internet of Nano Things (IoNT). 2015 *Internet Technologies and Applications (ITA)*, 219–224. https://doi.org/10.1109/ITechA.2015.7317398

Mirzaalian, F., & Halpenny, E. (2019). Social media analytics in hospitality and tourism: A systematic literature review and future trends. *Journal of Hospitality and Tourism Technology*, 10(4), 764–790. https://doi.org/10.1108/JHTT-08-2018-0078

Monino, J.L. (2021). Data value, Big Data analytics, and decision-making. *Journal of the Knowledge Economy*, 12(1), 256–267. https://doi.org/10.1007/s13132-016-0396-2

Morabito, V. (2015). Managing change for Big Data driven innovation. In *Big Data and Analytics* (pp. 125–153). Springer.

Morakanyane, R., Grace, A., & O'Reilly, P. (2017). Conceptualizing digital transformation in business organizations: A systematic review of literature. *30th Bled EConference: Digital Transformation – From Connecting Things to Transforming Our Lives, BLED* 2017, 427–444. https://doi.org/10.18690/978-961-286-043-1.30

Müller, V.C., & Bostrom, N. (2016). Future progress in artificial intelligence: A survey of expert opinion. In *Fundamental Issues of Artificial Intelligence* (pp. 555–572). Springer.

Murphy, J., Hofacker, C.F., & Gretzel, U. (2017). Robots in hospitality and tourism: A research agenda. In V.C. Müller (Ed.), *Fundamental Issues of Artificial Intelligence*. Synthese Library Vol. 376. Springer. https://doi.org/10.1007/978-3-319-26485-1_33

Nagle, T., Sammon, D., & Co-Directors (2013). Fast and flexible: Exploring agile data analytics. *Cutter Benchmark Review*, 13(2), 5.

Neuhofer, B., Buhalis, D., & Ladkin, A. (2014). A typology of technology-enhanced tourism experiences. *International Journal of Tourism Research*, 16(4), 340–350.

New York Times. (2021). Bitcoin uses more money than many countries. How is that possible? *New York Times*, 3 September. https://www.nytimes.com/interactive/2021/09/03/climate/bitcoin-carbon-footprint-electricity.html (accessed 23 June 2022).

Ng, I.C.L., & Wakenshaw, S.Y.L. (2017). The internet-of-Things: Review and research directions. *International Journal of Research in Marketing*, 34(1), 3–21. https://doi.org/10.1016/j.ijresmar.2016.11.003

Ngo, V.M., & Vu, H.M. (2020). Customer agility and firm performance in the tourism industry. *Tourism: An International Interdisciplinary Journal*, 68(1), 68–82.
Nikolskaya, E.Y., Zakharova, E.V., Galkin, D.V., Kovaleva, N.I., & Panova, N.A. (2021). The impact of digital technologies on the transformation of the tourism and hospitality industry. *Revista Geintec-Gestao Inovacao E Tecnologias*, 11(4), 623–632.
Noor, A.K. (2015). Potential of cognitive computing and cognitive systems. *Open Engineering*, 5(1), 75–88. https://doi.org/10.1515/eng-2015-0008
Noor, M.N., & Haneef, F. (2020). A review on big data and social network analytics techniques. *Researchpedia Journal of Computing*, 1(1), 39–49.
Nord, J.H., Koohang, A., & Paliszkiewicz, J. (2019). The Internet of Things: Review and theoretical framework. *Expert Systems with Applications*, 133, 97–108.
OECD. (2017a). Enhancing the contributions of SMEs in a global and digitalised economy. *Meeting of the OECD Council at Ministerial Level.* OECD Publishing.
OECD. (2017b). *Key Issues for Digital Transformation in the G20*. Report prepared for a joint G20 German Presidency/OECD Conference. https://www.oecd.org/g20/key-issues-for-digital-transformation-in-the-g20.pdf (accessed 30 July 2022).
OECD. (2019). *SME and Entrepreneurship Outlook 2019*. OECD. https://doi.org/10.1787/34907e9c-ene
OECD. (2020). Preparing tourism businesses for the digital future. In *OECD Tourism Trends and Policies* (pp. 59–85). OECD. https://doi.org/10.1787/6b47b985-en
Olszak, C. M., & Zurada, J. (2019). Big Data-driven value creation for organizations. *Proceedings of the Annual Hawaii International Conference on System Sciences*, January, 164–173. https://doi.org/10.24251/hicss.2019.021
Ozdemir, A.I., Ar, I.M., & Erol, I. (2020). Assessment of blockchain applications in travel and tourism industry. *Quality & Quantity*, 54(5), 1549–1563.
Palattella, M.R., Dohler, M., Grieco, A., Rizzo, G., Torsner, J., Engel, T., & Ladid, L. (2016). Internet of things in the 5G era: Enablers, architecture, and business models. *IEEE Journal on Selected Areas in Communications*, 34(3), 510–527.
Paraskevas, A. (2020). Cybersecurity in travel and tourism: A risk-based approach. *Handbook of E-Tourism* (pp. 1–24). https://doi.org/10.1007/978-3-030-05324-6_100-1
Parloff, R. (2016). From 2016: Why deep learning is suddenly changing your life. *Fortune*, 28 September. https://fortune.com/longform/ai-artificial-intelligence-deep-machine-learning (accessed 30 July 2022).
Pasca, M.G., Renzi, M. F., Di Pietro, L., & Guglielmetti Mugion, R. (2021). Gamification in tourism and hospitality research in the era of digital platforms: A systematic literature review. *Journal of Service Theory and Practice*, 31(5), 691–737. https://doi.org/10.1108/JSTP-05-2020-0094
Payne, A.F., Storbacka, K., & Frow, P. (2008). Managing the co-creation of value. *Journal of the Academy of Marketing Science*, 36(1), 83–96.
Pellizzoni, E., Trabucchi, D., & Buganza, T. (2019). When agility meets open innovation: Two approaches to manage inbound projects. *Creativity and Innovation Management*, 28(4), 464–476. https://doi.org/10.1111/caim.12337
Pesonen, J. (2020). Management and leadership for digital transformation in tourism. In Z. Xiang, M. Fuchs, U. Gretzel, & W. Höpken (Eds.), *Handbook of e-Tourism* (pp. 1–34). Springer. https://doi.org/10.1007/978-3-030-05324-6_68-1
Peterlin, J., Meško, M., Dimovski, V., & Roblek, V. (2021). Automated content analysis: The review of the Big Data systemic discourse in tourism and hospitality. *Systems Research and Behavioral Science*, 38(3), 377–385.
Petzold, B., Roggendorf, M., Rowshankish, K., & Sporleder, C. (2020). *Designing Data Governance That Delivers Value*. McKinsey & Company.

Pinheiro, A.B., Pinto, A.S., Abreu, A., Costa, E., & Borges, I. (2021). The impact of artificial intelligence on the tourism industry: A systematic review. *Smart Innovation, Systems and Technologies*, 208, 458–469. https://doi.org/10.1007/978-981-33-4256-9_42

Pogue, C. (2018). *The Black Report 2018. Decoding the minds of hackers*. https://www.nuix.com/sites/default/files/report_nuix_black_report_2018_web_us.pdf (accessed 23 June 2022).

Poslad, S. (2011). *Ubiquitous Computing: Smart devices, environments and interactions*. John Wiley & Sons.

Pramanik, P.K.D., Pal, S., & Choudhury, P. (2018). Beyond automation: The cognitive IoT. Artificial intelligence brings sense to the Internet of Things. In *Cognitive Computing for Big Data Systems Over IoT* (pp. 1–37). Springer.

prnewswire.com. (2021). 38% of Americans plan to travel more after the pandemic, nearly 1 in 4 plan to use crypto to pay for a part of their travel. 11 April. https://www.prnewswire.com/news-releases/38-of-americans-plan-to-travel-more-after-the-pandemic-nearly-1-in-4-plan-to-use-crypto-to-pay-for-a-part-of-their-travel-301267684.html (accessed 23 June 2022).

PWC. (2017). *2017 Global Digital IQ Survey: Blockchain*. http://usblogs.pwc.com/emerging-technology/2017-digital-iq-blockchain (accessed 29 October 2019).

Rao, B., Mohapatra, S., Saha, U., & Mitra, A. (2015). Applications of social network using quantum computing. *Conference on Inter Disciplinary Research in Engineering and Technology (ICIDRET 15)*, 144, 1144.

Reagin, M.J., & Gentry, M. V. (2018). Enterprise cybersecurity. *Frontiers of Health Services Management*, 35(1), 13–22. https://doi.org/10.1097/hap.0000000000000037

Reddy, S.K., & Reinartz, W. (2017). Digital transformation and value creation: Sea change ahead. *GfK Marketing Intelligence Review*, 9(1), 10.

Reinsel, D., Gantz, J., & Rydning, J. (2018). *The Digitization of the World – From edge to core* (November). https://www.seagate.com/files/www-content/our-story/trends/files/idc-seagate-dataage-whitepaper.pdf (accessed 23 June 2022).

Reis, J., Amorim, M., Melão, N., & Matos, P. (2018). Digital transformation: A literature review and guidelines for future research. *World Conference on Information Systems and Technologies*, 411–421.

Rialti, R., Marzi, G., Ciappei, C., & Caputo, A. (2018). Reframing Agile Organization: Do Big Data analytics capabilities matter? Paper delivered at conference, *Big Data and Managing in the Digital Economy*, University of Surrey Business School, 18–20 April.

Riasanow, T., Setzke, D.S., Böhm, M., & Krcmar, H. (2019). Clarifying the notion of digital transformation: A transdisciplinary review of literature. *Journal of Competences, Strategy & Management*, 10(1), 5–31.

Rigby, D.K., Sutherland, J., & Takeuchi, H. (2016). Embracing agile. *Harvard Business Review*, 94(5), 40–50.

Roxo, M.T., & Brito, P.Q. (2018). Augmented reality trends to the field of business and economics: A review of 20 years of research. *Asian Journal of Business Research*, 8(2). https://doi.org/10.14707/ajbr.180051

Ruíz, M.A.C., Bohorquez, S.T., & Molano, J.I.R. (2017). Colombian tourism: Proposal app to foster smart tourism in the country. *Advanced Science Letters*, 23(11), 10533–10537.

Russom, P. (2013). Managing Big Data. *TDWI Best Practices Report, TDWI Research*, 1–40.

Samara, D., Magnisalis, I., & Peristeras, V. (2020). Artificial intelligence and Big Data in tourism: A systematic literature review. *Journal of Hospitality and Tourism Technology*, 11(2), 343–367. https://doi.org/10.1108/JHTT-12-2018-0118

Santoro, G., Vrontis, D., Thrassou, A., & Dezi, L. (2018). The Internet of Things: Building a knowledge management system for open innovation and knowledge management capacity. *Technological Forecasting and Social Change*, *136*, 347–354.

Santos-Júnior, A., Almeida-García, F., Morgado, P., & Mendes-Filho, L. (2020). Residents' quality of life in smart tourism destinations: A theoretical approach. *Sustainability (Switzerland)*, 12(20), 1–24. https://doi.org/10.3390/su12208445

Scannell, E. (2021). Recent quantum computing advances point to brighter future. TechTarget Search Data Center. 10 December. https://www.techtarget.com/searchdatacenter/feature/Recent-quantum-computing-advances-point-to-brighter-future (accessed 1 March 2022).

Schwab, K. (2017). *The Fourth Industrial Revolution*. Crown Business.

Schwartz, E.I. (2002). *Digital Darwinism: 7 breakthrough business strategies for surviving in the cutthroat web economy*. Currency.

Schwertner, K. (2017). Digital transformation of business. *Trakia Journal of Sciences*, 15(1), 388–393.

Sebei, H., Hadj Taieb, M.A., & Ben Aouicha, M. (2018). Review of social media analytics process and Big Data pipeline. *Social Network Analysis and Mining*, 8(1), 1–28. https://doi.org/10.1007/s13278-018-0507-0e

Seigneur, J.M. (2018). Towards geneva crypto-friendly smart tourism. *Etats Généraux du Tourisme.* http://archive-ouverte.unige.ch/unige:103406 (accessed 4 October 2022).

Sestino, A., Prete, M.I., Piper, L., & Guido, G. (2020). Internet of Things and Big Data as enablers for business digitalization strategies. *Technovation*, 98, 102173. https://doi.org/10.1016/j.technovation.2020.102173

Setili, A. (2015). Does your leadership style destroy agility ... Or supercharge it? *Leader to Leader*, 2015(78), 56–61.

Shamim, S., Zeng, J., Shariq, S.M., & Khan, Z. (2019). Role of Big Data management in enhancing big data decision-making capability and quality among Chinese firms: A dynamic capabilities view. *Information and Management*, 56(6), 1–12. https://doi.org/10.1016/j.im.2018.12.003

Shamim, Saqib, Zeng, J., Shafi Choksy, U., & Shariq, S.M. (2020). Connecting Big Data management capabilities with employee ambidexterity in Chinese multinational enterprises through the mediation of Big Data value creation at the employee level. *International Business Review*, 29(6). https://doi.org/10.1016/j.ibusrev.2019.101604

Shim, J.P., Avital, M., Dennis, A.R., Rossi, M., Sørensen, C., & French, A. (2019). The transformative effect of the internet of things on business and society. *Communications of the Association for Information Systems*, 44(1), 5.

Shoval, N., Isaacson, M., & Chhetri, P. (2014). GPS, smartphones, and the future of tourism research. In A.A. Lew, C.M. Hall, & A.M. Williams (Eds.), *The Wiley Blackwell Companion to Tourism* (pp. 251–261). John Wiley & Sons. https://doi.org/10.1002/9781118474648.ch20

Singh, A., & Hess, T. (2020). How chief digital officers promote the digital transformation of their companies. In *Strategic Information Management* (pp. 202–220). Routledge.

Singh, M., Halgamuge, M.N., Ekici, G., & Jayasekara, C.S. (2018). A review on security and privacy challenges of Big Data. In *Cognitive computing for big data systems over IoT* (pp. 175–200). Springer.

Skinner, H., Sarpong, D., & White, G.R.T. (2018). Meeting the needs of the Millennials and Generation Z: gamification in tourism through geocaching. *Journal of Tourism Futures*, 4(1), 93–104. https://doi.org/10.1108/JTF-12-2017-0060

Soeparno, H., & Perbangsa, A.S. (2021). Cloud quantum computing concept and development: A systematic literature review. *Procedia Computer Science*, 179, 944–954. https://doi.org/https://doi.org/10.1016/j.procs.2021.01.084

Stylos, N., Zwiegelaar, J., & Buhalis, D. (2021). Big Data empowered agility for dynamic, volatile, and time-sensitive service industries: The case of tourism sector. *International Journal of Contemporary Hospitality Management*, 33(3), 1015–1036. https://doi.org/10.1108/IJCHM-07-2020-0644

Tableau. (2022). What Is data visualization? Definition, examples, and learning resources. https://www.tableau.com/learn/articles/data-visualization (accessed 12 February 2022).
Tabrizi, B., Lam, E., Girard, K., & Irvin, V. (2019). Digital transformation is not about technology. *Harvard Business Review*, 13(March), 1–6.
Tajeddini, K., Ratten, V., & Merkle, T. (2019). *Tourism, Hospitality and Digital Transformation: Strategic management aspects*. Routledge.
Teece, D.J. (2007). Explicating dynamic capabilities: The nature and microfoundations of (sustainable) enterprise performance. *Strategic Management Journal*, 28(13), 1319–1350.
Teichert, R. (2019). Digital transformation maturity: A systematic review of literature. *Acta Universitatis Agriculturae et Silviculturae Mendelianae Brunensis*, 67(6), 1673–1687. https://doi.org/10.11118/actaun201967061673
Theoharidou, M., Kokolakis, S., Karyda, M., & Kiountouzis, E. (2005). The insider threat to information systems and the effectiveness of ISO17799. *Computers & Security*, 24(6), 472–484.
Thrassou, A., Vrontis, D., Chebbi, H., & Yahiaoui, D. (2012). A preliminary strategic marketing framework for new product development. *Journal of Transnational Management*, 17(1), 21–44.
Torchia, M., & Calabrò, A. (2019). Open innovation in SMEs: A systematic literature review. *Journal of Enterprising Culture*, 27(02), 201–228. https://doi.org/10.1142/s0218495819500080
TravelPort. (2017). *The Global Digital Traveler Research*. https://marketing.cloud.travelport.com/gdtr-en-uk-download (accessed 30 July 2022).
Trenkle, J. (2019). Survival in the digital age – A framework for formulating a digital transformation strategy in SME. *Proceedings of the 19th International Conference on Electronic Business*, 428–442.
Tribe, J., & Mkono, M. (2017). Not such smart tourism? The concept of e-lienation. *Annals of Tourism Research*, 66, 105–115. https://doi.org/10.1016/j.annals.2017.07.001
Trustwave. (2020). *2020 Trustwave Global Security Report*. https://www.trustwave.com/en-us/resources/library/documents/2020-trustwave-global-security-report (accessed 30 July 2022).
Tsaih, R.H., & Hsu, C.C. (2018). Artificial intelligence in smart tourism: A conceptual framework. *Proceedings of the International Conference on Electronic Business (ICEB)*, 2018(December), 124–133.
Udovita, P. (2020). Conceptual review on dimensions of digital transformation in modern era. *International Journal of Scientific and Research Publications*, 10(2), 520–529.
Ullah, R. (2019). Agile Analytics: How organizations can benefit from the agile methodology for smoother delivery of data-driven analytics projects. MSc thesis, Universität Koblenz-Landau. https://hbz.opus.hbz-nrw.de/opus45-kola/frontdoor/deliver/index/docId/1918/file/RafiUllah_Masterarbeit.pdf (accessed 30 July 2022).
UNCTAD. (2018). *Technology and Innovation Report 2018: Harnessing frontier technologies for sustainable development*. United Nations. https://unctad.org/en/PublicationsLibrary/tir2018overview_en.pdf (accessed 23 June 2022).
Unhelkar, B. (2017). *Big Data Strategies for Agile Business: Framework, practices, and transformation Roadmap*. CRC Press. https://doi.org/10.4324/9781315120232
UNWTO. (2021). *UNWTO World Tourism Barometer*. https://doi.org/https://doi.org/10.18111/wtobarometereng
Vaska, S., Massaro, M., Bagarotto, E.M., & Dal Mas, F. (2021). The digital transformation of business model innovation: A structured literature review. *Frontiers in Psychology*, 11, 3557.
Veit, D., Clemons, E., Benlian, A., Buxmann, P., Hess, T., Kundisch, D., Leimeister, J.M., Loos, P., & Spann, M. (2014). Business models: An information systems research agenda. *Business & Information Systems Engineering*, 6, 45–53.
Vial, G. (2019). Understanding digital transformation: A review and a research agenda. *The Journal of Strategic Information Systems*, 28(2), 118–144. https://doi.org/https://doi.org/10.1016/j.jsis.2019.01.003

Viñan-Ludeña, M.-S. (2019). A systematic literature review on social media analytics and smart tourism. In V. Katsoni & M. Segarra-Oña (Eds.), *Smart Tourism as a Driver for Culture and Sustainability* (pp. 357–374). Springer. https://doi.org/10.1007/978-3-030-03910-3_25

Webster, C., & Ivanov, S. (2020). Future tourism in a robot-based economy: A perspective article. *Tourism Review*, 75(1), 329–332. https://doi.org/10.1108/TR-05-2019-0172

Wei, W. (2019). Research progress on virtual reality (VR) and augmented reality (AR) in tourism and hospitality: A critical review of publications from 2000 to 2018. *Journal of Hospitality and Tourism Technology*, 10(4), 539–570. https://doi.org/10.1108/JHTT-04-2018-0030

West, J., & Bogers, M. (2013). Leveraging external sources of innovation: A review of research on open innovation. *Journal of Product Innovation Management*, 31(4), 814–831.

World Economic Forum. (2017). *Digital Transformation Initiative: Aviation, travel and tourism industry.* No. 060117; White Paper. http://reports.weforum.org/digital-transformation/wp-content/blogs.dir/94/mp/files/pages/files/wef-dti-aviation-travel-and-tourism-white-paper.pdf (accessed 30 July 2022).

Xu, F., Nash, N., & Whitmarsh, L. (2020). Big Data or small data? A methodological review of sustainable tourism. *Journal of Sustainable Tourism*, 28(2), 144–163.

Yarali, A., Joyce, R., & Dixon, B. (2020). Ethics of Big Data: Privacy, security and trust. *Wireless Telecommunications Symposium, 2020, April.* https://doi.org/10.1109/WTS48268.2020.9198734

Yin, C.Z.Y., Jung, T., tom Dieck, M.C., & Lee, M.Y. (2021). Mobile augmented reality heritage applications: Meeting the needs of heritage tourists. *Sustainability (Switzerland)*, 13(5), 1–18. https://doi.org/10.3390/su13052523

Yoo, Y., Boland, R. J., Lyytinen, K., & Majchrzak, A. (2012). Organizing for innovation in the digitized world. *Organization Science*, 23(5), 1398–1408. https://doi.org/10.1287/orsc.1120.0771

Yung, R., & Khoo-Lattimore, C. (2019). New realities: A systematic literature review on virtual reality and augmented reality in tourism research. *Current Issues in Tourism*, 22(17), 2056–2081.

Zapata, M.L., Berrah, L., & Tabourot, L. (2020). Towards the definition of an impact level factor of SME features over digital transformation. *IFIP International Conference on Advances in Production Management Systems*, 123–130.

Zeng, J., & Glaister, K.W. (2018). Value creation from Big Data: Looking inside the black box. *Strategic Organization*, 16(2), 105–140.

Zine, P.U., Kulkarni, M.S., Chawla, R., & Ray, AK. (2014). A framework for value co-creation through customization and personalization in the context of machine tool PSS. *Procedia CIRP*, 16, 32–37.

Zobel, A.-K., Balsmeier, B., & Chesbrough, H. (2016). Does patenting help or hinder open innovation? Evidence from new entrants in the solar industry. *Industrial and Corporate Change*, 25(2), 307–331.

Zsarnoczky, M. (2018). The digital future of the tourism & hospitality industry. *Boston Hospitality Review*, 6, 1–9.

Index

Note: The page numbers in italics and bold represents figures and tables respectively.